SEWAGE SLUDGE TREATMENT AND DISPOSAL

SEWAGE SLUDGE TREATMENT AND DISPOSAL

Richard W. James

NOYES DATA CORPORATION
Park Ridge, New Jersey, U.S.A.
1976

Library of Congress Catalog Card Number: 76-17939
ISBN: 0-8155-0630-9
Printed in the United States

Published in the United States of America by
Noyes Data Corporation
Noyes Building, Park Ridge, New Jersey 07656

FOREWORD

The detailed, descriptive information in this book is based on U.S. patents issued since the late 1950s that deal with sewage sludge treatment and disposal.

This book serves a double purpose in that it supplies detailed technical information and can be used as a guide to the U.S. patent literature in this field. By indicating all the information that is significant, and eliminating legal jargon and juristic phraseology, this book presents an advanced, commercially oriented review of modern sludge treatment as depicted in U.S. patents.

The U.S. patent literature is the largest and most comprehensive collection of technical information in the world. There is more practical, commercial, timely process information assembled here than is available from any other source. The technical information obtained from a patent is extremely reliable and comprehensive; sufficient information must be included to avoid rejection for "insufficient disclosure." These patents include practically all of those issued on the subject in the United States during the period under review; there has been no bias in the selection of patents for inclusion.

The patent literature covers a substantial amount of information not available in the journal literature. The patent literature is a prime source of basic commercially useful information. This information is overlooked by those who rely primarily on the periodical journal literature. It is realized that there is a lag between a patent application on a new process development and the granting of a patent, but it is felt that this may roughly parallel or even anticipate the lag in putting that development into commercial practice.

Many of these patents are being utilized commercially. Whether used or not, they offer opportunities for technological transfer. Also, a major purpose of this book is to describe the number of technical possibilities available, which may open up profitable areas of research and development. The information contained in this book will allow you to establish a sound background before launching into research in this field.

Advanced composition and production methods developed by Noyes Data are employed to bring our new durably bound books to you in a minimum of time. Special techniques are used to close the gap between "manuscript" and "completed book." Industrial technology is progressing so rapidly that time-honored, conventional typesetting, binding and shipping methods are no longer suitable. We have bypassed the delays in the conventional book publishing cycle and provide the user with an effective and convenient means of reviewing up-to-date information in depth.

The Table of Contents is organized in such a way as to serve as a subject index. Other indexes by company, inventor and patent number help in providing easy access to the information contained in this book.

15 Reasons Why the U.S. Patent Office Literature Is Important to You –

1. The U.S. patent literature is the largest and most comprehensive collection of technical information in the world. There is more practical commercial process information assembled here than is available from any other source.

2. The technical information obtained from the patent literature is extremely comprehensive; sufficient information must be included to avoid rejection for "insufficient disclosure."

3. The patent literature is a prime source of basic commercially utilizable information. This information is overlooked by those who rely primarily on the periodical journal literature.

4. An important feature of the patent literature is that it can serve to avoid duplication of research and development.

5. Patents, unlike periodical literature, are bound by definition to contain new information, data and ideas.

6. It can serve as a source of new ideas in a different but related field, and may be outside the patent protection offered the original invention.

7. Since claims are narrowly defined, much valuable information is included that may be outside the legal protection afforded by the claims.

8. Patents discuss the difficulties associated with previous research, development or production techniques, and offer a specific method of overcoming problems. This gives clues to current process information that has not been published in periodicals or books.

9. Can aid in process design by providing a selection of alternate techniques. A powerful research and engineering tool.

10. Obtain licenses — many U.S. chemical patents have not been developed commercially.

11. Patents provide an excellent starting point for the next investigator.

12. Frequently, innovations derived from research are first disclosed in the patent literature, prior to coverage in the periodical literature.

13. Patents offer a most valuable method of keeping abreast of latest technologies, serving an individual's own "current awareness" program.

14. Copies of U.S. patents are easily obtained from the U.S. Patent Office at 50¢ a copy.

15. It is a creative source of ideas for those with imagination.

CONTENTS AND SUBJECT INDEX

INTRODUCTION

To an increasing degree municipalities throughout the world are beset by difficulties in disposing of the ever larger quantities of refuse that are being produced. On the one hand, the quantity of refuse being produced per capita increases, whereas on the other hand the requirements made of disposal facilities are becoming more and more strict. In many instances, particularly in metropolitan areas, it is no longer possible to merely cart refuse to a designated dump, because space availability for such dumps is limited and in many instances has been exhausted or very nearly so.

The per capita production of sewage in the United States varies from less than 100 gallons per day for a residential area to well over 300 gallons per day for a highly industrialized area. While the original need for sewage-purification plants stemmed largely from the prevention of disease and the general concern for public health, the growing shortage of water and the pollution of our lakes, rivers and streams has greatly accelerated the demand for efficient, low cost treating facilities in all but the smallest rural communities.

Sludge is the major by-product of industrial and domestic water and waste treating processes. In fact, one of the major problems in a well-run water pollution control program is the sludge conditioning method used in dewatering sludges. The term sludge dewatering process as used in the art means any process which reduces the water content of the sludge from its usual value of 93 to 99 percent by weight to about 90 percent by weight or less. That is, it concentrates the sludge solids to about 10 percent by weight or greater.

Wastewater sludge is basically characterized according to three factors which are: (1) sludge source, (2) sludge processing, and (3) degree of treatment. By sludge source is meant whether the sludge is from municipal (domestic) wastewater or industrial wastewater or a combination thereof. The sludge processing characteristic defines whether the sludge is raw untreated sludge, anaerobic or aerobic digested sludge, air flotation sludge or digested elutriated sludge.

The degree of treatment indicates whether the sludge is primary sludge, activated sludge, waste activated sludge, chemically precipitated sludge, trickling filter humus or a combination of one or more of these such as waste activated sludge combined with primary sludge. It is generally accepted that each individual sludge has a different characterization and this sludge character more or less dictates the dewatering process used.

The various mechanical dewatering processes now commonly used in the art are gravity filtration, vacuum filtration, centrifugation, flotation, and sedimentation. However, regardless of the mechanical process used for dewatering, it has become standard practice in the art to chemically condition the sludge prior to dewatering. This chemical conditioning of the sludge enhances the mechanical dewatering process dramatically. The water content of the sludge can be reduced from concentrations in the neighborhood of 93 to 99 percent water to those of about 60 to 90 percent or less by proper chemical conditioning prior to mechanical dewatering.

A desirable and convenient way of disposing of organic and other combustible waste materials is by complete combustion of these materials, in such a manner that the only residue is a relatively lightweight ash, which may be trucked away and disposed of as landfill and the like. Any particulate materials which would otherwise exit the incineration process through the flue system may be suitably wet scrubbed with water. This serves to reduce the temperature and remove entrained particles from the effluent stream. Alternately these particulates may be electrostatically precipitated in order that the gaseous discharge of the incineration process does not produce atmospheric pollution.

Since these incineration processes are nonproductive, and do not usually produce commercially useful by-products, it is most desirable to maintain the construction cost of incineration systems, and the cost of operating these incineration systems, as low as possible, in order to achieve optimum economy of operation, while maintaining optimum efficiency of waste disposal within the limits of air pollution emission standards established by federal and state agencies.

In some instances, services may be provided for collecting liquid, semiliquid and solid organic and other combustible wastes from plant sites and other waste-producing sites at which it is not economical to build incinerators, and for trucking these collected wastes to central incinerator facilities at which the wastes from many industrial plants and/or other waste-producing facilities are incinerated, with each manufacturing plant and other facility paying a fee for the hauling away and incineration of its wastes.

In other instances, large industrial plant complexes produce wide varieties of combustible wastes in large quantities which can economically justify the construction and operation of on-site waste incineration facilities. In these circumstances it is desired to achieve optimum combustion efficiency of combustible materials at the minimum initial cost for incineration equipment, and minimum operating cost for day-to-day operation of these incinerator units.

ACTIVATED SLUDGE PROCESS

The activated sludge process uses a heterogeneous microbial population to oxidize soluble and colloidal organics to carbon dioxide and water in the presence of molecular oxygen. Turbulent mixing holds the organisms in suspension and also provides the necessary oxygen required by the cells for respiration. During the oxidation process a portion of the organic material is synthesized to form new cells while the other organic material oxidized by the microorganisms is used to provide energy for synthesis and motility. Part of the synthesized mass undergoes endogenous respiration, and the remainder forms excess sludge. In a continuous process the excess sludge flows out of the aeration tank and settles under quiescent conditions in a final clarifier. A portion of the solids is recycled to the aeration tank, and the remainder must be further treated for final disposal.

Much of the development of the activated sludge process is a result of solving operational problems rather than applying fundamental concepts. Activated sludge originated in 1882, when Angus Smith aerated sewage and discovered that the oxidation of ammonia to nitrate was accompanied by the removal of organic matter. The basic process was actually developed and named in London by Ardern and Lockett in 1914.

Initially, the studies were performed on a batch basis using a six-hour retention time, which is still used as a set design parameter by some engineers today. Experimentation with mechanical aeration resulted in the design of the first complete mix plant at Bury, England in 1919. However, the advantages of the complete mix regime were not apparent at that time. Most of the plants being built consisted of long, rectangular aeration units which produced a plug flow regime and what is known as the conventional activated sludge process.

The importance of microbiology to the design and operation of a complete mix plant was not completely realized by design engineers until the early 1950's. In 1923 a study at the University of Illinois established the microbial population dynamics of the activated sludge process as it is known today. Unfortunately, there was a lack of communication between the biologists and engineers resulting in omission of the application of biological principles to the design of activated sludge processes. Even though the importance of microbiology is relatively well established now, some engineers continue to ignore this truth and prefer to simply follow what has been done before.

So that one may fully understand the mechanisms involved in an activated sludge process, a brief description of the microbiology of the process will be presented. The stabilization of organic material and the flocculation of the sludge are related to the growth of microorganisms. Bacteria, which are the primary consumers of organics in an activated sludge process, metabolize their food to grow and obtain energy via a variety of complex cycles and pathways.

To facilitate this conversion, the bacterial cells are equipped with enzymes, organic catalysts known for their efficiency of conversion and their specificity for a particular substrate. Enzymatic reactions are affected by environmental conditions such as temperature and pH; therefore, these parameters are important in operation of an activated sludge process.

In addition to enzymes, bacterial cells require energy for their biochemical processes. Energy is released from oxidation of the organic material, and this energy is stored by the compound adenosine triphosphate (ATP). After this stored energy is released for cell synthesis, maintenance, and motility, it is reduced to adenosine diphosphate (ADP). The ADP molecule is then available to accept the energy released by oxidation of organic matter and is converted to ATP to complete the cycle. Thus, energy is obtained by assimilation of the organic material, and this energy is used for cell synthesis.

Other microorganisms important for the successful operation of activated sludge are protozoa and rotifers. These animals establish a prey-predator relationship which ensures good settleability of the activated sludge floc. The protozoa and rotifers consume bacteria as their food source, reducing the number of bacteria. However, because of the increase in the food to microorganism ratio, bacterial growth is accelerated, and hence mcre organic matter is removed from solution.

The growth of microorganisms in an activated sludge system occurs in three phases: log growth phase, declining growth phase, and endogenous phase. In the log growth phase there is always an excess of food for the microorganisms, and the growth rate is at a maximum, limited only by the ability of the microorganisms to process the substrate. When the microorganisms need more food than is available, their rate of growth decreases, which is the declining growth phase. As the food concentration reaches a minimum, growth ceases, the microorganisms metabolize their own protoplasm, and the food concentration is slowly decreased. This is known as endogenous respirat on.

The following figure illustrates this relationship between food and mass in an activated sludge system. Whereas the conventional activated sludge operates somewhat in each of the three phases of growth from the head of the tank to the end of the tank, complete mix activated sludge operates at a point on the growth curve, preferably between the log growth and declining growth phases for optimum substrate removal and good sludge characteristics.

Metabolic Relationships of Activated Sludge

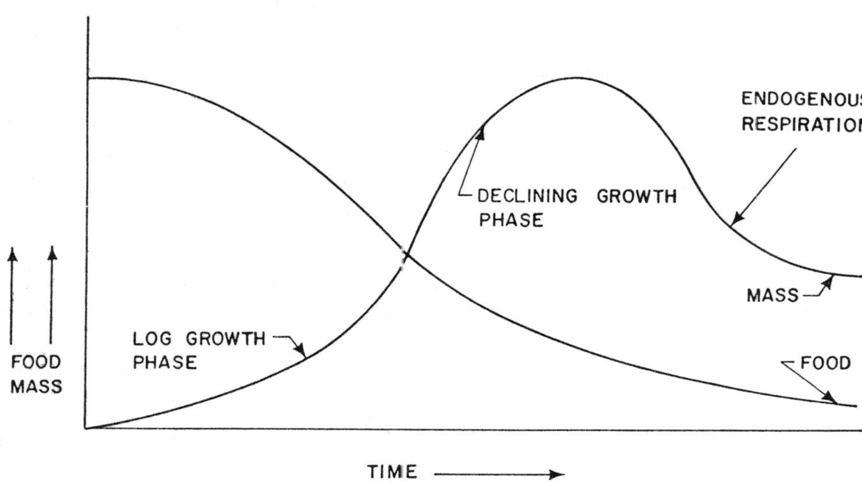

The activated sludge process has many modifications and flow regimes. The varied modifications will not be discussed here, but there are two distinct flow regimes, or mixing models, used in activated sludge processes which are of primary importance. The first of these mixing models is the plug flow model which is associated with the conventional activated sludge process. In a plug flow system the individual particles of the influent pass through the reaction vessel in the same sequence in which they entered, and there is no intermixing or interaction between the particles. This type of flow regime is difficult if not impossible to actually attain in an activated sludge system, but the long, narrow tanks typical of a conventional activated sludge system have traditionally associated it with plug flow. There is always some mixing or interaction between the particles because of the turbulence required for aeration.

On the other hand, the complete mix flow regime, in which the influent particles are completely mixed with the liquid and suspended matter in the reactor, can be attained in laboratory as well as field scale systems. A complete mix activated sludge plant offers many advantages over the conventional process. An understanding of the operation of each process will reveal the advantages of CMAS. It is important to this study to apply these advantages to possible additional advantages of CMAS units in series.

The conventional process mixes incoming waste with return activated sludge at the head of the aeration tank creating a high food to microorganism ratio. In the presence of excess food the bacteria grow rapidly demanding a large amount of oxygen to complete their metabolism. Providing adequate oxygen transfer at the beginning of the tank is often difficult or impossible. As the waste moves down the length of the tank, biodegradation of the organics approaches completion, and the bacterial mass undergoes endogenous respiration; that is, the cells consume their own protoplasm.

By the time the sludge reaches the end of the tank it has become inactive. It is this largely inert mass which is recycled to be mixed with the high strength waste at the head of the tank. The microbial population never reaches an equilibrium, but is in a continual cycle of growth and starvation. The resulting imbalance between the microorganisms and the organic load is the reason for the irregular operation of the system.

The complete mix activated sludge system alleviates many problems experienced by the conventional activated sludge process. The operational advantages of the CMAS process are accomplished by equalizing the organic load over the entire mass of microorganisms in the aeration tank. Complete mix means that the feed particles and recycled sludge become completely intermixed in the aeration chamber, thereby losing their individual identity.

Attainment of complete mixing is assured when the oxygen uptake rate is uniform throughout the aeration volume, provided that there is sufficient mixing to prevent settling of suspended solids in the aeration tank. An important added advantage of CMAS is its ability to act as a surge tank when the system is subjected to shock loads. The increased load is distributed uniformly over the entire contents of the tank, thereby effecting better use of the dissolved oxygen throughout the tank. In summary, operation of a CMAS system is smoother than the conventional system because of CMAS's stable microbial growth, uniform oxygen demand rate, uniform distribution of food, and reduced sensitivity to shock loadings.

More information on the efficacy of the complete mix activated sludge process in modular mode appears in PB 211 156 (July 1972) which is available from the National Technical Information Service.

The Water Pollution Control Act of 1972 calls for nationwide achievement of the installation of secondary treating facilities in all communities in the country by mid-1977. Recent estimates by the Environmental Protection Agency, using a 200 million dollar estimate per plant for conventional equipment, indicate that the total cost of compliance would amount to 13 billion dollars in the United States.

This book describes over 190 patented processes relating to all phases of industrial and domestic water purification. These processes include large scale community systems as well as special compact units which have been developed for remote locations and watercraft waste disposal.

INCINERATION

FLUIDIZED BED

Nozzle Feed System

L.R. Van Gelder; U.S. Patent 3,709,170; January 9, 1973; assigned to Chicago Bridge & Iron Co. describes a sludge feed system for a reactor having a nozzle, and a conduit communicating with the outside of the reactor and the nozzle for feeding sludge to the nozzle under pressure, the nozzle having a mouth in communication with the interior of the reactor for forming the sludge into a thin layer which breaks up in the reactor.

Referring to Figures 1.1a and 1.1b, enclosed reactor **10** has a lower cylindrical portion **11**, a central conical section **12** and an upper cylindrical portion **13**. The described reactor portions comprise an exterior metal shell with a suitable refractory material lining the interior surface thereof, particularly if the reactor is to be employed in high-temperature reactions or treatments. The bottom **14** of the reactor and the top **15** are of similar construction to the reactor vertical walled portions already described. The reactor is typical of the type used in fluidized bed treatments and particularly in the combustion of waste materials.

A constriction plate **16** spans the horizontal interior area defined by the lower cylindrical portion of the reactor support. The space between the constriction plate and the reactor bottom constitutes air box **18**. A plurality of vertical tubes **17**, in the constriction plate, communicate at their lower ends with the air box. The upper ends of the tubes, which extend above the constriction plate, terminate in horizontally located small tubular sections **19** through which air is expelled from the air box under pressure to fluidize particulate bed **20** located in the reactor. Conduit **21** feeds air or hot gases or whatever gas is appropriate to the air box. Conduit **22** in the upper part of the reactor is used to withdraw gases and products of combustion from the reactor.

Nozzle **30** is positioned so that its mouth is in communication with the interior of the reactor. As shown in Figures 1.1a and 1.1b, the nozzle mouth is so positioned in the lower portion of the reactor as to be slightly above the top surface of the fluidized bed **20**. However, the nozzle can be so located as to deliver sludge being fed from the mouth directly into the fluidized bed.

With reference to Figures 1.1c and 1.1d, nozzle **30** has a tubular portion **31** which communicates with conduit **32**. The tubular section is circular in cross section and its inner end joins inlet end **32** of nozzle transition element **33** which terminates in nozzle transition element mouth **34**. The inlet end of the transition element is circular in cross section and progressively becomes oval and finally terminates in the mouth which is shaped in the form of an elongated rectangle which is much wider than it is high.

As shown in Figure 1.1c, the top and bottom walls of the transition element converge towards the nozzle mouth while as shown in Figure 1.1d the opposing side walls of the transition element diverge towards the nozzle mouth. The smooth contour of the interior surface of the transition element serves to progressively shape or form a sludge passing through it into a wider and thinner layer as it moves forward so that it is ultimately expelled from the nozzle mouth in a thin disintegratable layer which breaks up into pieces in the reactor. The nozzle mouth is advisably positioned to be approximately flush with the reactor interior surface to avoid undue abrasion of it by the dynamic activity of the fluidized bed.

FIGURE 1.1: SLUDGE FEED SYSTEM

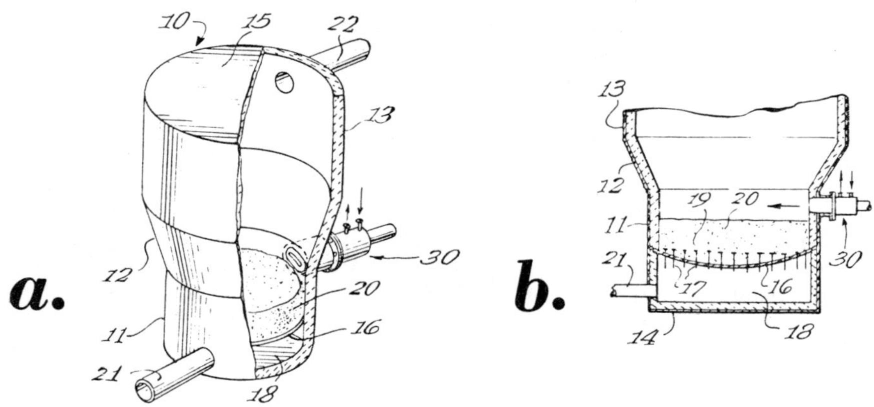

(a) Isometric View, Partially Broken Away, Showing a Reactor Provided with a Nozzle for Feeding Sludge in a Thin Layer to the Interior

(b) Vertical Sectional View Through the Lower Part of the Reactor of Figure 1.1a

(continued)

FIGURE 1.1: (continued)

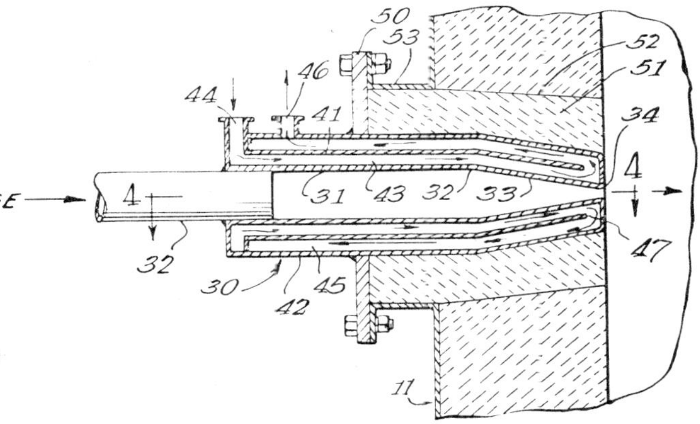

c.

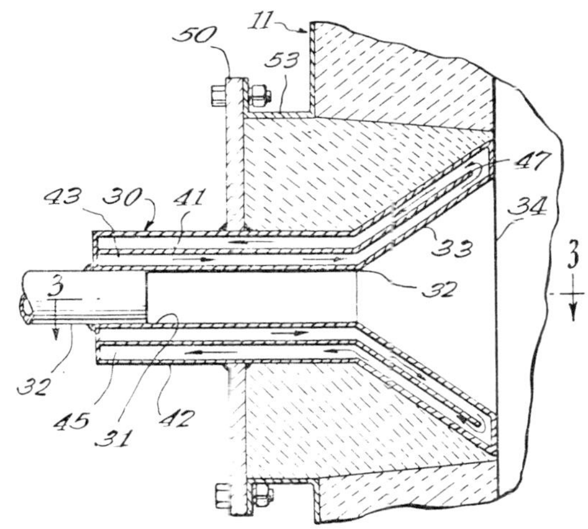

d.

(c) Vertical Sectional View Through the Nozzle and Adjacent Reactor Wall Taken on the Line **3–3** of Figure 1.1d

(d) Top Sectional View Taken Along the Line **4–4** of the Nozzle Shown in Figure 1.1c

Source: U.S. Patent 3,709,170

It is advisable, although not essential in all cases, to provide means to regulate the temperature around and in the nozzle. This can be accomplished by suitable jacket means provided adjacent to the nozzle through which a suitable fluid can be circulated for temperature control of the nozzle. The nozzle embodiment as shown in the drawings, with reference to Figures 1.1c and 1.1d in particular, is provided with an integral jacket. Interior wall **41** is spaced away from the wall-defining tubular portion **31** of the nozzle and the wall of transition element **33** to provide a path **43** for inward flow of cooling fluid supplied to the jacket through inlet opening **44**.

Jacket interior dividing wall **41** is surrounded by spaced apart exterior wall **42** thereby defining a path **45** for return flow of cooling fluid to exit opening **46**. The forward end **47** of the inner wall terminates just short of the end of the nozzle to provide a pathway for the cooling fluid to flow from the path **43** to path **45**. Water is the preferred coolant and in the case of sewage sludge serves to prevent the sludge from drying into a hard cake before it is expelled into the reactor. In addition, the coolant prevents the nozzle from becoming damaged through contact with high temperatures in the reactor.

To facilitate insertion of the nozzle in the reactor and to aid its removal for servicing and repair, an opening is provided in the reactor wall to receive the nozzle and suitable attaching means. Thus, as shown in Figures 1.1c and 1.1d, a plate **50** is positioned normal to and joined to the outer wall of the cooling jacket. Insulation material is positioned around the inner end of the nozzle and is suitably secured thereto and to the plate.

The inner end of insulation **52** is suitably tapered so that when it is placed in the reactor opening it will fit tightly in contact with the opening surface in the reactor wall. A cooperating flange **53** is provided around the reactor opening for bolting plate **50** thereto so that the entire nozzle and auxiliary supporting structure can be securely fixed in a removable arrangement in and to the reactor.

It is considered advisable to dimension the nozzle mouth to have a width at least five times, and better at least ten times, greater than the mouth height to shape the sludge into a suitably thin layer. More specifically, it is usually appropriate to eject the sludge as a layer of no more than 2" thickness, and advisably of no more than 1" thickness. For optimum results it presently appears advisable to have the layer no more than ½" thick and about 12" to 15" wide.

A specific nozzle for feeding 1,700 lb/hr of wet sewage sludge of about 28 to 33% solids in water to a fluidized bed reactor can have a transition element circular inlet of 3" diameter and a rectangular mouth ½" high and 14" wide. The sludge layer ejected from such a nozzle is not self-supporting and breaks up as it leaves the nozzle mouth. The scrubbing action of the fluidized bed will break the sludge pieces into smaller particles which will completely burn.

Pressure-Cooked Sludge as Feed

A process described by *R.P. Logan and O.E. Albertson; U.S. Patent 3,580,193; May 25, 1971; assigned to Dorr-Oliver Inc.* relates to disposing by combustion of sewage or other waste sludges, by first heat-treating the sludge at an elevated temperature and pressure for an extended period of time as described in British Patents 1,129,362 and 1,129,361. This treatment results in a sludge which may be mechanically dewatered to a cake having a much higher solids content and which is much more friable and easily handled than conventionally thickened and mechanically dewatered sludge.

By combining the pretreatment of sludge with means by which the sludge is fed continuously to dewatering means such as a solid bowl centrifuge or a rotary filter and with continuous feed to a fluid bed combustion chamber an extremely compact and efficient thermal disposal system results. This system requires minimum space, no preheating of combustion air, and requires no auxiliary fuel during normal operation.

The product of the sludge pressure-cooked as set forth in British Patent 1,129,361 after mechanical dewatering is unlike other dewatered sludges which are plastic and cohesive. This pressure-cooked sludge after mechanical dewatering, is of a dry friable nature, substantially free from odor, which readily crumbles into small pieces which are especially suited to extremely rapid and substantially complete combustion in a fluidized bed of inert granular material. Such material may despite its high solids content be fed by a screw feed, and due to its crumbly nature may be also broken up into small pieces and blown into the combustion chamber.

Referring to Figure 1.2a, shown below, the thickened sludge is fed from a thickener **11** through conduit **12** and pump **13** which feeds the sludge under pressure to a heat exchanger **14**. Since a doubling of the solids content of the sludge without further significant change doubles the capacity of the sludge pressure-cooking system, it is important that a highly thickened sludge be obtained such as supplied by the thickening process described in U.S. Patent 2,850,449 in which a solids content from 5 to 10% or over is secured.

FIGURE 1.2: PRESSURE-COOKING OF SLUDGE PRIOR TO COMBUSTION

a.

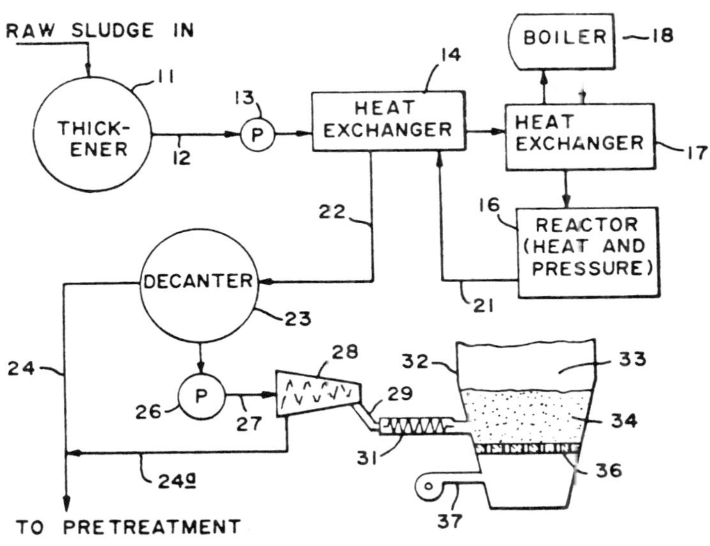

Process Schematic

(continued)

FIGURE 1.2: (continued)

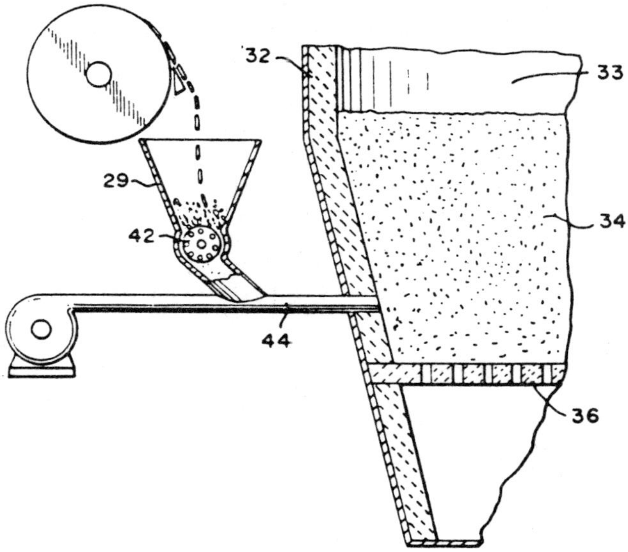

b.

Feed and Fluidized Bed Chamber

Source: U.S. Patent 3,580,193

This process is characterized by controlling the operation of a clarifier having an upper body of liquor for effluent and a lower bed of sludge in which (a) the ratio of sludge volume in the bed compared with the sludge withdrawn in a day is in the range between 0 and 3, (b) the depth of the bed is maintained between 1 and 5 feet, (c) the dry weight of solids fed to the clarifier is between 4 and 40 lb/ft^2 of sludge bed area, (d) the rate of feed is between 250 and 1,500 gal/ft^2 of sludge bed area and (e) the dilution of the feed to the clarifier is between 1,000 and 6,000 parts of solids per million parts of liquid.

The heat exchanger **14** heats the thickened incoming sludge by heat exchange from the hot outgoing sludge coming from the reactor or pressure cooker **16**. The sludge from the heat exchanger is raised in temperature in booster heat exchanger **17**, heated by boiler **18** to between 300° and 450°F, preferably in the range of 360° to 400°F, for a period between 20 and 40 minutes, preferably 30 minutes at a corresponding pressure of 15 to 17 atm, sufficient to prevent evaporation.

From the reactor the sludge passes through conduit **21** to the heat exchanger again where much of its heat is given up to the incoming sludge and the temperature is brought down to between 80° and 90°F. From the heat exchanger, through conduit **22** the sludge goes to a continuous settling and decanting tank and sludge reservoir **23**, from the upper part of which relatively clear liquid freed by the pressure-cooking from the biological cells of the sewage sludge is decanted as overflow and passes to pretreatment or clarification through conduit **24**. The underflow sludge from the bottom of the decanting tank **23** is pumped by pump **26** through conduit **27** to a mechanical dewatering unit **28** preferably a solid long bowl centrifuge, though a conventional rotary drum filter may also be used. The pressure-cooking system process and apparatus employed is of the continuous feed type described in British Patents 1,129,362 and 1,129,361.

The cake which issues from the combined pressure-cooking and dewatering unit is, from most sewage sludges, of an entirely different physical character from the cake resulting from treatment at room temperature and pressure. The sludge cake resulting from mechanically dewatering most thickened sewage sludge without heat treating results in a cohesive, plastic material having the consistency of thick semidry mud and a rather low solids content particularly as to secondary or mixed secondary and primary sludges, on the order of 15 to 25%.

There has been much difficulty in feeding this cake, and after feeding it, getting it distributed evenly in a fluid bed or other combustion chamber, particularly in larger installations, resulting in expensive and elaborate combined paddle feeds, pumps and screw feeds and a multiplication of feed points into the combustion chamber.

This pressure-cooked sewage sludge, in contrast, after mechanical dewatering as by a solid bowl centrifuge, has a solids content of 35 to 50% or more, readily permitting autogenous combustion, without preheating of combustion air as hitherto common when burning biologically treated sludges. Preheating is troublesome and expensive because of corrosion problems and thermal expansion, and the requirement of a constriction or support plate for the fluidized bed, consisting of refractory material because of the high wind box temperatures resulting from preheated combustion air.

But the most striking characteristic of the cake resulting from heat-treating most sewage sludges in accordance with the process is the friable crumbly nature of the cake which renders it easily broken into small pieces, somewhat granular in structure, which greatly aids feeding, and permits a choice of feeding methods such as milling and air conveying or blowing to the combustion chamber or screw feeding, and which promotes distribution and even burning in the combustion chamber and is particularly advantageous when fed to a fluidized bed combustion system because of its rapid and substantially uniform dispersal in the fluidized bed.

Also, because of the greatly increased solids content, approximately double that of untreated sludge, much smaller combustion chambers are needed, calling for fewer feed points and reducing capital costs of the thermal reactor. This pressure-cooked sludge has proved in tests to produce particularly favorable results as to high solids content and high recovery of solids when dewatered by a solid long bowl centrifuge such as the trademarked MercoBowl.

The results of a series of MercoBowl centrifuge tests on sewage sludge pressure-cooked as above-described, conducted on a variety of primary and secondary sludges proved that:

(1) The centrifuge will dewater all types of primary plus secondary sewage sludges to concentrations in excess of 40% total solids at

high recoveries.

(2) The capacity of the centrifuge is doubled by heat treating the sludge. This doubling capacity of the centrifuge is partly due to the higher concentration of settled sludge achieved by heating, and by the elimination of slimes.

(3) Chemical cost, if used to achieve high recoveries at high feed rates, would be only 1/6 that currently employed.

(4) Solids concentration achieved from the pressure-cooking and centrifuging of raw activated sludge will be sufficient to sustain autogenous conditions in the reactor, that is, will exceed 35% total solids on activated sludge and 40 to 50% total solids on mixtures of primary and activated sludge.

As shown in Figure 1.2a, the mechanically dewatered sludge is continuously fed from the centrifuge **28** into a conduit **29** from which a screw feed **31** delivers the cake through the sidewall **32** of the combustion chamber **33** directly into the fluidized bed **34** of incandescent inert granular material such as sand resting on an air-permeable constriction plate **36**.

A blower **37** delivers combustion air at room temperature in a cold wind box below the constriction plate at a space rate sufficient to fluidize the bed. Due to the highly mobile nature of the fluidized sand and its grinding effect on the readily crumbled material particles, distribution and burning are very rapid and substantially complete in the bed. Because the ash is finely divided and passes out through an exhaust outlet with the combustion gases to be removed by a cyclone gas cleaner, there is no ash removal problem.

However, the nature of this pressure-cooked sludge is such that it makes practical other methods of feeding. In Figure 1.2b, the dewatered sludge issuing from the filter to hopper **29** is milled by cage mill **42** into small pieces and delivered into the path of a stream of air from a blower in conduit **44** passing through the wall **32** of the combustion chamber to flow the sludge into the chamber.

K. Ishii; U.S. Patent 3,559,596; February 2, 1971; assigned to Kurita Water Industries, Ltd., Japan describes a method of incinerating sludge which comprises the steps of subjecting watery organic sludge to pressure and heating concurrently and then jetting the thus-treated sludge into a combustion chamber so as to evaporate most of the water contained in the sludge instantaneously with the jetting and incinerating the solid matters taking place within the combustion chamber. The chamber is provided with a packed layer of refractories designed to catch the solid matters.

Cascade Addition of Sludge

R. Menigat; U.S. Patent 3,736,886; June 5, 1973; assigned to Metallgesellschaft AG, Germany has found that it is possible to vastly increase the efficiency of a fluidized bed furnace for the combustion of sludge, especially a furnace operating with an expanded fluidized bed, in which a distillation of sludge is permitted in cascade or free fall onto the bed and an afterburning of the volatile components of the sludge is performed above the inlet of the latter but within the fluidized bed chamber.

The clarifier sludge is supplied to a fluidized bed furnace in the form of a liquid, paste or solid (generally a semisolid or a flowable solid), in which a fluidized bed is maintained in a burning state with the aid of oxygen-containing fluidizing gases. At least a portion of the sludge (or other waste to be burnt) is introduced into the chamber at a point above the highest combustion zone of the expanded fluidized bed (i.e., above the flame at the top of the fluidized bed), the point at which the sludge is fed to the furnace being disposed below or at a sufficient spacing from the exhaust gas outlet, that within the fluidized bed chamber, there is effected an afterburning of volatile constituents of the sludge.

Surprisingly, the simple free fall or cascade of the sludge, or a portion thereof, to the expanded fluidized bed from above may result in a distillation of volatile constituents regardless of the original consistency of the sludge, these volatile constituents interacting with the gases rising from the fluidized bed to provide the afterburning mentioned earlier. The fact that afterburning is carried out completely within the fluidized bed chamber and between the inlet for the sludge and the gas outlet, provides maximum utilization of the heat value of the materials.

The fluidized bed furnace illustrated in Figure 1.3 comprises a refractory-lined, steel-encased shaft **1** provided with a grate **2** defining the bottom of a fluidized bed chamber **1a**. Above the grate there is maintained an expanded fluidized bed **3** of inert fluidized solids, e.g., quartz sand, at which combustion of volatile components of a sludge and burnable solids is carried out.

Below the grate, there is formed a windowbox **2a** into which an oxygen-containing fluidizing gas stream is introduced from a combustion chamber **5**. The latter may be fed with atmospheric air from a compressor or blower **4** through an inlet **5a** when only air is to be introduced or when air is to be supplied to the fluidized bed in combination with combustion gases. A valve **4a** controls the proportion of air supplied for this purpose. Another valve **4b** controls the air supplied to a burner **6** opening into the combustion chamber and supplied with fuel at a line **6a**. The fuel introduced here may be gas or oil.

It will be apparent that heat can be supplied to the fluidized bed by sustaining combustion in chamber **5** and introducing a mixture of hot combustion gases and air to the fluidized bed. At any rate the fluidizing gas must be rich in oxygen or must contain oxygen in a stoichiometric excess over any combustible materials introduced by the fluidizing gas. Another portion of the fluidizing air, controlled by a valve **4c** may be fed at **7a** through a heat exchanger **7** and returned at **7b** to the windowbox. The heat exchanger is located above the gas outlet **13** at the upper end of the combustion chamber **1a**.

It has been found to be advantageous, from time to time, to introduce unheated air or oxygen-enriched air to the combustion chamber above the fluidized bed or into the fluidized bed by lances represented at **8a, 8b** and **8c**, the latter lances serving to supply preheated air to the system. The gases distributed to these lances via supply lines **8** through valves are represented at **4d, 4e** and **4f**. Fuel may be introduced through the lances **8c** and **8d** as well. Spaced above the top **3a** of the fluidized bed, which is maintained approximately in line with the primary sludge inlet **9**, by a distance **11**, is the secondary sludge inlet **10**.

The inlet is located 6 meters above the grate and about 4.5 meters above the fluidized bed which may have a high ranging between 1 and 1.5 meters. The pipe **10**, moreover, should be located somewhat above (say 0.5 to 1 meter), the top of the combustion zone. Above this pipe, at least at its mouth **10a**, there is located the exhaust gas outlet **13** at a spacing of at least 3 meters and represented by the dimension **12**. Within this afterburning zone, the gas velocity is maintained at or below 3 m/sec. A temperature control **13a** at the outlet regulates the operating parameters of the burner and gas feed to maintain the exhaust gas temperature in the range of 800° to 900°C.

FIGURE 1.3: FLUID BED COMBUSTION PROCESS

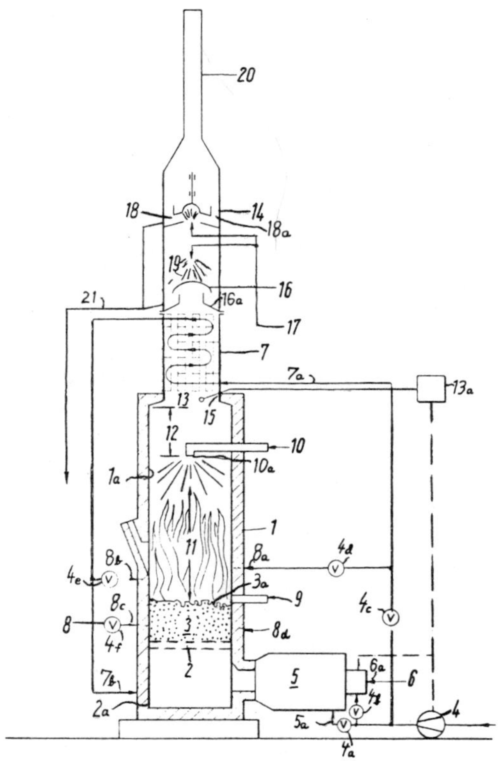

Source: U.S. Patent 3,736,886

The hot exhaust gases discharged through the outlet **13** pass at **15** into the heat exchanger **7** in which sensible heat is transferred to gases which are used to feed the bed. Thereafter, the exhaust gases traverse a scrubber **14**, e.g., a radial flow scrubber as shown in Figure 1.3, and then enter a chimney or stack **20** which is advantageously mounted directly upon the furnace. Water is sprayed into the scrubber as shown at **17, 18** and **19** and a shield is provided at **16** of the bell type. The shield cooperates with a cone **16a** and an annular gap **18a** is provided at the scrubber **18** for additional separation of impurities. The shield **16** and the gas **18a** are designed so that the scrubbing water is properly drained at **21** with the resulting seals acting as an explosion vent. The annular gap is adjusted in response to pressure differential to maintain a superatmospheric pressure within the combustion chamber and insure a constant scrubbing effect even when the combustion rate and the exhaust gas rate vary. The scrubbing of the exhaust gas effectively removes sulfur trioxide and hydrogen chloride where alkaline scrubbing water is employed.

As a specific example of the process, with the outlet **10a** located 3 meters below the gas outlet **15** but 6 meters above the grate **2**, with a fluidized bed of quartz sand maintained at a height of about 1.5 meters and a gap of 1 meter between the top of the combustion zone and the secondary waste inlet **10a**, the temperature maintained at the gas outlet **15** is about 800° to 900°C. The system is used to burn refinery sludge consisting of 95% by weight water and 5% oil and industrial solids.

Without a secondary combustion as illustrated, 300 to 500 kg of refinery sludge could be processed per hour per square meter of the grate area. Using a secondary flow of 30 to 60% of the total flow, it was possible to increase the combustion rate to 600 to 1,000 kg of sludge per hour per square meter of the grate area. The fluidizing air is heated to a temperature of 450° to 600°C and the temperature can be controlled to maintain a level of 800° to 900°C at outlet **13** by feeding fuel into the system at the lances **8c** and **8d**.

Secondary Bed Elutriation Arrestor Labyrinth Construction

R.G. Reese; U.S. Patents 3,882,798; May 13, 1975; and 3,818,846; June 25, 1974; both assigned to Combustion Power Co., Inc. describes a fluid bed disposal method and apparatus including a secondary bed/elutriation arrestor obstruction of labyrinth construction in the fluid bed reactor having lower and upper vertically spaced-apart perforate retention plates between which are positioned at least first and second vertically spaced-apart grates with the grate bars of the grates staggered with respect to one another. The lower surface of the bars of the lower perforate plate is described as stepped between at least two horizontal surface locations. At least two fluid bed/elutriation arrestor labyrinth obstructions spaced apart above the bed of granular material are used.

While the process is well suited for the pollution-free disposal of solid waste with the possible additional simultaneous disposal of various liquid materials, one configuration of the process is useful in a municipal solid waste disposal plant where the solid waste can be used as a fuel to dispose of high water content municipal sewage sludge from a given population segment. Accordingly, the process will be described with reference to such a system.

Referring to Figure 1.4a, the combustion system includes three principal assemblies, first the waste delivery assembly **10** where solid waste material is collected, shredded and separated for introduction of the principal consumable portion into a combustion assembly, second the combustion assembly **50** where the solid waste is consumed and where the solid waste can also serve as a fuel to dispose of sewage sludge and third the combustion gas particle cleanup assembly **90** where entrained particles are removed from the combustion gases which are then either exhausted or conveyed on to perform work.

FIGURE 1.4: FLUID BED REACTOR FOR LIQUID DISPOSAL

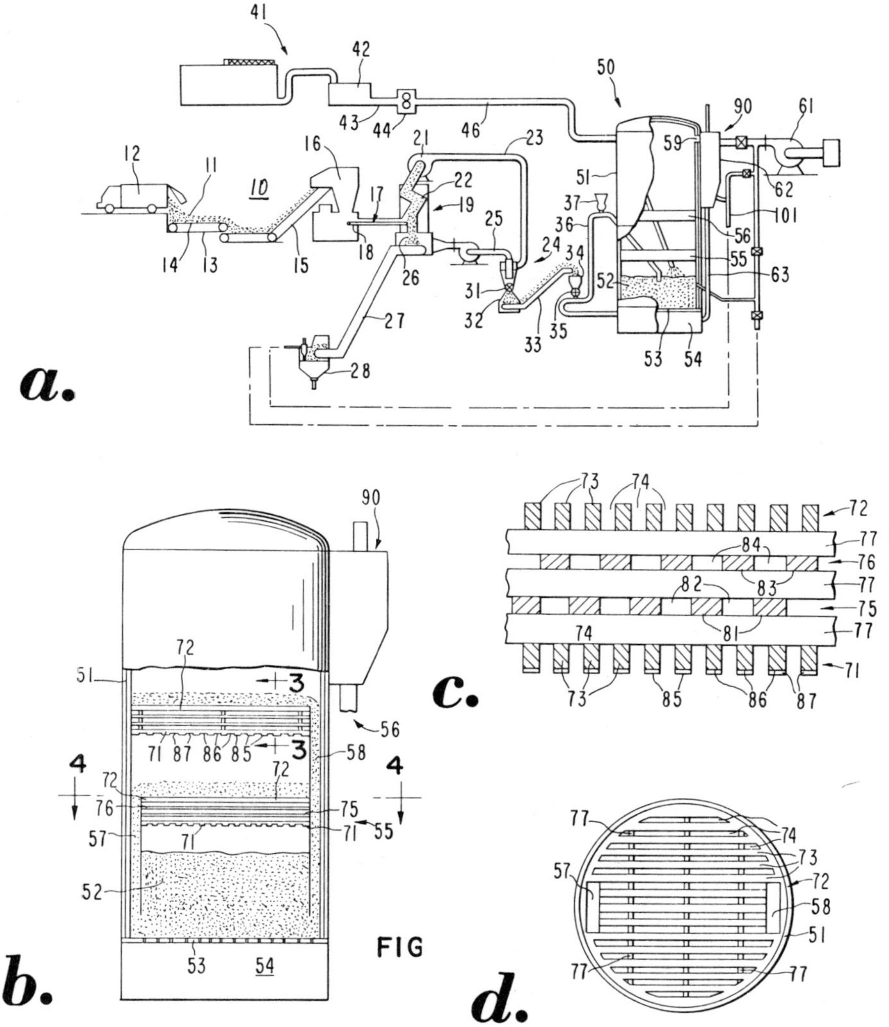

(a) Side-Elevational Schematic View Illustrating the Process

(b) Enlarged Side-Elevational View of a Combustion Housing Constructed in Accordance with the Process with the Waste Material Feed Deleted for Clarity

(c) Elevational, Sectional View of a Portion of the Structure Shown in Figure 1.4b, Taken Along Line **3–3** in the Direction of the Arrows

(d) Horizontal, Sectional View of a Portion of the Structure Shown in Figure 1.4b, Taken Along Line **4–4** in the Direction of the Arrows

Source: U.S. Patent 3,882,798

In the waste delivery system **10**, solid waste **11** is received, typically in refuse or garbage collection trucks **12** which deliver the waste in either compacted or uncompacted form into a receiving pit **13** sized to act as a storage area so that the system can operate continuously while receiving solid waste only over a specified period of the day. From the pit the waste is delivered first by a pit conveyor **14** and then a shredder feed conveyor **15** into the top of a shredder **16** such as a belt-driven vertical, rotor, shredder grinder. The waste is gradually reduced in size and ejected into an exhaust chute **17** where it is conveyed with the aid of a conveyor **18** to an air classifier **19**.

The effluent from the shredder is a homogenous mixture with particle size depending upon the shredder setting and the feed rate, varying from a very fine to 1" by 2" and larger. The steel and other ductile metal is in small balls with some coinage. Wood is toothpick to pencil size. The more flexible material, such as soft plastic and soft material such as rubber, are the largest particles.

High-density, noncombustible particles coming out of the shredder are removed from the process by the air classifier. This is done primarily to protect subsequent feed processes since the material removed has little heating value and hence would be little altered in volume or weight by the combustion process. Also, the air classifier provides a means for separating the reclaimable metal, glass, etc., for recycling to the appropriate industries for reuse if desired.

In the air classifier a large blower **21** is used to suck the refuse up through an inclined treacherous path **22** and out through a tube **23** to a velocity reducer/solid waste accumulator system in the form of a cyclone inertial separator **24**. The exhaust stack **25** of the inertial separator containing the major portion of the air used during the air classifying cycle returns this air to the lower inlet of the air classifier.

The size of the heavy particles that are dropped out by the air classifier can be regulated by adjustment of the velocity of air going through the system. The air classifier/solid waste feed subsystem is a completely enclosed system with the result that all odors, dust and potential pollution problems are completely contained. The heavy noncombustible reject material **26** from the air classifier is carried via a conveyor **27** to a residue storage bin **28** from which it can be disposed of periodically by landfill or ultimately by further reclamation of some or all of the materials contained in the residue.

The light particle fraction of the solid waste which has been carried from the air classifier via conduit **23** to the inertial separator drops preferably into a mulcher **31** which breaks apart any clods, chunks or large pieces that may have passed the shredder or formed after shredding. This mulcher and mulching process eliminate possible clogging in pneumatic feed portions of the rest of the system. From the mulcher the solid waste drops into a storage accumulator **32** which serves as a reservoir to smooth out the flow of solid waste. From the storage bin the solid waste is carried such as via a conveyor **33** into a hopper **34** for introduction into the entrance of an air-lock feed valve **35**.

This air-lock feed valve introduces the solid waste into a fluid conduit **36** where the material is pneumatically conveyed into the fluid bed. A bed additive feed system **37** is provided for automatically feeding, upon demand, fresh limestone or other bed additive chemical reactive materials to the fluidized bed during operation. These additives react with the corrosive or other pollutant gases or materials within the bed and reduce or prevent noxious fumes from escaping to the atmosphere.

Where the combustion system also serves to consume liquid waste from a sewage treatment plant **41**, sludge from sludge storage tanks **42** is removed via one or more pipes **43** to one or more sludge pumps **44** and via one or more pipes **45** to the combustion chamber where it is injected as a spray on top of the bed of granular material or injected above or below secondary beds in the combustion system.

It has been discovered that with the waste delivery assembly **10** the air-volume-reducing cyclone inertial separator **24** reduces the volume of air from the air classifier **19** to the air-lock feed valve **35** so that solid waste is fed through this valve with minimum turbulance and as dense as possible while the inertial separator and storage bin **32** act as a plenum or reservoir to smooth out the flow of solid waste.

In the combustion assembly **50** a hollow, cylindrical combustion chamber **51** is provided to contain a bed of granular material **52** which, when not being fluidized, is supported on a distributor plate **53** which provides even distribution of air across the bottom of the bed from a plenum chamber **54**. Positioned within the combustion housing **51** above the fluid bed **52** are a lower secondary bed/elutriation arrestor **55** and then upper secondary bed/elutriation arrestor **56** constructed in a manner described in greater detail below, each to provide an obstruction of labyrinth construction.

A return conduit or downcomer **57** is provided open at the top of the lower arrestor and extending through the lower arrestor to a position below the surface of the bed of granular material. A second return conduit or downcomer **58** is provided open at the top of the upper arrestor and extending through the second and first arrestors to a position below the surface of the bed of granular material.

Exhaust gases from the combustion chamber are conveyed out of a number of exhaust ports **59** below a cover over the top of this chamber and passed through the gas particle cleanup assembly typically incorporating inertial separators for separation and return of granular material carried out of the bed with the exhaust gases for later return to the bed. Combustion air is directed to the combustion chamber from a blower **61**, through a housing **62** surrounding the inertial separators for heat exchange to heat up the blower air and delivery via a conduit **63** to the plenum.

When burning or consuming more volatile liquid or solid waste materials which are difficult to oxidize or thermally break down in a fluid bed without the process, afterburning, unburned hydrocarbons, incomplete thermal breakdown of toxic materials, and the like can occur. The addition of multiple secondary bed/elutriation arrestors, each forming an obstruction of labyrinth construction reduces or eliminates these more difficult combustion problems.

The secondary bed/elutriation arrestors are mounted in the housing **51** in a horizontal plane or at a slight angle with the downcomers. Other secondary bed/elutriation arrestors may be added above arrestor **56** with a downcomer or return conduit penetrating to within a short distance of the distribution plate. It has been discovered that different configurations of the secondary bed/elutriation arrestors result in different performances in the disposal of various solid wastes.

One of the more difficult materials to burn in a fluid bed arrestor is aluminum-laden solid waste with the aluminum in the form of aluminum foil, flip-top can lids and the like. During combustion of this type of material, molten nodules of aluminum are formed and upon impact with one another will oxidize resulting in slagging, clinkering and ash buildup.

The addition of the secondary bed/elutriation arrestor provides a means of breaking apart these nodules as they flow through the labyrinth obstruction, depending upon the configuration of the obstruction. It has been found during tests that a labyrinth obstruction in the form of a plurality of large spheres such as of ceramic positioned between a pair of spaced-apart perforate plates results in aluminum oxide buildup and consequent unacceptable use when burning aluminum-laden solid waste.

The particular configuration of the preferred labyrinth obstruction is illustrated in Figures 1.4b through 1.4d. As illustrated there, the secondary bed/elutriation arrestor includes the lower and upper vertically spaced-apart perforate retention plates **71** and **72**. These plates include a series of parallel bars **73** spaced apart to provide spaces **74** between the bars for passage of the gases moving upwardly in the combustion chamber. Between the retention plates at least first and second vertically spaced-apart grates **75** and **76** are spaced apart from each other and from the retention plates by spacer bars **77** extending transversely thereof.

Each of the first and second grates include a plurality of spaced-apart grate bars, the bars **81** of the lower grate **75** being spaced apart by spaces **82** and the bars **83** of the upper grate **76**, being spaced apart by spaces **84**. Bars **81** and **83** of grates **75** and **76**, respectively are staggered with respect to one another so that grate bars **83** of upper grate **76** are positioned above spacings **82** between bars **81** of lower grate **75**.

Similarly, the bars **81** of lower grate **77** are positioned below the spaces **84** between the bars **83** of the upper grate. These spaced-apart bars have been found to effectively prevent the buildup of aluminum oxide in the obstruction of labyrinth construction so that strips of aluminum oxide which begin to form are broken up and eventually reduced to bed material size particles.

It has been found additionally advantageous to prevent the buildup of aluminum oxide strips on the lower surface of the lower perforate retention plate by stepping this lower surface between at least two horizontal surface locations such as shown in Figures 1.4b and 1.4c where the lower surface of retention plate **71** contains surface segments **85** in one horizontal plane and stepped from surfaces **86** in another horizontal plane by vertical steps **87** It has been discovered that any aluminum oxide that begins to build up on lower surfaces of a perforate retention plate of the latter configuration will flake off in small pieces and not lodge within the secondary bed/elutriation arrestor.

The fluidizing air is caused to flow through the bed material particles under carefully controlled conditions, chief among these conditions being the requirement that the air velocity through the bed, and hence the pressure drop, be greater than the value required to support the bed weight and less than the value required to sweep the particles out of the bed. In addition, the bed must consist of particles within a suitable range of size, shape and density. When these conditions are all satisfied, the stationary bed of particles will have expanded and the bed particles will exist in a fluidized state.

If the movement of one specific particle could be observed, it would be seen to undergo a continuous, turbulent motion and would wander throughout the bed in a random manner. Viewed as a whole, the dynamic condition of the fluidized bed resembles a tank of boiling water in the sense that there is considerable turbulence and bursting of bubbles at the surface. It is this dynamic characteristic which imparts to a fluidized bed its unique advantages.

The inert or chemically reactive bed materials used with the fluid bed combustor are chosen to withstand the combustion temperatures without melting or slagging. The specific bed materials selected can fall within a wide variety of inert particles, ranging from the inexpensive silica beach sand to the more exotic ceramic materials (usually alumina or silicon carbide) and to chemically reactive materials such as ground limestone, dolomite, and others. The bed material is preheated to a predetermined temperature and then fluidized for injection and combustion of the solid waste material.

The cross-sectional area of the downcomers **57** and **58**, the angle of the support plates **71** and **72**, the thickness, configuration and size of the openings in the support plates, the configuration of the labyrinth obstruction established by the grates **75** and **76**, and the height relationship of the secondary bed/elutriation arrestor assemblies **55** and **56** and their relationship to one another are all chosen in the particular application to provide and maintain a shallow bed of fluidizing material on top of the secondary bed/elutriation arrestors.

These shallow bed depths along with the labyrinth obstruction in the arrestors whose voids are filled with small particles during fluidization of the granular bed **52**, provide a very even flow of gases through the plurality of arrestors, and the action of the small particles filling the voids in the labyrinth obstructions allows each arrestor to prevent large bubble formations and the associated elutriation from the top surface of the primary bed and excess elutriation from the top surface of the lower arrestor **55**. Each arrestor acts as a separate turbulence generator providing additional and more thorough mixing of air and fuel and provides additional high heat transfer zones for complete oxidation, decomposition, destruction, etc., of the products being consumed in the system.

Distributor Plate Arrangement

C.S. Miller, Jr., H.K. Staffin and R. Staffin; U.S. Patent 3,772,999; November 20, 1973; assigned to AWT Systems Inc. describe an improved fluidized bed apparatus for combusting or reacting mixtures of combustible and noncombustible waste matter. The improvement resides in a distributor plate having sides sloping toward the bottom of the apparatus and having an inlet for solid feed in the side of the distributor plate. A pipe extends downwardly from the base of the distributor plate and functions to capture noncombustibles in the solid feed. A lock is provided which permits noncombustibles to be removed from the apparatus without interruption of operation.

P.R. Evans and D.H. Graham; U.S. Patent 3,776,150; December 4, 1973; assigned to AWT Systems, Inc. describe a fluidized bed system for pyrolysis or incineration of solid wastes. The solids feed is force-fed to a fluidized bed apparatus having a conically shaped distributor plate and a first internal chamber above the distributor plate and a smaller second internal chamber connected to the first chamber and positioned directly below the first chamber.

Solid waste is fed into the apparatus by a feeder which seals the feeder from the apparatus at a point spaced from hot-zone of the apparatus. The system has fluidizing gas inlets which are designed so that the fluidized bed in the first internal chamber is more highly fluidized than the fluidized bed in the second internal chamber. A lock in the second internal chamber permits removal of nonreacted materials without interruption in operation of the system.

Multistage Dryer Above Furnace

R. Menigat; U.S. Patent 3,772,998; November 20, 1973; assigned to Metallgesellschaft AG, Germany describes a sludge-burning method and apparatus in which a fluidized bed forms the combustion zone and is surmounted by a multistage drying and preheating zone such that the exhaust gases of the combustion zone rise directly through the multistage zone above the fluidized bed.

The multistage drying and preheating zone is in the form of a multiple-hearth furnace with annular supporting surfaces, each associated with one or more rotary brakes or agitators which continuously break up the layer on the respective stage and keep the sludge moving downwardly from stage to stage until a dried, preheated, comminuted solid material is distributed over the entire cross section of the fluidized bed. Hence it is essential that the fluidizing gases contain oxygen to effect a combustion in the bed, that the wastes are comminuted and transformed into a particulate or other subdivided form during drying and preheating, and that the comminuted matter is distributed over the grate area of the furnace.

In the multiple-hearth furnace the annular, vertically spaced, axially superposed hearths are cantilevered from the inner wall of the housing of the drying zone and define a central space in which a shaft is rotatable, this shaft carrying a plurality of radial arms at each stage, the arms having depending plate-like rake teeth for scraping the deposit on the annular platform overhung thereby. The teeth may be provided at various inclinations to the rotation circles defined by the arms, i.e., lie along tangents or chords to induce a movement of the comminuted waste spirally inwardly or outwardly depending upon the location of the outlet for each platform.

The platforms are alternately provided with outlets along the inner periphery or the outer periphery so that, for example, the comminuted drying material may be induced to flow inwardly at one annular platform, is caused to cascade downwardly onto an intercepting inner peripheral portion of the next lower platform, is induced to move outwardly along this latter platform by the associated rake or comminution arms, and is caused to cascade downwardly along the outer periphery of this platform onto still another platform underlying it.

In addition, the apparatus comprises a fluidized bed furnace which forms the combustion zone and is surrounded by the multiple-hearth furnace, while a transfer conduit is provided for recycling the waste gases of the multiple-hearth furnace into the fluidized bed furnace. The action of the rake teeth produces a comminuted material which is highly combustible and easily strewn across the grate of the fluidized bed, or, more specifically, across the full cross section of the fluidized bed furnace with uniform distribution.

Dryer and Heat Exchange System

H.K. Staffin and R. Staffin; U.S. Patent 3,779,181; December 18, 1973; assigned to AWT Systems, Inc. describe an improved heat exchange system which permits the economic disposal of a slurry of combustible waste materials by drying the combustible waste materials prior to reaction (i.e., incineration or pyrolysis), utilizing off-gases from the reactor in which the combustible materials are reacted to effect the drying.

The fluid bed dryer comprises a drying chamber and a heating jacket contiguous with the drying chamber. The drying chamber and the heating jacket are each adapted to contain a bed of fluidizable particulate heat transfer material positioned in their lower portions and both the drying chamber and the heating jacket have a fluid gas inlet and an exhaust gas outlet. The drying chamber also has a feed inlet and a feed outlet. Preferably, the feed outlet is vertically spaced from the feed inlet and a conveyor means extends from the feed outlet into the drying chamber for transferring dried solids from the drying chamber to the feed outlet.

The heat exchange system comprises in combination: (a) a fluid bed dryer such as described above, (b) means for conveying a slurry comprising liquid and combustible waste solids into the drying chamber of the dryer, (c) a reactor or incinerator having a feed inlet, a reaction section, and a vapor exhaust port, (d) conveyor means for transferring solids from the fluid bed dryer feed outlet to the feed inlet of the reactor, (e) conduit means connecting the vapor exhaust port of the reactor to the gas inlet of the heating jacket of the fluid bed dryer, and (f) means for providing fluidizing gas to both the reactor and the drying chamber of the fluid bed dryer.

The heat exchange system may include a conduit communicatively connecting the exhaust gas outlet of the drying chamber of the fluid bed dryer with the reactor. The reactor or incinerator is preferably a fluidized bed reactor. The heat exchange system is more fully described with reference to the following detailed description and Figure 1.5.

Referring to Figure 1.5a, a slurry feed **10** is held in storage bin **12**. The feed is transferred through bin conveyor **14** to the slurry transfer pump **16**. The slurry feed is pumped through line **18** into dryer manifold **20**. The slurry feed passes from the manifold directly into the base of drying chamber **22** of dryer **21**, illustrated in Figures 1.5b and 1.5e. The slurry feed is contacted with hot fluidized solids, such as sand, aluminum oxide, etc., in the drying chamber. Fluidizing gas, usually air, is forced into the base **24** of the drying chamber by blower **26**. As the slurry feed is heated in the drying chamber water and volatile constituents in the feed are vaporized and the solids in the feed are slowly lifted by the buoyancy forces and the flow of fluidizing gas to the discharge level **28** of the drying chamber.

The vaporized gas from the drying chamber may be discharged through exhaust opening **76** and line **102** into the incinerator **48**. Line **102** feeds this vaporized stream into the top of the incinerator only when the material being dried contains odorous volatiles. In the incinerator, the vaporized gas is heated to above 1200°F and thereby deodorized. If the material being dried and incinerated is not odorous, line **102** may simply go to the atmosphere.

A first solids screw conveyor **30** extends across the diameter of the dryer and passes through an opening **32** in the side wall **34** of the drying chamber **22** and through a conduit **36** in heating jacket **38** to the exterior of the heat-exchanger dryer. A drive means **40** is employed to impart rotation to the solids screw conveyor. The dried solids removed from the drying chamber are discharged into a solids receiver **42**. A second conveyor screw **44** extends from the solids receiver to the inlet **46** of reactor **48**. This reactor is principally operated as an incinerator.

The dryer is more completely described with reference to Figures 1.5b, 1.5c, 1.5d and 1.5e. As clearly shown in Figure 1.5b, the dryer is comprised of an inner vertical cylindrical drying chamber **22** which is surrounded throughout a substantial portion of its length by a cylindrical heating jacket **38**. At the base of the heating jacket is gas manifold **50** which distributes gas to the base of the heating jacket through inlets **52, 54, 56** and **58** as shown in Figure 1.5d. A gas distributor plate **121** may extend across the heating jacket at the base of the jacket and directly above the gas inlets. The distributor plate may be eliminated if desired; the use of a distributor plate provides more efficient fluidizing, but has a greater chance of plugging.

FIGURE 1.5: FLUID BED DRYER AND HEAT EXCHANGE SYSTEM

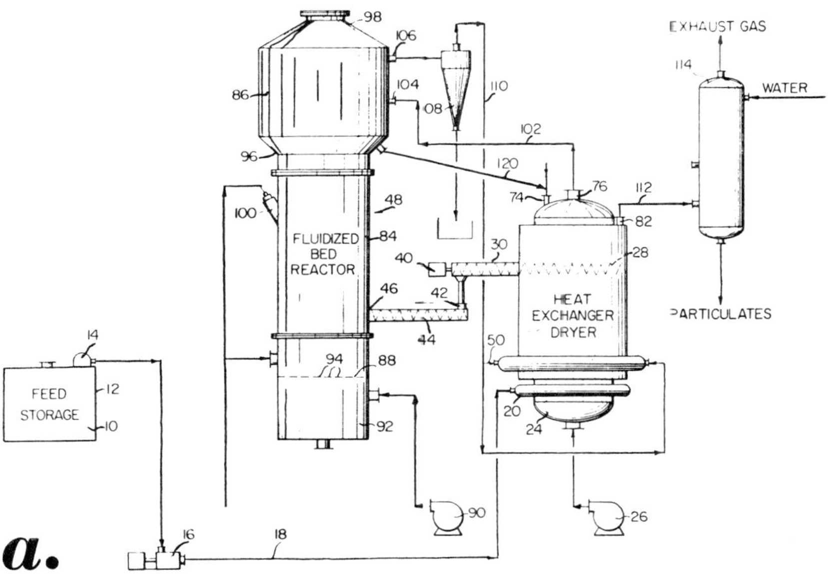

a.

Schematic Diagram Illustrating the Heat Exchange System in Which the Reactor Employed Is an Incinerator.

(continued)

FIGURE 1.5: (continued)

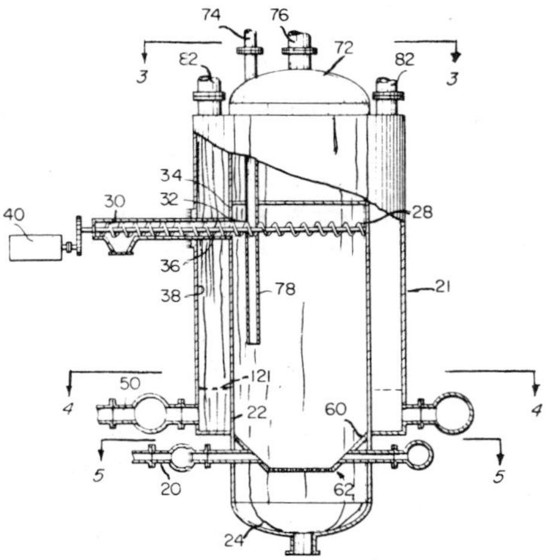

b.

c.

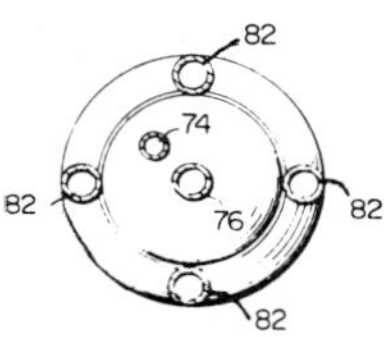

d.

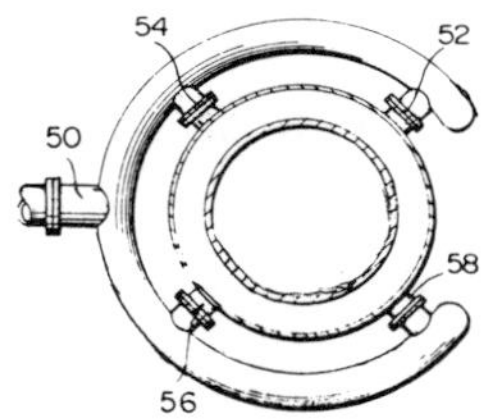

e.

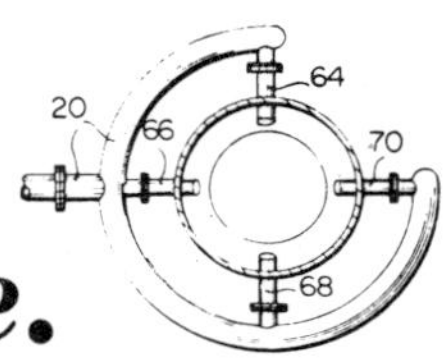

(b) Side View of the Fluidized Bed Heat Exchanger-Dryer Illustrated in Figure 1.5a, Partly Broken Away and Partly in Section

(c) View Taken Along Line **3–3** of Figure 1.5b Illustrating Various Outlets from the Heat Exchanger-Dryer

(d) View Taken Along Line **4–4** of Figure 1.5b Illustrating the Flue Gas Inlet to the Heating Jacket of the Heat Exchanger-Dryer

(e) View Taken Along Line **5–5** of Figure 1.5b Illustrating the Slurry Feed Inlet to the Heat Exchanger-Dryer

Source: U.S. Patent 3,779,181

The combustible waste slurry is fed to feed inlets in the sloping side walls **60** of dryer distributor plate **62** through feed manifold **20**. The slurry is distributed about the distributor plate through feed inlets **64, 66, 68** and **70** as shown in Figure 1.5e. Instead of the sloping side wall distributor plate shown in Figure 1.5b fed by a plurality of inlets, a flat distributor plate fed by one or more slurry inlet ports may be used.

The top of inner vertical cylindrical drying chamber **22** is capped with a convex head **72** which has a fluidized solids return inlet **74** and an exhaust port **76** communicating the drying chamber to the exterior of the dryer. The fluidized solids return inlet communicates with a conduit **120** (Figure 1.5a) which extends from the upper portion of the incinerator **48**. A down pipe **78** extends downwardly from the fluidized solids return inlet into drying chamber **22** to the area of the chamber below the discharge screw **30**. At the upper end of heating jacket **38** are a number of gas outlets **82**. The arrangement illustrating the various outlets and inlets on the upper-end of dryer **21** is further described in Figure 1.5c.

Referring again to Figure 1.5a, it is seen that a preferred incinerator is comprised of a vertical cylindrical incinerator body **84** and an incinerator head **86** extending directly above the incinerator body. The incinerator is adapted to be operated as a fluidized bed incinerator. The incinerator head has a larger diameter than incinerator body **84** to permit reduction in fluid gas velocity thereby providing a vapor disengagement section for the incinerator.

A gas distributor plate **88** extends across the entire cross-sectional area of the incinerator body near the bottom. The incinerator is equipped with an air blower **90** which pressurizes a plenum chamber **92** directly below the distributor plate. The pressurized air passes through a multiplicity of openings **94** in the distributor plate fluidizing the fluid bed medium employed. This air also provides the necessary oxygen for burning of the dried waste material.

The portion of the incinerator body extending from the distributor plate to the base **96** of the incinerator head is referred to as the incineration or reaction section. The portion of the incinerator extending from the base of the incinerator head to the top **98** of the head is referred to as the vapor disengaging section. Fluidizable solids may be returned to the drying chamber from the lower portion of the vapor disengaging section through conduit **120**.

In operation, dried combustible waste feed from the dryer is fed into the incinerator in which the feed is contacted with the hot fluidized bed particles and air. The solids in the feed are ignited by burner **100** and are burned. The gases of combustion pass upwardly into the vapor disengaging section of the incinerator. Vapors resulting from the drying of the slurry feed in the dryer may be also transferred through line **102** and gas inlet **104** into the incinerator head. The combined gaseous stream (off-gas) is withdrawn from the incinerator through gas outlet **106** and into a cyclone type separator **108**.

Uncombustible feed solids and a small amount of fluidized bed particles are separated from the gas in the separator. The separated off-gases are forced through line **110** to the inlet of the gas manifold **50** of dryer heating jacket **38**. These hot off-gases are employed to fluidize fluid bed heat transfer material supported within the heating jacket. The gases pass through the heating jacket and are then discharged through gas outlet **82** and line **112** to a water scrubber **114** in which particulates and water-soluble contaminants are removed. Substantially pollutant-free air is then exhausted to the atmosphere.

In the system described, the hot off-gases resulting from incineration of combustible waste solids are used to fluidize heat transfer material in the heating jacket of the dryer. The cross-sectional area of the heating jacket must be sized so that the incinerator off-gas is sufficient to form a well-agitated fluidized solids bed. Typically 0.5 to 1.5 ft/sec superficial velocity of gas in the jacket is suitable.

A high heat transfer coefficient results between the wall of the fluidized bed heating jacket and the wall of the fluidized bed drying chamber.

The system can result in a heat transfer coefficient of about 50 Btu/hr-ft^2-°F on each side of the drying chamber wall (assuming the fluidized bed in the dryer is well agitated). The heat transfer system also eliminates heat exchanger fouling because of the lower temperature at the heated walls of drying chamber, as a result of increased heat transfer, and reduced contact of the solids being dried with the heat exchange wall. Another advantage to the process is that odorous volatiles can be deodorized by afterburning in the vapor-disengaging section of the incinerator and made substantially odor-free before being discharged into the atmosphere.

The operating temperature of the fluidized bed dryer for drying waste feeds containing high water concentrations, i.e., up to about 90% water is 212°F. Air is provided to the fluid bed dryer from a blower in an amount to maintain a state of minimum fluidization (about 0.5 ft/sec superficial vapor velocity) in the bed of the dryer without the presence of waste feed. This fluidization level insures that the waste feed material can be pumped into the drying chamber without difficulty. Vaporization of water and volatile constituents in the waste feed provides additional fluidizing gas and results in a well-agitated fluidized solids bed (about 2 ft/sec superficial vapor velocity is desirable).

In actual tests performed on drying of a paunch manure feed containing about 82% by weight water, steam, air and other gases are discharged from the dryer exhaust and piped to the vapor disengagement section of an incinerator, where these gases are combined with the incinerator combustion gases which are at elevated temperature on the order of about 1500°F. The combined off-gas equilibrates to a temperature of about 1350°F. The off-gas enters the base of the heating jacket at about 1350°F and is discharged from the heating jacket at about 700°F. The water content of the slurry is reduced by about 60%. The cooled off-gas is piped to a water scrubbing system for removal of particulates and soluble contaminants and discharged to the atmosphere.

High-Melting Inert Particles

In a process described by *T. Shirai; U.S. Patent 3,411,465; November 19, 1968* moist waste material is incinerated in fluidized beds of higher-melting inert particles in an approximately cylindrical vessel where a slow stirring is applied to the fluidized bed, and if desired, dehydration of feedstock is effected in that vessel by passing combusted gas upward through a dehydrating bed charged with moist feed material while applying slow stirring, the gas being flowed upward at a sufficiently high rate relative to that of the particles to prevent downflow of water, thereby causing free water to become present at the upper end of the vessel, and withdrawing free water.

In Figure 1.6a on the following page, an example is shown where the apparatus of the process is applied to the incineration of fowl excretions. The apparatus comprises an upright cylinder. Air for combustion is supplied, at a temperature in the vicinity of normal temperature or after being preheated, from the bottom portion of the apparatus through an inlet **101** and through a perforated plate **102** to the fluidized bed **103** in which solid particles such as sand grains are fluidized. **104** represents the peripheral wall of the incineration furnace. **105** represents an ignition burner or fluid inlet which is used to heat the interior of the apparatus to the combustion temperature.

FIGURE 1.6: FLUIDIZED BED INCINERATOR

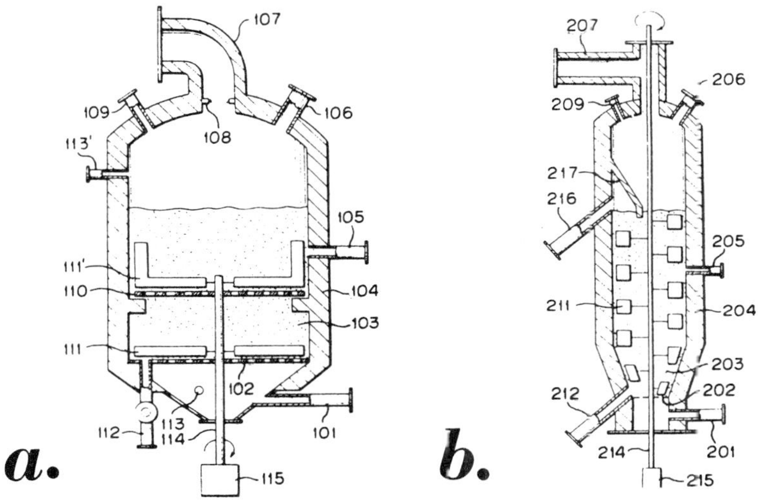

Incineration of Fowl Excretions   Incineration of Sludge from Pulp Factory

Source: U.S. Patent 3,411,465

The fowl excretions which are to be incinerated are fed through a supply inlet **106** into the apparatus. While the excretions descend through the fluidized bed, they are contacted by the high-temperature combusted gas and by the sand grains and thus they are heated and dried, eventually combusted to ashes.

The fluidized bed is held at a high combustion temperature by virtue of the combustion temperature which is produced during the foregoing process. In the event that there is want of calories, a combustion gas may be directly introduced into the bed through a gas inlet **105** by means of a burner or any other appropriate means. **107** represents a combusted gas exhaust pipe. **108** represents an afterburner for depriving the combusted gas of its odor. **109** represents a supply inlet of sand grains to the fluidized bed.

It is preferred that the fowl excretions which constitute the material to be incinerated are continuously fed into the apparatus in their dispersed state. Since it may occur, however, that the excretions are fed into the apparatus in a mass of considerable size dropping directly on the perforated plate **102** where they may accumulate to block the openings of the plate, there is provided obstructing means **110** in the form of a coarse grate while gentle stirring is performed in the area immediately above this obstructing means by stirring vanes **111'** to thereby disperse the masses and to avoid the occurrence of the aforesaid blocking of the openings of the perforated plate.

The masses of excretions which have passed through the spaces in the grate **110** downward and which have not dispersed into smaller clods are dried, destroyed and dispersed in the highly heated fluidized bed and are eventually combusted to ashes. The area immediately above the perforated plate **102** is also gently stirred to avoid adherence of the particles to the plate by the stirring vanes **111** which are similar to those described above, and these vanes are mounted to the periphery of the common rotary shaft **114**.

Depending on the type of the material to be incinerated, the combination of the grate and the stirring vanes **111'** or the meshes of the grate may be appropriately modified. Furthermore, the combination of the grate and the stirring vanes may be provided in plural sets, one set being located above the other. In some instances, one or both of them may be omitted as required. The grate, the perforated plate, and the stirring vanes are constantly polished by being contacted by the fluidized particles. Therefore, no blocking of the meshes of the grate after the particles of waste materials have dropped downward therethrough will occur.

Such materials as fowl excretions which are to be incinerated, often contain small shells and small pebbles which would heap up on the preforated plate. These obstacles are removed together with a part of sand grains through a discharge outlet **112**. This removal of pebbles or the like may be performed jointly through a withdrawing outlet provided at an appropriate portion of the apparatus such as the region where the grate is located. The sand grains which have been taken out from the apparatus are screened to remove the foreign elements, and the segregated sand grains may be mixed with fresh sand grains and supplied to the fluidized bed **103** through the supply inlet **109** of sand grains.

Numeral **115** represents a low-speed motor which is used to rotate the stirrer shaft **114** at low speed. The vanes **111** and **111'** which are fixed to the periphery of this shaft are preferably so angled as to work to take up particles and send them upward since such shape of the vanes will serve to effect smooth operation of the apparatus continuously for an extended period and to thereby save the power.

113 and **113'** represent holes for measuring the pressure. Besides these, a plurality of thermocouples for the measuring of temperature are inserted in the wall of the apparatus at appropriate sites thereof. Those fine powders and ashes which scatter outside the system from the outlet **107** are collected by a powder separator such as cyclone. For the sake of recovery of heat, a boiler or other appropriate means may be provided.

Figure 1.6b illustrates an example of the apparatus which is applied to the treatment of condensed sludge from the liquid-form waste discharged from a pulp factory. The reference numerals appearing on this figure whose last two digits are the same as the last two digits of the reference numerals appearing in Figure 1.6a identify corresponding parts. Such sludge contains small pieces of wood, barks, wood meal and fibrous materials or organic substances such as dissolved lignin or inorganic materials.

Inorganic materials are represented by sodium sulfite and sodium carbonate. This example shows how such inorganic materials are collected as solid particles. This is effected by feeding condensed sludge into an upright cylindrical apparatus in which solid particles of inorganic particles of appropriate size are fluidized and which has a conically shaped bottom structure, whereby perfect combustion of all that is combustible takes place.

In case the material to be treated is in the form of suspension in a liquid, it may be supplied in atomized state into the apparatus through a supply inlet located at the top, **206**, or it may be injected directly into the fluidized bed at the inlet **205** through the peripheral wall **204**. Coarse particles are drained from the draining outlet **212** located at the bottom of the apparatus, while relatively fine particles are drained from a draining outlet **216** provided near the surface of the fluidized bed.

In addition to these outlets, there may be provided a particle-draining outlet at an arbitrary level of the fluidized bed. A part of the sludge introduced into the apparatus is immediately dried and combusted and reduced to fine particles of inorganic substances alone, which will either stay in the bed by being involved in the fluidized bed or will be discharged together with the combusted gas outside the system from the outlet **207** and will be collected by cyclone so as to be returned to the bed again.

These fine particles of inorganic substances will act as cores and collect the sludge onto their surfaces where the deposited sludge is dried and combusted. Thus, the particles of inorganic substances will gradually grow into larger particles and they are discharged locally outside the system from the particle-draining outlets.

By arranging so that a small air current flows near the outlets against the direction in which the particles are discharged, this air current will serve to cool and at the same time to classify the particles, so that particles of inorganic substances having relatively uniform size will be collected easily. **217** is a partition wall to prevent by-passing of particles around the bed.

At a temperature lower than the melting point of particles, sintering and fusing of particles are completely avoided by the stirring of the fluidized bed. Therefore, by employing as high a temperature as possible and by thus holding the bed in the state of reducing atmosphere it is possible to collect the inorganic substances in the form of reclaimed chemical particles for being used again. In some cases however, consideration and care may have to be given to hold the structural materials of the apparatus below the permissible temperature by passing water or the like to cool the stirrer shaft and the inside portions of the vanes.

Fluidized beds are usually of an average density of 1.0 g/cc or more, so that even when wood pieces or the like, are introduced into the apparatus they are dried and combusted while they are floating on the surface of the bed. Thus, wood pieces or like materials will never sink deep into the bed. If small pebbles or nails, which have a larger specific gravity, are mingled into the apparatus, they will be taken up by the current produced by the vanes rotating at low speed immediately above the perforated plate **202** located in the bottom portion, and thus they are transferred to the outlet **212** wherefrom they are discharged together with coarse particles.

Example: Incineration of Fowl Excretions — Fowl excretions usually contain 80% of water and also pieces of shells, small pebbles and sand grains. It is generally quite difficult to incinerate them smoothly when they are in such a state.

The excretions are incinerated in the apparatus of the type shown in Figure 1.6a. The apparatus comprises a cylinder made of mild steel, with the inner peripheral wall face being lined with refractory cement. The inner diameter of the cylinder is 340 mm, and the height is 600 mm.

In the bottom portion is disposed an air-distribution plate which is comprised of a perforated plate having an appropriate ratio of opening (in this instance, the ratio is about 1%). Through this perforated plate, air is passed at the rate of 30 to 50 m^3/hr STP so that the sand grains are fluidized to the level which is about the middle of the height of the apparatus. The velocity of starting fluidization (minimum fluidization velocity, U_{mf}) of the sand grains used is 14.3 cm/sec at 28°C, and this velocity decreases with the increase in the temperature of the fluidized bed.

Into a bed of sand grains already fluidized by air at normal temperature was fed hot combustion gas of LPG directly from the side of the bed to heat the entire apparatus. After a predetermined temperature was reached, fowl excretions containing 70 to 85% of water were continuously introduced from the upper portion of the apparatus for incineration. Checked temperature of the fluidized bed was t_B = 370° to 850°C. Fowl excretions are fed at the average rate of 40 kg/hr. LPG was supplied at the rate of 0.35 to 0.7 m^3/hr STP. The excess air rate was 20 to 40%.

Stirring vanes were disposed immediately above the upper face alone of the perforated plate, and were 40 mm wide, 320 mm long and 2 mm thick, and were angled at 45° in the direction in which the particles were taken up and hurled. In this instance, the omission of the grate **110** and the stirring vanes **111′** in Figure 1.6a caused no disorder of operation. The motor used in this example was of the capacity of 5 to 28 rpm on 75 watts.

However, only several watts or less were sufficient for running the main shaft of the apparatus. When the apparatus was operated by suspending the stirring, sintered blocks of sand grains containing small pebbles and pieces of shells were always noted. These sintered foreign materials grew into a mass of 2 to 5 cm in thickness and accumulated on the upper face of the perforated plate in several hours after the operation was started.

It required a considerable amount of labor to dismantle the apparatus and to remove the mass from the perforated plate. When the stirrer was rotated even at a low speed, the materials did not become sintered on the surfaces of the sand grains into larger masses, provided that the vanes were running. Operation was conducted intermittently for over 2 months without replacing or substituting the sand. Thereafter, the sand was drained and screened. It was found that there were sintered sand particles of several millimeters in size mingled with small pebbles and pieces of shells.

These were measured to be only 0.5% of the total volume of the sand. These small masses were considered to be produced when the sand grains were exposed directly to the flame and heat from the LPG burner. The formation of these masses are considered to be reduced further by improving the nozzle portion of the burner. Masses of the aforesaid size were however, fluidized well together with sand grains and did not settle on the plate and, therefore, they practically did not interfere with normal operation. Most of the ashes were carried outside the fluidized bed with the combusted gas and were collected by the cyclone-separator. Hardly any ashes were noted to remain in the apparatus. In case the temperature of the combusted gas was 700°C or above, no odor was sensed.

Because air was supplied at normal temperature from the bottom of the apparatus, the temperature of the perforated plate **102**, the stirring vanes **111** and the stirrer shaft **114** was considerably lower than that of the combustion zone **103**. Ordinary mild steel materials with which the perforated plate and the stirrer are constructed stood the use of over several consecutive months, and still they were suitable for further use.

The inner wall faces of the apparatus containing the fluidized bed, the surfaces of the perforated plate and the vanes were clean and no disorder was noted. Only the wall faces in the upper portion and those portions other than the fluidized bed were slightly covered with soot and with ashes.

P. Foch; U.S. Patent 3,515,381; June 2, 1970; assigned to Charbonnages de France, France describes a method for burning sludge by introducing the sludge in dispersed form from the top of the combustion chamber into a fluidized layer of inert particles traversed by a constant quantity of oxidizing fluidizing agent.

The temperature of the fluidized layer of inert particles is maintained at constant value to insure the combustion of the product to be treated and the temperature of the zone above the fluidized layer is maintained at a constant value less than the sintering temperature of the ash formed by regulating the speed of introduction of the dispersed sludge. A temperature-controlling fluid is injected into the system over the whole section of the chamber immediately under the fluidized layer.

Inert Fluidized Bed

A process described by *O.E. Albertson and W.M.H. Kilmer; U.S. Patent 3,319,587; May 16, 1967; assigned to Dorr-Oliver Incorporated* is concerned with direct combustion of organic waste material having a high moisture content, under specially controlled conditions. The waste matter is fed continuously into a body of hot inert fluidized granular material having heat-storing and heat-radiating capabilities, e.g., sand.

The fluidized or teeter state of this granular material is maintained in the combustion chamber by the combustion air being forced upward therethrough. Thus, the air pressure should be sufficient to keep the sand particles in teeter which in turn depends upon the inventory of inert particles to be thus maintained in depth. The devices for introducing the wet waste material into the combustion chamber should provide an adequate seal against that pressure.

Under the foregoing conditions, and with certain controls of the combustion process itself, direct and complete combustion of the wet waste is attainable with the result that the stack gases discharging from the combustion chamber comprise only inert and innocuous combustion gases along with water vapor, carrying with them the residual ash and inert fines.

To this end, the controls of this combustion process should maintain a high enough combustion temperature in the body of fluidized material, while maintaining a low excess of combustion air. However, optimum efficiency of such a combustion operation further depends upon achieving the combustion of the organic waste within the confines of the body of hot inert fluidized material, which presents the problem of minimizing or eliminating the afterburning of waste material and/or of its distillation gases that might occur in the freeboard space above the body of the fluidized granular material.

According to the process, under the above conditions afterburning is minimized by extruding the high moisture waste material tightly compacted into the lower zone of the body of hot inert fluidized material maintained in the combustion chamber.

Thus, the waste material is delivered from an extrusion device in the form of plugs or chunks compacted to a density approximately equal to the density of or greater than the surrounding body of fluidized material. Gradually these chunks become more and more fragmented, finally disintegrated, and destroyed by combustion, with evaporation of the water taking place concurrently within the confines of the body of inert fluidized material.

In related work *O.E. Albertson and W.E. Budd; U.S. Patent 3,319,586; May 16, 1967; assigned to Dorr-Oliver Incorporated* describe a process which involves effecting the combustion of the moist raw sludge in a bed of inert hot granular material maintained in a state of fluidization which is the formation of a dense and turbulent suspension of particles in an upflowing steam of treatment gas termed a "turbulent layer" or fluidized bed.

Combustion thus conducted under the conditions of the process will reduce the organic solids in the sludge to stable gaseous products, with a large portion of the heat liberated by the combustion of the sludge fuel used to flash off the moisture in the sludge while also heating the combustion gases to the bed temperatures ensuring complete odorless combustion.

There is hereby provided a method where a concentrated sludge, for example, in the range of 25 to 35% solids concentration, is fed directly into a combustion chamber where, in the presence of the suspended inert particles of hot inert auxiliary material, a complete and odorless combustion is maintainable and readily controllable so that the combustion gases will be kept odor-free. To that end, this process provides a combustion chamber wherein a quantity of inert granular material having high heat storage, as well as high heat-radiating capability, for example, graded sand, is maintained in a heated fluidized state.

Screen Deliquefier

A process described by *O.E. Albertson and W.M.H. Kilmer; U.S. Patent 3,375,794; April 2, 1968; assigned to Dorr-Oliver Incorporated* relates to an internal screen deliquefier for pressure conduits carrying moist fibrous material, and is particularly useful with screw or progressive cavity feeding of dewatered waste organic material into an incandescent fluidized inert material bed reactor as described above in U.S. Patent 3,319,587.

Fluidized Solids Burner

D.L. Campbell; U.S. Patent 3,306,236; February 28, 1967; assigned to Esso Re-

search and Engineering Company describes an improved burner for waste materials, i.e., for wastes which may contain gases, solids and/or liquids. The burner operates on the fluidized-solids principle. Fine particles of refractory material are used as heat carriers. A multiple bed arrangement is employed with one dense fluid solid bed preferably arranged above another. Elastic fluids, which may be either gases or vapors or both, are used to keep the dense beds fluidized.

In Figure 1.7 below, there is shown in elevation a single burner structure or vessel which involves a lower burning zone **11** and an upper burning zone **13**. Both of these zones are fluidized by passing gases, preferably air, upward through them. A suitable blower for the air, driven by a motor not shown, is indicated at **15**. It supplies compressed air through a line **17**. A substantial part of this air flows upward into a bottom zone **19** of the lower burner and into the fluid solids bed through a perforated grid **21**.

FIGURE 1.7: FLUIDIZED SOLIDS BURNER

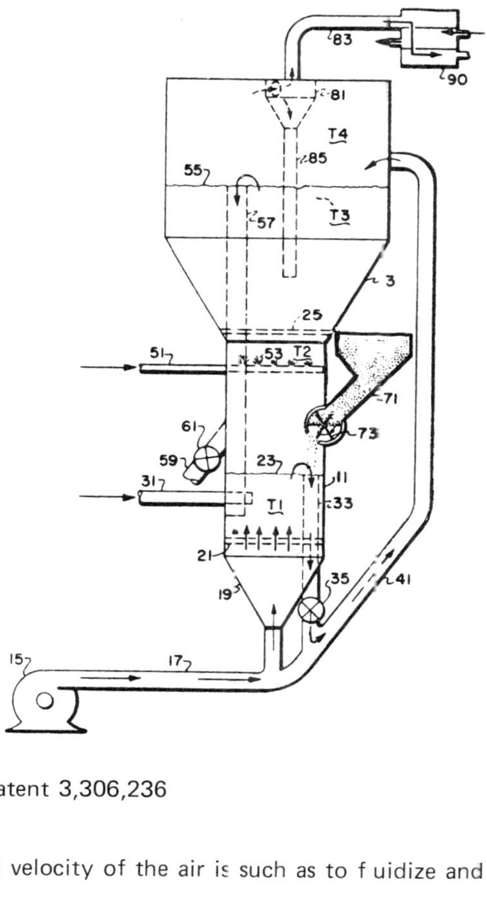

Source: U.S. Patent 3,306,236

The volume and velocity of the air is such as to fluidize and maintain fluidity in the solids in the lower burner in a well-known manner. Gases passing through or generated in the bed continue to rise from the bed through the intersurface **23**. The latter defines essentially the top of the dense phase fluid solids in the lower vessel. The gases, and vapors which may be included, pass on up into the bottom of the upper burner vessel **13** and through a perforate grid **25**, on which another bed of fluidizable solids is supported. The action of the rising gases of course fluidizes the latter solids in the same general manner as in the lower bed.

Fluid materials to be burned, which may comprise gases, or liquids, or solid matter suspended in gas or liquid, etc., will be introduced to the lower zone and preferably directed into the lower bed through a line **31**. As the vapors and liquids are consumed or vaporized, etc., residual solids such as ash, or carbon in some cases, will be deposited upon the more refractory solids, which comprise the major part of the original solids in the lower vessel.

As the total volume of these deposited and original solids increases, the excess will flow over a downcomer or standpipe **33**. This standpipe may be partly under control of a valve **35**. By suitable adjustment of this valve and especially by suitable aeration to maintain fluidity in the line **33** above it, which will usually be desirable, the flow rate of solids down the standpipe can be controlled.

A portion of the air stream from blower **15** also passes into branch line **41**. The descending solids from the interface pass down through the standpipe and are picked up by the gas or air passing upward and to the right in line **41**. By this means, the solids are carried up into the top of the upper vessel **13**. They fall into and augment the upper dense bed and add their heat to it. A fuel, which may be either liquid or gas, or a mixture, is introduced into the gas space above the interface in the lower vessel through a line **51**. This line conveys the fuel to a burner **53** where it is consumed, adding substantial heat to the rising gases from the interface.

The latter, plus the products of fuel combustion, pass upward through the grid **25** into the upper burning zone. By this means the temperature of the solids in the upper vessel or compartment may be made substantially higher than that of the solids in the lower vessel or compartment. These solids form a dense phase having an interface as indicated at **55**. As they build up, they overflow through the downcomer **57**, thus being returned to the lower bed. A branch withdrawal line **59** is provided in the downcomer, a valve **61** being installed therein to control the withdrawal of solids from the system. The withdrawal rate will, of course, depend on the amount of ash and other incombustibles coming into the system from the waste products being burned.

Solids, either waste products or replenishment refractory particles, may be introduced into the lower bed from a hopper **71**. This is under control of a star wheel or lock hopper arrangement **73**, as will be understood by those skilled in the art. Several such devices are known and used by which the solids may be added without letting gases escape from the vessels.

These solids may be waste matter which itself contains the needed refractory materials to replenish those which are removed by attrition or by mixture with ash which is withdrawn, or they may consist merely of solid waste products to be consumed in the burner. In the latter case, it may be necessary to add small amounts of refractories from time to time or even continuously, to keep the bed composition of desired consistency.

The vessel may be constructed as a single unit with walls of suitable material, such as steel, lined with refractory brick or cement, etc. Alternatively, separate vessels may be superimposed. The temperature T_1 in the lower bed is sufficiently high, after the system gets into operation, to insure gasification in some way, e.g., at least vaporization or partial cracking, etc., of all the waste products except their ash or perhaps their carbon content.

For example, a good working temperature is about 1000°F. It may be less, but preferably will not be below 800°F nor more than 1200°F in most cases. At these temperatures in the lower fluidized bed, any waste material supplied thereto, whether in liquid or solid form, will either be vaporized, or cracked or burned, at least to some extent.

Generally speaking, ash will remain as the chief solid component and it is deposited on or mixed with the refractory particles constituting the original bed. Heat for the lower fluidized bed is supplied primarily by the stream of solids flowing down the downcomer **57** from the upper level. Gases passing upward from the interface **23** of the lower bed ordinarily will not be completely consumed or oxidized at this point. They will contain highly undesirable atmospheric pollutants of at least a nuisance type. They may even be dangerous in some situations, if they were released in such condition.

They may contain odorous compounds and/or incomplete combustion products such as carbon monoxide or hydrogen, H_2S, etc. The additional fuel injected through line **51** is intended to burn these materials completely. It is designed also to heat the rising gases in the vapor zone to a temperature T_2 substantially higher than T_1, about 2000°F. At this temperature all the gaseous and vapor products passing to the zone above are essentially completely converted to their final oxidation products, such as CO_2, H_2O, SO_2 or SO_3. The hot gases including the complete combustion products then pass through the upper bed.

They impart some but not all of their heat. Therefore, the upper bed will have a temperature T_3 substantially higher than that of the lower bed, though less than T_2. Finally, the gases emerging from the upper bed have a temperature T_4 at which they are passed through the solids separating device such as a cyclone **81** with its outlet line **83**. Separated solids return to the upper bed through downspout **85** in the conventional manner. The gases passing out through line **83** are completely innocuous, or substantially so. They preferably go through a heat exchanger **90** where they are cooled, their heat content being used to generate steam or for such other useful purposes as may be desired.

The arrangement, in essence, is such that the gases rising from a lower zone through the upper fluidized bed give up a certain amount of heat, which heat equals, of course, the sensible heat of the stream of solids circulating back to the lower bed. The solids which rise through line **41**, leaving the lower bed at a temperature T_1, for example, of 1000°F will reach the higher temperature T_3 in the upper vessel. The difference is essentially the heat applied to vaporization, combustion, cracking, etc., of the waste products fed into the lower bed through line **31**.

The primary point of control of the system is the temperature of the lower bed. It is controlled eventually of course, by firing supplementary fuel above the bed in suitable quantity, the heat being imparted first and mainly to the solids in the bed above, which reach the temperature T_3. In many cases the supplementary fuel added through line **51** need be only very small in quantity. When there is enough fuel value in the waste products being consumed to substantially meet all heat requirements, fuel will not be required at all. Generally speaking however, with waste products such as sewage which contain a great deal of water, often around 90%, fuel for line **51** will be required. The fuel feed rate through this line is generally controlled in accordance with the water content of the waste products being consumed.

The rate of air flow from the pump **15** is generally substantially constant and is sufficient to insure good fluidization and proper mixing in the two beds as well as solids circulation between them. The rate of refractory solids cycling between the two beds also is preferably maintained substantially constant.

Surplus solid materials, if and as they accumulate, may be withdrawn through line **59**. It may be necessary, under some conditions, to supplement or occasionally replenish the solids which constitute the fluidized beds. For this purpose refractory solids of suitable type such as granules of metal oxides, metallic particles, ceramic particles, etc., and mixtures thereof may be added to the hopper **71** and its feed wheel **73**.

Since the temperature in the lower bed is generally rather moderate, the grid or grate **21** at the bottom usually can be made of relatively inexpensive materials. Preferably an alloy is used which is reasonably resistant to oxidation at the operating temperatures. In cases where the waste solids include pieces of metal or other solids too large to fluidize, these may accumulate on the grate and may require periodic removal therefrom.

This may be done either through a suitable cleanout or by shutting down the system and removing the grid, etc. Generally speaking, the upper bed will not require a grate in the sense of a support for massive solids. Normally, the first dense fluid solids bed is arranged beneath the second and is at lower temperature, the temperature between the two beds being higher than either. In some cases the beds need not be superimposed in this manner. The second bed supplies heat to the first by solids circulation.

Production of Active Carbon

R.D. Nickerson and H.C. Messman; U.S. Patent 3,887,461; June 3, 1975; assigned to Combustion Engineering, Inc. describe a process of removing solids from sewage, subjecting these solids to pyrolysis, thereby producing active carbon and char, and using both of these materials as adsorbent agents in the treatment of the sewage or other wastewater.

Referring to Figure 1.8, **10** is a chamber for incoming raw sewage, where floating material is screened out, and heavy inert solids are settled out, and additives may be added to the sewage. The sewage then progresses to the settling tank **12**, where some of the additives along with the absorbed matter and other solids are allowed to settle out of the sewage water to form a bottom sludge. The sewage water moves on to a subsequent treatment tank **14**, where it can be treated with further additives, and further settling occurs.

FIGURE 1.8: PROCESS FLOW SHEET FOR ACTIVE CARBON PRODUCTION

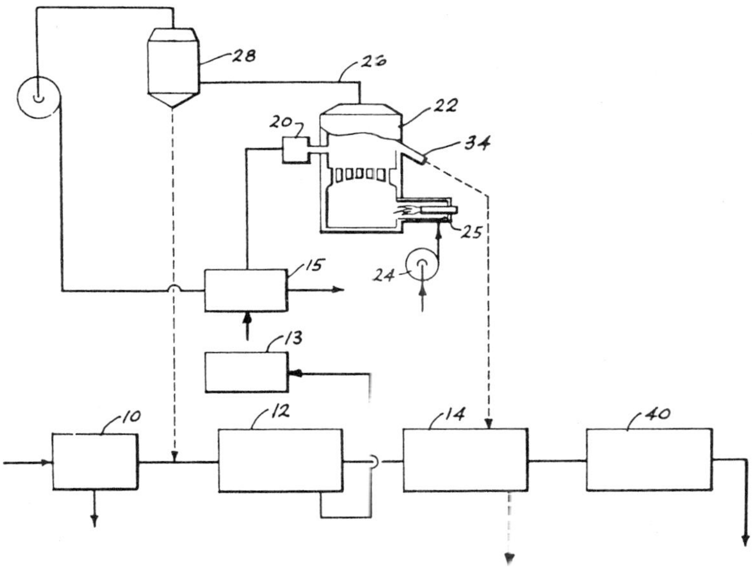

Source: U.S. Patent 3,887,461

From tank **12**, the sludge flows to mechanical dewatering device **13**, where it is substantially dewatered by vacuum filtration or other means. Filtrate water from **13**, taken together with weir overflow from the settling tank **12**, is commonly said to have had primary treatment and, as such, is often discharged from a sewage treatment plant, even though it is still highly polluted, except where a substantial amount of active carbon has been used in the primary system.

The solids which are filtered out of the sewage in the form of cake, are conveyed to a dryer **15**, where they are at least partially dried by hot gases. A screw feeder **20** introduces the sludge cake into a fluidized bed reactor **22**, where it is flash-heated and carbonized and where it simultaneously disintegrates into small particles, most of which will pass through 16 mesh (U.S. Standard Sieve) and some of which will pass through 325 mesh.

Hot flue gas is supplied to the reactor by means of blower **24** and the velocity of the gas flowing through the reactor is sufficient to maintain the carbonizing sludge particles in a fluidized state. Any suitable fuel can be combusted in burner **25**. The temperature within the reactor is held within the range of 500° to 1000°C, and the gas contains little free oxygen, so that the sludge is subjected to pyrolysis.

The only oxygen contained in the combustion gases is the excess air supplied to the burner in order to maintain optimum combustion within this burner (optimum combustion being accomplished with a slight excess of air). Thus the amount of free oxygen contained in the combustion gases flowing through the reactor would be within the range of 0.1 to 2% by volume. Surprisingly, the sludge particles, after having their residual moisture flashed off and most of their pyrolitic volatile products driven off, are in the form of active carbon and char, which does not require further conventional activation (selective oxidation) over extended periods, typically one to several hours, at elevated temperature, to exhibit useful adsorptive capacity.

The gases discharged from the fluidized bed through duct **26** pass through a wet scrubber **28**, (or other dry collector) where the fine carbon particles carried along in the gas stream are separated out. The separated carbon particles are relatively pure active carbon, typically having (BET) surface area of 300 m^2/g or more. The carbon slurry from the scrubber is recycled to the sewage system and introduced therein, to adsorb impurities from the sewage and to markedly improve the quality of primary effluent. The activated carbon can be introduced downstream of tank **10**, as shown, or can alternatively be introduced upstream of, or directly into the tank, and still be effective.

An overflow line **34** is used for continuously removing the heavier char solids that have not gone into dispersed phase above it, from the fluidizing bed **22**. These solids are not recycled to the primary system because they would cause an undue ash buildup in it. Nevertheless, this high ash char has some useful adsorbent capacity, typically 100 m^2/g. It may therefore be employed to finish primary effluent, or as an aid in secondary or tertiary treatment, in various manners readily apparent to those skilled in the art of treating sewage and other wastewater.

Particularly where lime has been a primary additive, this high ash active char has useful capacity for chemisorbing phosphates by what appears to be an ion-exchange mechanism. As shown, the char is introduced into the secondary treatment tank **14**. This char is allowed to settle out of this tank, and is then disposed of. It is not conveyed to the fluidized bed for recycling, because of its high ash content.

From the above, it can be seen that through the use of a fluidized bed, a separation of the fines from the heavier char and ash particles is accomplished. The fine active carbon particles, with a density of 0.3 to 0.6 g/cm^3, are entrained in the gases flowing through the reactor **22**, whereas the char and ash particles, which are of higher density, are removed through overflow line **34**.

The amount of fines entrained in the gases leaving reactor **22** will vary, depending on the original makeup of the sewage. Typically, the fines might comprise 10 to 35% of the dry sludge on an ash-free basis. For example, if a yield from a sludge is 25% on the above basis and the sludge assays 30% ash, actual yield of gas entrained fines is 0.25 (1.00 to 0.30) or 17.5% of the gross dry sludge.

The above applies only to new, first-cycle carbon being made from sludge. Also, it should be noted that although the majority of the airborne fines are active carbon, there will be some percentage, typically 10 to 25%, ash content. The ash-forming ingredients contribute to adsorptive capacity and, to some extent, constitute a portion of all commercial carbons, the percentage previously mentioned being applicable.

The effluent from subsequent treatment tank **14** is next subjected to chlorine treatment in chlorination tank **40**, prior to its being discharged from the sewage treatment plant. For the initial start-up of the fluidized bed, fluid coker petroleum coke or coal coke breeze may be used as a bed material. After desired bed temperature and dense-phase fluidization are established, sludge cake may be fed to the bed. The rate of gas flow, and the rate of sludge feed should thereafter be adjusted to obtain the optimum yield of powdered active carbon. These rates will vary, depending on the moisture and the fixed carbon content of the sludge cake being fed to the fluidized bed, as well as on other factors.

It should be noted that the adsorbent capacity of the powdered active carbon increases with recycling. While powdered active carbon loss may amount to 5 to 30% per cycle, there will be a net gain of active carbon in the primary system with every cycle, until the active carbon loading in the primary system exceeds sludge filtration capacity. Thus primary effluent of high quality can be achieved. The abovementioned system can be utilized on any municipal or other sewage system. It will also have application for any wastewater cleanup operation, so long as the wastewater contains combustibles.

Organic Nitrogen Compounds

K. Kishigami, H. Kobayashi, T. Sente and K. Sugiyama; U.S. Patent 3,888,194; June 10, 1975; assigned to Babcock-Hitachi KK, Japan describe a method for incinerating industrial wastage containing organic nitrogen compounds, by using a fluidized bed furnace, in which industrial wastage together with materials containing a great amount of carbon, such as pulverized coal, heavy oil or the like is charged in a furnace to thereby fluidize the wastage and the carbon-abundant material together with fluidizing medium with air.

The industrial wastage is then burnt, while the interior of the fluidized bed (or fluidized layer) is maintained in a reductive atmosphere, whereby the nitrogen oxides (NO_x) produced from the organic nitrogen compounds contained in the wastage are reduced during the combustion of the wastage, to thereby obtain nitrogen gas (N_2) thus preventing the production of nitrogen oxides.

As is known, the fluidized bed is provided by blowing air or gas into a furnace from its bottom and fluidizing the fluidizing medium in the form of solid particle layer which is supported on a perforated plate within a furnace. In general, silica-base particles (primary constituent SiO_2) such as river sand, which is chemically inactive, are used as a fluidizing medium of a solid particle layer, the particle size thereof being in the range from 0.1 to 2.0 mm. In case sulfur and chlorine are contained in the wastage and desired to be directly fixed and absorbed within the fluidized bed, it is preferable that chemically active cement clinker particles be used as a fluidizing medium. The composition of cement clinker particles is, for instance, 60 to 70% CaO, 6 to 8% Fe_2O_3 and 3 to 6% SiO_2.

The fluidized bed incineration furnace utilizes a fluidized bed for incineration. The temperature of the fluidized bed is normally in the range from 750° to 850°C. In this respect, such a temperature is extremely lower than the temperature (over 1000°C), at which nitrogen oxides are formed by the reaction of nitrogen with oxygen contained in air, as is the cases with industrial furnaces of a general type, such as a boiler. Furthermore, at such a low temperature, the quantity of nitrogen oxides produced according to the reaction of nitrogen with oxygen in air is negligible.

What is of consequence in this respect is the production of nitrogen oxides due to the combustion of organic nitrogen compounds contained in the wastage, because as has been described earlier, nitrogen oxides may be produced in a great quantity even at a temperature as low as 750° to 850°C.

Vigorous reaction may be achieved in the fluidized bed furnace, because the fluidizing solid particle layer consists of particles of a size from 0.1 to 2.0 mm and thus presents extremely large contact surface area at the time when the particle layer stands still (e.g., the height of the layer is 500 to 1,000 mm). Accordingly, even the wastage such as those having poor combustibility, like a dehydrated mudcake (water content, 70 to 90%) may be burnt at a very low excessive air ratio (1.1 to 1.3). In case a dehydrated mudcake having a high water content and containing materials having a high carbon content is burnt in the fluidized bed, the following reaction formulas are considered to take place, because the combustible solid matter in the mud consists essentially of carbon, hydrogen, sulfur and nitrogen:

$$C + O_2 \longrightarrow CO_2$$

$$4H + O_2 \longrightarrow 2H_2O$$

$$S + O_2 \longrightarrow SO_2$$

$$2N + O_2 \longrightarrow 2NO$$

However, in practice, the wastage particles are maldistributed microscopically to some extent, despite that muddy wastage has been dried, crushed and placed within the fluidized bed. Furthermore, the wastage particles are subjected to combustion at a very small excessive air ratio, such that a reductive atmosphere short of oxygen encircles around the small particles of crushed wastage.

One example of the fluidized bed muddy wastage incineration apparatus is given in detail with reference to Figure 1.9. An incineration furnace **1** includes a perforated plate **2** and an air chamber **3** in the lower portion thereof, while the dehydrated mud, which contains pulverized coal or fine granular coal, is supplied from the top portion of the incineration furnace therein from a mud supply device **4**.

The air which is required for fluidizing and combustion is supplied through a blower **8**, damper **15** and duct **9** to the air chamber, thereby fluidizing by passing through the perforated plate the fluidizing medium particles such as cement clinker and the like as well as the mud which contains pulverized coal and fine granular coal, the particles and mud being positioned on the perforated plate, and thus the fluidized bed is provided.

The operation of the incineration furnace begins with heating the fluidized bed by using a burner **6a**, after which the incineration of the wastage proceeds due to the combustion of the pulverized coal (or fine granular coal) mixed with the muddy wastage. If the heat accruing from the combustion of the pulverized coal is insufficient, a burner is used as an auxiliary means. The waste gas is discharged through a duct **7** out of the furnace. When the increase in the quantity of fluidizing gas is required for forming a fluidized bed **5**, the waste gas is introduced from duct **7**, to thereby supply the waste gas through a damper **14** and blower **16** into duct **9**.

FIGURE 1.9: FLUIDIZED BED INCINERATION FURNACE

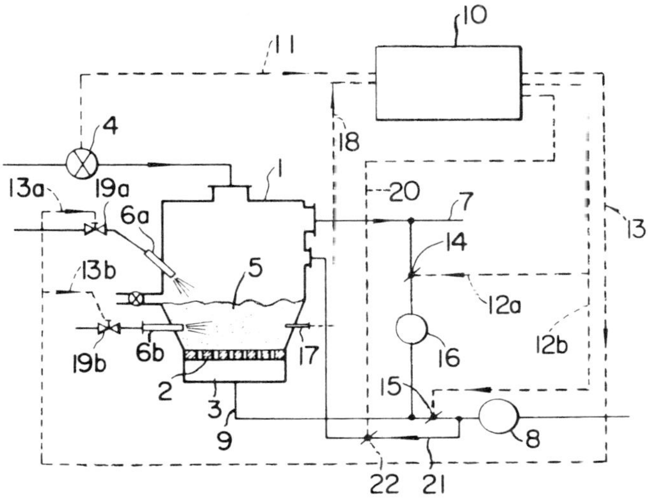

Source: U.S. Patent 3,888,194

For operation of the incineration furnace **1**, it is required to control the quantity of air and waste gas to be mixed. To this end, a signal **11** which represents the supplied amount of muddy wastage and is obtained in connection to rpm of a rotary feeder of the muddy wastage supply device **4**, and a temperature signal **18** from a thermometer **1** adapted to measure temperatures at the fluidized bed **5**, are fed into a control box **10**, whereby inlet dampers **15** and **14** of the blowers **8** and **16** are each controlled by means of signals **12a** and **12b**, while the control valves **19a** and **19b** for controlling the amount of oil for burners **6a** and **6b** are controlled by means of signals **13a** and **13b**.

The burner **6a** is primarily used for starting a furnace or for an auxiliary purpose, while the burner **6b** is used for supplying combustible materials abundant in carbon content, such as heavy oil, to the fluidized bed and is normally used in case the muddy wastage to be charged in the furnace **1** contains no pulverized coal or the like. On the other hand, a damper **22** provided in the secondary air supply duct **21** is controlled by means of a signal **20**.

The features of the process reside in that, for incineration of combustible industrial wastage containing organic nitrogen, particularly muddy industrial wastage of a high water content, there is used a fluidizing bed which permits the incineration at an extremely low excessive air ratio (1.1 to 1.3); carbon components of no less than 20% by weight, based on the weight of the wastage, are mixed with the solid matter such as muddy wastage to thereby form a reductive atmosphere within the fluidized bed, while the temperature at the fluidized bed is maintained to be not less than 500°C to incinerate the wastage; nitrogen oxides produced are reduced by CO, H_2, CH_4 and the like which are thus produced, whereby nitrogen oxides may be removed from the waste combustion gas from the wastage in the form of nitrogen gas.

According to another aspect of the process, there are used, as fluidizing medium, cement clinker particles containing a great amount of Fe_2O_3, Cr_2O_3, MnO, Al_2O_3, chrominum ore particles and manganese ore particles for the purpose of accelerating the reductive reaction, thereby effectively reducing and removing nitrogen oxides.

The reactions which take place within the fluidized bed incineration furnace according to the process are summarized as follows. The following reactions take place for the combustible matter and the added carbon components, within the reductive fluidized bed which is maintained at a temperature of not less than 500°C.

$$2C + O_2 \longrightarrow 2CO$$
$$C + H_2O \longrightarrow CO + H_2$$
$$C_nH_m \longrightarrow aCH_4 + bC_{n'}H_{m'}$$

where C_nH_m represents hydrocarbon groups contained in the combustible matters, part of which is subjected to dry distillation and decomposition within the fluidized bed under a reductive atmosphere to thereby produce methane (CH_4) and the like.

Carbon monoxide, hydrogen and methane which are produced from the reactions maintain the fluidized bed in a reductive atmosphere, whereby nitrogen oxides produced simultaneously with the production of the gases are reduced according to the following reactions:

$$2NO + CO \longrightarrow N_2O + CO_2$$
$$N_2O + CO \longrightarrow N_2 + CO_2$$
$$2NO + 2H_2 \longrightarrow N_2 + 2H_2O$$
$$CH_4 + 4NO_2 \longrightarrow 4NO + CO_2 + 2H_2O$$
$$CH_4 + 4NO \longrightarrow 2N_2 + CO_2 + 2H_2O$$

The quantity of CO produced within the fluidized bed according to the above reactions is larger than that of nitrogen oxide. Excessive CO, H_2, $C_{n'}H_{m'}$, CH_4 and the like are oxidized by means of the secondary air in a space over the upper surface of the fluidized bed and are converted into CO_2 and H_2O, harmless gases.

When incinerating sewage mud having a water content of 78%, solid matter of 22% and nitrogen content in solid matter of 5.5%, by using the above fluidized bed incineration furnace, the concentrations of nitrogen oxides (NO_x) in the waste gas are as follows:

> when pulverized coal is added the concentration of nitrogen oxides in the waste gas is 65 ppm and
>
> when pulverized coal is not added the concentration of nitrogen oxides in the waste gas is 1,230 ppm.

INCINERATION AND FURNACE DESIGN

Multihearth Furnace

J.C. Eck; U.S. Patent 3,777,680; December 11, 1973; assigned to Wilputte Corp. describes a furnace of the multihearth type, comprising a lower combustion zone having a plurality of vertically spaced annular hearths and an upper drying zone having a plurality of vertically spaced annular hearths. The central area of the drying zone hearths is open and provides a passage for rubbish and garbage to fall directly from a hopper at the top of the furnace upon the uppermost hearth in the combustion zone. Rabbling apparatus moves the rubbish and garbage progressively from hearth to hearth and then in ash form to a removal chute.

The sewage sludge is deposited on the uppermost of the drying zone hearths and moved progressively downward by rabbling apparatus in a zigzag path from hearth to hearth and then falls in a dry state to the uppermost hearth of the combustion zone, where it is incinerated along with the garbage and rubbish, and then removed in ash form via the removal chute. A forced air system for cooling the rabbling apparatus is also provided.

Referring to Figure 1.10, the multihearth furnace **10** shown comprises a structure or tower of essentially circular form. The walls of the furnace consist of an outer protective layer **11** of sheet metal and an inner layer **12** of ceramic or refractory material. Omitted, for simplicity, are the structural support members by which the walls, roof and floor of the furnace **10** are supported above the surface of the ground.

The furnace is divided vertically into a combustion zone in the lower part thereof and a drying zone in the upper part thereof. Structurally supported within the combustion zone are a plurality of superposed hearths **13, 14** and **15**, suitable oil or gas burners **16** extending into the combustion chambers above each hearth. Supported above each hearth by a central rotary shaft **17** are a plurality of radial rabbling arms **18, 19** and **20** respectively, the rabbles of which serve, by rotation of the rabble arms to progressively move the refuse material on the hearth toward the outer or inner periphery of the hearth in characteristic manner.

As shown, hearths **13, 14** and **15** have central openings **22, 23** and **24** through which the rotary shaft **17** extends. Hearth **13** also has openings **25** at the outer periphery thereof through which the refuse material falls to the hearth **14** immediately below. The central opening **23** of hearth **14** is substantially larger in diameter than the rotary shaft thereby providing passage via which the refuse material falls to the hearth **15** immediately below. At the outer periphery of hearth **15** is opening **26**, communicating with a chute **27** through which the ashes remaining from the combustion of the refuse material are deposited in a vehicle or on a conveyor for removal.

The furnace walls in the combustion zone are provided with a plurality of openings **28** having damper doors **29** which multiply serve to inspect the operation, replace or adjust rabbles or rabble arms and to adjust the amount of air admitted to the combustion zone in consequence of suction created by a blower (not shown) connected to the exhaust duct or pipe **30** opening out of the top or roof of the furnace. The arrow adjacent the duct **30** indicates the direction of exhaust flow of the combustion gases.

FIGURE 1.10: MULTIHEARTH FURNACE FOR COMBINED INCINERATION OF RUBBISH, GARBAGE AND SEWAGE SLUDGE

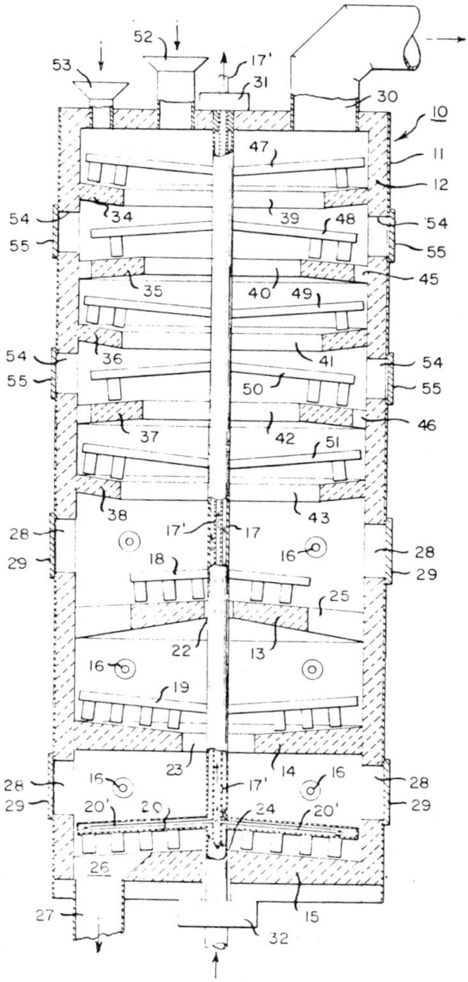

Source: U.S. Patent 3,777,680

Rotary shaft **17** is rotationally supported by suitable bearing units **31** and **32** in the roof and floor structure of the furnace **10**. The shaft is rotated by a suitable motor through a gear or chain drive in the manner exemplified in the prior art (U.S. Patents 2,108,997; 2,121,661; and U.S. Patent Reissue 20,046). Also, the shaft, which is hollow, communicates at its lower end with a blower which provides flow of cooling air through the shaft for cooling the rabbling arms **18, 19** and **20** and other rabbling arms, as later to be described. Structurally supported in superposed relation within the upper or drying zone are a plurality of hearths **34, 35, 36, 37** and **38**. These hearths have central openings **39, 40, 41, 42** and **43** respectively, corresponding substantially in diameter and area to that of the uppermost hearth **13** in the combustion zone.

The hearths **35** and **37** are provided at the outer periphery with a plurality of openings **45** and **46**, respectively, through which refuse material may drop to the hearth below. Associated respectively with each of the hearths **34, 35, 36, 37** and **38** are radially extending rabble arms **47, 48, 49, 50** and **51** attached to and rotated by the shaft **17**.

A cooling arrangement is provided for the rabbling arms **18, 19** and **20** in the combustion zone and for the rabbling arms **47** to **51** of the drying zone. Each of the rabbling arms is hollow, as illustrated in connection with rabbling arm **20**, and contains a coaxially extending tube (for example, **20'**) communicating at its inner end with the tube **17'** and open at its outer end to the interior of the rabbling arm.

The flow of cooling air from the lower end of the shaft is closed off at the top of the shaft. Thus under forced draft, cooling air is supplied into the rabbling arms and then out through the inner tube in each arm back to the central tube **17'**, whence it exits at the upper open end, as indicated by the arrows.

Mounted on the roof structure of the furnace and communicating with openings in the roof itself are two hoppers **52** and **53**. The hopper **52** is substantially larger in cross-sectional area than hopper **53** and is located within a radius from the shaft such that rubbish and garbage deposited therein fall gravitationally through the central openings **39** to **43** of hearths **34** to **38** directly upon the uppermost hearth **13** of the combustion zone.

Hopper **53** is located at a greater radius from the shaft than hopper **52** and such that sewage sludge deposited therein falls directly on the uppermost hearth **34** of the drying zone. The diameter of the central opening **39** in hearth **34** is somewhat larger than that of the central opening **40** of the hearth **35** immediately below. Thus, the sludge material is moved by the rabbles on rabble arms **47** toward the central opening and then falls directly upon the hearth **35** beneath.

The rabbles on rabble arms **48** associated with hearth **35** thus move the sludge toward the outer periphery of the hearth, whence it falls through the openings **45** directly upon the hearth **36** beneath. Rotation of rabble arms **49** in turn moves the sludge inwardly over hearth **36** until it drops through opening **41** upon hearth **37** beneath.

Rotation of rabble arms **50** associated with hearth **37**, in turn, causes radially outward movement of the sludge on the hearth **37** from which the sludge falls through openings **46** directly upon the hearth **38** beneath. The rabbles on rabble arms **51**, in turn, cause movement of the sludge on hearth **38** toward the central opening **43**, through which the sludge falls upon the hearth **13** beneath or through openings **25** in hearth **22** upon the intermediate hearth **14**. Similarly to the combustion zone, the drying zone has a plurality of openings **54** in the furnace wall covered by inspection doors **55**, used to inspect the operation and replace or adjust rabbles or rabble arms.

The method of operation of the furnace **10** should be apparent in general from the foregoing description. However, a summary description of the operation will offer additional clarification. Assuming that the furnace is conditioned for operation, that is with the combustion zone fired by burners **16** and with shaft **17** and the attached rabble arms rotating, garbage and other rubbish may be deposited in hopper **52** from which it falls directly upon hearth **13** within the combustion zone. From hearth **13** the waste material is progressively advanced by operation of the rabbling arms **18, 19** and **20** until it falls, in ash form through chute **27** and is thence removed.

Sewage sludge, which may have about 75 to 80% water content, may be deposited in hopper **53** separately or concurrently with deposit of rubbish and garbage in hopper **52**. In either event, the forced current of hot air from the combustion zone, initially created partly by heat supplied by the burners **16** and partly by the heat of combustion of the waste material itself, rising in the furnace through the drying zone and traveling over the hearths **34** to **38,** to the exhaust duct **30** effectively evaporates, that is dries, the moisture from the sludge on the hearths in the drying zone.

By appropriate adjustment of damper doors **29**, the moist sludge entering the drying zone is thoroughly dried, the water evaporated being carried along with the hot combustion gases out through exhaust duct **30**. Accordingly, at the time the sludge drops from the lowermost hearth **38** in the drying zone upon the uppermost hearth **13** or hearths **13** and **14** of the combustion zone, the sludge has attained such a dry condition that quenching of the rubbish and garbage waste material on the hearths in the combustion zone does not occur.

Consequently, the sludge is incinerated along with the waste material directly deposited through hopper **52**, and its ash commingled with the ashes forced out through chute **27**. Moreover, since the sludge is introduced into the combustion zone in a dry state, the residue from the combustion zone is entirely free of combustible material. Of special importance is the fact that the quenching of the rubbish and garbage in the combustion zone is avoided, thereby avoiding the production of noxious odors and air pollution caused by quenching.

In the operation of the furnace, combustion heat must be supplied initially from the burners **16**. However, as the waste material in the combustion zone becomes fired, the heat of combustion of the waste material is sufficient to continue the furnace operation without supplementary heat from the burners.

Cyclone Hearth

G. Quesnel; U.S. Patent 3,521,581; July 21, 1970; assigned to Stein & Roubaix, France describes an installation that permits the performance of drying and incineration of the sludges with a deodorization of the gases, giving residues that are bacteriologically neutral.

For this purpose, the installation comprises essentially a dryer and a cyclone hearth, the latter serving to burn the sludges dried in the dryer and giving combustion gases at high temperatures of the order of 1400° to 1600°C, a first circuit leading the moist gas from the dryer to the outlet of the cyclone hearth where it mixes with the combustion gases from the hearth, a second circuit leading the mixture to the inlet of the dryer and having a sufficient length, taking into account the temperature of the mixture of gases for the latter to be deodorized at the inlet to the dryer, and a third circuit taking off part of the deodorized gases at the inlet to the dryer in order to discharge it to the atmosphere. In a preferred practical form, the third circuit comprises a reheater traversed by the first circuit, for lowering the temperature of the gases intended to be discharged to the atmosphere, while reheating the moist gas from the dryer before it is mixed with the exhaust gas from the cyclone furnace.

A second reheater may be provided in the third circuit for lowering further the temperature of the gases intended to be discharged to the atmosphere, while effecting preheating of the combustion air intended for the cyclone hearth. Complete details of the apparatus are provided.

In related work, *C. Maille; U.S. Patent 3,628,473; December 21, 1971; assigned to Stein Industries, France* describes a cyclone furnace for an installation for drying and incinerating mud or sludge which comprises a vertical cylindrical chamber closed at its upper end and extended downward by an internal frustoconical wall. The fuel and air for combustion are introduced, optionally with the material to be incinerated, tangentially into the cylindrical chamber. The waste gases are removed by a duct at the lower end of the combustion chamber. The temperature is sufficiently high to enable the ash to drop from the lower end of the combustion chamber in liquid form.

Rotating Cylindrical Carbonizing Oven

A process described by *J. Von Klenck; U.S. Patent 3,859,933; January 14, 1975; assigned to Mannesmann AG, Germany* involves feeding problem waste to a carbonizing oven for indirect heating while circulating under exclusion of air, so that the waste is continuously agitated during progressive carbonization.

Upon removal from the coking facility on a continuous basis acidized sewage slush is heated to a temperature from 60° to 80°C and mixed with the coke for draining. The mixture is preferably neutralized with lime, drained and the mixture is then burnt together with the problem free waste. The coke-slush mixture is drained, that is, partially dehydrated to a solid content of at least 30% prior to burning.

The gas developed during carbonizing the problem waste contains hydrogen, carbon oxide, some carbohydrates, H_2S, HCl and, for example, some PVC waste. Accordingly, this gas is cleaned by washing, whereby watery hydrochloric acid and either H_2S or just sulfur, depending on the mode and manner of washing, is processed to make sulfuric acid. Alternatively, the sulfur can be burnt to obtain SO_2 and that oxide, in turn, can be used to decontaminate slush or liquids containing chromium. After such purification, the coke gas can be used for indirectly heating the waste in the coking furnace.

The coke produced upon degassing problem waste is very well suited for draining the slush developed in sewage treatment plants. The slush is preferably thickened first to have about 7% dry substance. Draining of the mixture may result simply by settling and centrifuging continuously the slush-coke mixture so as to remove moisture until combustion can be sustained. About 30% and up dry substance is required. The method is preferably practiced in a carbonizing oven constructed as a turning cylindrical oven chamber, which rotates on its longitudinal axis in horizontal or slightly inclined disposition and having an acid proof inner lining. The front ends of that furnace are closed except for the waste charging opening and the coke removal opening. The furnace is indirectly heated in that the inner lining is provided with ducts for passage of fuel gas. These ducts are connected to receive the cleaned gas developed by the furnace as raw gas. Additionally, the furnace is provided with discharge openings of the burnt smoke gas.

High Velocity Cyclonic Motion in Reactor

B.R. Ankersen; U.S. Patent 3,884,161; May 20, 1975; assigned to Ankersen Resource Systems, Inc. has developed a reactor and process for the pollution free thermal disposal of all types of organic waste materials and for the recovery of all usable resources from such waste including the metallic and glassy values in solid form and the latent fuel values as steam energy. The process is based upon the concept of depositing into the interior of the main reactor, solid waste materials as typically generated by a modern community.

Liquid organic wastes such as oils, paint thinners, solvents and other combustible liquids received from industrial sources are stored and injected at selected points in the reactor. High calorific fuel in liquid form from these sources is preferably injected and ignited through the first four nozzles at a control end of the reactor, remote from the stacks. Combustion through these first four nozzles will be preferably maintained at all times serving in effect as a large pilot flame for the entire reactor system and assuring ignition of all other fuels introduced into the reactor at a time when such ignition is desired.

Due to the continuously sustained operation of an induced draft exhaust fan at the discharge end of the return flue, the entire reactor system will at all times operate under a substantial negative pressure. This negative pressure will cause the hot products of combustion from the liquid fuel injectors to proceed through, above and around the still unburned solid waste materials which have been deposited in the downstream portion of the main reactor. As these products of combustion come into intimate contact with the solid waste materials these solids will dry and their moisture will be drawn out through the hot gas flue exhaust duct through the air pollution control equipment and thence exhausted by the induced draft fans through the stacks.

As the moisture content of the solid waste materials is thus gradually removed the solid waste will approach a condition of complete absence of moisture. Since the hot products of combustion from the pilot fire are continuing to penetrate the solid waste materials their temperature will gradually increase to the point where the entire bulk of solid waste will proceed to pyrolize.

Since heating of the solid waste is being provided entirely by completely combusted products of combustion from the pilot fire and since no oxygen is being supplied at all to the portion of the reactor in which the solid waste is stored no combustion can take place and only pyrolysis will occur releasing combustible gases from the solids. These combustible gases, primarily CO and methane, will travel through and around the stored solid waste and issue through the end of the pile toward the downstream end of the reactor. At this portion of the reactor where no solid waste is stored a high temperature cyclonic combustion is maintained by the injection of liquid high calorific fuel through the tangential air nozzles as described previously.

This flame, which is of similar nature to that maintained in the upstream end of the reactor for pilot purposes, is maintained at a much higher temperature of operation, that is, between 2800° and 3000°F. The combustible gases issuing from the stored solid wastes into this high intensity cyclonic combustion chamber will burn to absolute completion. From this very high temperature cyclonic combustion only completely combusted gases approaching 3000°F will proceed on into the return gas duct through the air pollution control equipment and induced draft fan and out the stack.

As the pyrolysis process throughout the entire volume of stored solid waste within the reactor approaches completion and the emission of combustible gases from its process decreases below a desirable level the high velocity air injection nozzles at the immediate end of the solid waste volume within the reactor are energized and simultaneously those nozzles in the downstream section of the reactor farthest removed from the solid waste are deenergized providing a very violent agitation and oxidation of the predried pyrolyzed and preignized bulk of solid waste contained within the extreme end of the stored volume.

The combustion reaction resulting from this very high intensity oxidation results in temperatures approaching 3000°F which will further stabilize very rapid ignition and combustion of the portions of the solid waste volume which will successively become exposed to this very high intensity reaction. Since the temperature thus maintained within the combustion section of the reactor is far in excess of the melting point of any of the noncombustible components of the solid wastes contained within the reactor these noncombustibles will melt and form a glassy molten slag.

The air which is injected through the high pressure air nozzles enters the interior of the reactor at approximately 500 fps and therefore imparts to the gases within the combustion zone a very high velocity angular cyclonic motion within the cylindrically shaped reactor chamber. This very high velocity swirling motion of the combusting gases will lift the solid combustible and noncombustible materials resting on the bottom of the reactor chamber and carry these pneumatically in suspension in the hot swirling gases.

As the combustible portions of the solid wastes complete their combustion thus suspended in the hot gases and surrounded by sufficient oxygen to complete their combustion and release their heat energy for subsequent use in stream generators elsewhere in the plant the noncombustible components will melt and remain swirling in cyclonic fashion in the hot gases. The noncombustibles which are largely SiO_2 and traces of metallic oxides etc., have a density of approximately 5 to 7 times that of the combustible portions.

Due to their high density these molten noncombustible components will form red hot molten droplets swirling cyclonically in the gases and due to the high angular velocity these molten droplets will be forced outward in their circular motion against the hot refractory walls of the reactor. Due to their viscosity and sticky consistency in their molten state they will adhere to the hot refractory walls and after having collected in sufficient quantities will proceed to run down the refractory walls to the bottom and subsequently out through the slag discharge outlet at the bottom of the reactor and into an adjacent water quench tank. As the molten slag drops into a flowing stream of cold water within the quench tank the slag will immediately solidify and due to the severe thermal shock and its glassy nature the slag will shatter upon cooling and form small crystallized glass granules which will subsequently be hydraulically conveyed by the quench water to the discharge end of the quench tank where the granulated slag will be separated from the water in a conventional clarifying system. This granulated glass material after having been separated from the quench water may be sold to road building contractors for use as aggregate black top road material, or to building block manufacturers for use as aggregate in building blocks or the so-called cinder blocks.

Due to this useful and economically feasible application of the only solid residue from this operation the process can truly be termed a resource recovery system. When the granulated glass is conveyed out from the clarifier a magnetic separating system is employed to separate the small agglomerated iron particles from the slag bulk. This iron stems from all iron components which were part of the original solid waste charged to the reactor such as refrigerators, hot water heater tanks, and automobile parts of various descriptions.

This iron when melted together with the slag will tend to agglomerate into separate little iron droplets which are readily separable magnetically from the bulk of granulated glass as it is conveyed on the loading conveyor to the trucks. This iron by virtue of the melting process by which it was recovered is generally completely pure and free of any foreign materials and therefore immediately saleable and applicable to the steel melting industry as scrap.

It should be pointed out that if for any reason an insufficient quantity of high calorific fuel should be available for the operation of the pilot fire at the upstream end of the reactor as previously described the operation can be maintained in exactly the same manner as described by igniting and burning the upstream end of the solid waste pile for purposes of piloting the entire operation.

If in some instances some of the combustible components of the waste materials be so large and heavy that they will not initially become airborne in the high velocity injected air they will be subject to an extremely high rate of combustion while remaining stationary in the bottom of the reactor due to the extremely high velocity of the oxygen and hot gases passing over them far in excess of that required for immediate ignition and providing an overabundance of oxygen right at the surface of these materials while they are burning.

In order to economically utilize the enormous quantities of thermal energy released by the combustion of these waste materials a number of high pressure steam generators are installed in the immediate vicinity of the return gas duct. Spaced at regular intervals of approximately 40 feet along the length of the return flue gas duct cross ducts are arranged at right angles to the return flue gas duct and of similar cross section for the purpose of conveying a proper quantity of the hot flue gases from the return flue gas duct into the entrance opening of high pressure steam generators located at these positions for the utilization of the heat energy contained in the hot flue gases by the generation of high pressure steam.

Each steam generator is provided with its own gas exhaust fan and high energy wet scrubber for the purpose of cleaning the thus exhausted flue gases prior to the emission of these into the surrounding atmosphere. The steam generator exhaust fan by creating a negative pressure within the steam generator in excess of the negative pressure already present within the hot flue gas exhaust ducts below will draw a proper quantity of hot flue gases from the reaction process below through the steam generator and thereby extract the heat energy contained in these gases by transferring the heat from the hot gases through the boiler tube walls into the steam and simultaneously cool the hot flue gases and thereby make them suitable for cleaning by the high energy wet scrubbers for air pollution control purposes.

Drying and Incineration Furnace

Y. Yamato, K. Terada and B. Nakamura; U.S. Patent 3,824,935; July 23, 1974; assigned to Yamato Sanko Mfg. Co., Ltd., Japan describe a drying and incinerating furnace containing means for effecting desiccation and incineration, in a single process, of a muddy material such as an activated sludge.

The furnace comprises a housing, having a vertical hollow-cylindrical wall and a circular hearth, the centers of the wall and hearth being aligned with each other. An inlet port is provided in the wall for introducing the material to be treated on the peripheral portion of the hearth and the outlet port is located in the center of the hearth for removing the treated material from the housing. A device for treating the material is mounted in the housing and comprises a rotary tube vertically inserted through the outlet port from the underside of the hearth into the housing.

An air supply system communicates with the bottom of the rotary tube, and a drive system is associated with the rotary tube for rotating the tube. The rotary tube is rotatably supported by the housing. A plurality of nozzles and blades are provided on the underside of each branch pipe. A burner is provided in the wall and a gas exhauster is mounted on the top of the housing. A plurality of radial branch pipes are secured to and communicated with the top of the rotary tube and are rotatable therewith.

In an example of the process, the nozzle is disposed with a predetermined angle about the axis of the branch pipe from the vertical plane toward the rotary direction of the branch pipe, so as to blow the material in the direction of rotation of the branch pipe. The blade is secured to the branch pipe in a manner that its flat face is inclined with a preselected angle to the direction of rotation of the branch pipe with respect to the vertical plane passing through the axis of the branch pipe. As the rotary tubes rotate, the blades rake and transport the material centripetally and vortically to the center of the hearth or the outlet port.

Referring to Figures 1.11a through 1.11c, the furnace illustrated comprises a housing **1** having a vertical, hollow, cylindrical wall **1a** and a circular hearth **1b**, the centers of the cylindrical wall and circular hearth being aligned with each other. The housing has an outlet port **2** provided in the center of hearth **1b** for discharging the treated material from the housing.

A rotary tube **3** has its lower portion rotatably supported by the housing and its upper portion vertically inserted from the underside of hearth **1b** into the inside of the housing through and concentrically with outlet port **2**. A plurality of branch pipes **4** have their bases fixed to and internally communicated with the top of the rotary tube, each branch pipe extending radially from the rotary tube. The rotary tube is operatively connected with a geared motor **20** through a bevel

FIGURE 1.11: DRYING AND INCINERATING FURNACE

a.

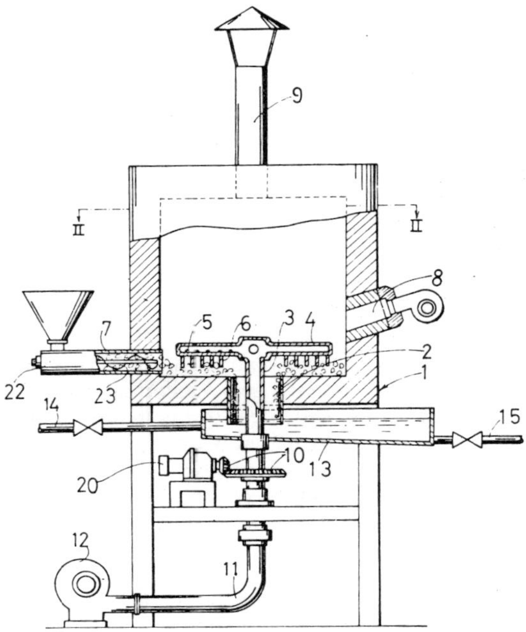

Elevation, Partially in Section, of the Furnace

b.

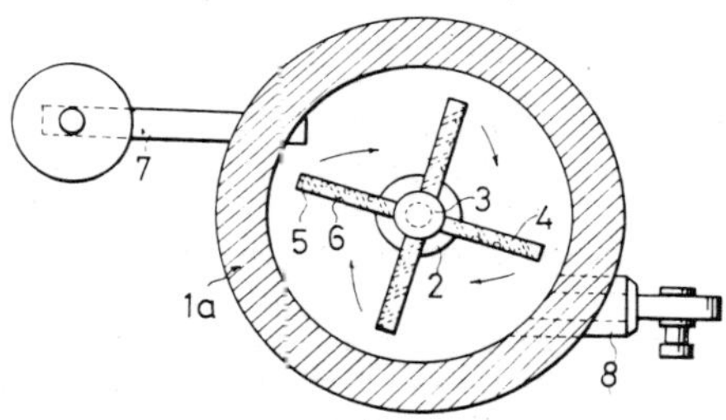

Section Taken Along Line II–II of Figure 1.11a

c.

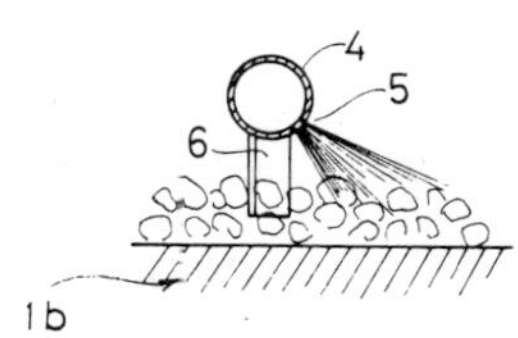

Cross-Sectional Enlarged View Illustrating Radial Branch Pipe

Source: U.S. Patent 3,824,935

gear transmission **10**, whereby the rotary tube as well as the branch pipes is rotated in the direction shown by the arrows of Figure 1.11b. The rotary tube **3** has its bottom end internally communicated with a blast tube **11** connected to the delivery side of a fan **12**. A plurality of nozzles **5** and blades **6** are provided on the underside of each branch pipe **4**. The nozzles are preferably disposed with their axis inclined with respect to the vertical plane containing the axis of the branch pipe, as seen in Figure 1.11c, in order to blow the material in the direction of rotation of the branch pipe.

As shown in Figures 1.11b and 1.11c, blade **6** is secured to the branch pipe with its flat face inclined relative to the vertical plane containing the axis of the branch pipe, so that, while the branch pipe rotates, the blades rake the material on the hearth in a centripetally vortical way, thereby gathering the material in the central portion of the hearth. An inlet device **7** including a material supply tube **22** and a screw conveyor **23** mounted therein is disposed in the cylindrical wall **1a** in a manner such that the material to be treated is deposited on the peripheral portion of the hearth **1b**.

An exhaust means **9** is mounted on the top of housing **1** for the exhaust of gas from the housing. A gas or oil burner **8** is provided in the wall **1a**. In order to quickly heat up the temperature in the housing and exhaust the dust-free gas after separation of the dust, the axis of the burner is preferably arranged to direct its products tangentially to an imaginary circle spaced a predetermined distance from the center of the cylindrical wall and the burner is also disposed in a position slanted to the surface of the hearth.

A water bath **13** is disposed beneath outlet port **2** for chilling the treated material or ash that drops through the outlet. A water pipe **14** is connected with the bath for supplying the water and the ash is removed through a drain pipe **15** together with the water.

The manner of operation of the furnace is as follows. First, the burner is fired to flame and heat the inside of the housing up to a desired temperature for treatment. Then, the fan is started to cause air to blow out through nozzles **5** against the surface of hearth **1b**, while the rotary tube **3** is rotated. Meanwhile, the material to be treated is deposited on hearth **1b** through material supply tube **22** by screw conveyor **23**.

The material located on the peripheral part of the hearth is gradually and centripetally and vortically moved toward the central part or the outlet port by blades **6**, while it is heated and dried by the air blowing through the nozzles. The blades penetrate deeply into the material deposit which serves for quickly drying and insuring complete combustion of the material.

Until the material is dried to a critical moisture content, it still possesses a wet bulb temperature. But, after exceeding the critical moisture content, the material soon reaches its firing temperature and makes a violent complete combustion through the help of fresh air supply through the nozzles. Thus, when the material arrives in the central part of the hearth, it is completely converted into ash and drops through outlet port **2** into the bath therebeneath, where the ash is chilled and then removed from the apparatus through drain pipe **15**. Some ash is rolled up by the rotational air flowing in the housing. However, it is separated from the exhausting gas by centrifugal motion and returned into the material on the hearth along the wall. It is advantageously simple to operate the furnace because the material to be treated can be easily transported from the inlet or the peripheral portion of the hearth to the outlet port by the rotation of the rotary tube **3**, as described above. Another advantageous effect of the furnace is that it is very high in heat-efficiency since the material is burned after being dried to the proper moisture content.

The nozzles and blades provided on the branch pipes can supply air and heat energy deep into the material all over the surface of the hearth, so that a complete combustion following the rapid drying of the material is achieved. This results in neither discharging incompletely burned products nor exhausting injurious gases. The housing as well as other components, for example, the rotary tube, the branch pipes, and the blades disposed in the high temperature environment are endurable for many years service, because the fan causes fresh air to pass through and about these elements thereby protecting them from heat damage.

Pyrolysis of Refuse and Sewage Treatment

In a process described by *J. Osterman; U.S. Patent 3,472,186; October 14, 1969; assigned to Combustion Engineering Inc.,* refuse is subjected to pyrolysis or decomposition by the action of heat in an agitated reaction chamber. The combustible volatiles that are driven off during this process can be used as a fuel, for example, in a steam generator. The remaining solids from the process, in the form of char, can be conveyed to a sewage treatment plant and used as an adsorbent and filter aid in separating the solids from the effluent.

After the solids, in the form of a sludge, have been concentrated to a sufficiently high percentage, they are transferred to an incinerator and burned. If an auxiliary fuel is necessary to maintain ignition of the wet sewage sludge, a portion of the combustible volatiles driven off from the refuse can be used for this purpose. The heat released from the combustion of the sewage sludge and char mixture is used for the pyrolysis of the refuse.

Referring to Figure 1.12, refuse is first introduced into separation device **10**, where large objects which cannot be easily reduced in size, for example, refrigerators, are separated from the rest of the refuse. Noncombustible items, such as cans, can also be separated out at this point. The rest of the refuse moves onto grinding or shredding apparatus **12**, where it is reduced to a size which can be easily reacted on in the pyrolysis chamber **14**.

The shredded or comminuted refuse is then fed into the fluidized bed chamber **14** where it is agitated and subjected to a decomposing atmosphere by the gaseous fluid flowing upwardly through the bed. The fluidized bed chamber is maintained at approximately 1200°F and contains an atmosphere with insufficient oxygen to permit combustion so that pyrolysis drives off the combustible volatiles contained in the refuse.

These combustible volatiles are exhausted through upper duct **22**, flowing to a dry cyclone separator **26**, where any solid particles are separated out and discharged into mixing chamber **30** where water is introduced to form a slurry. The volatiles exiting from the separator then are conveyed to fractionation and condensation apparatus **28**. Here the volatiles are broken down, with the gaseous fuel being discharged through line **27**.

FIGURE 1.12: COMBINATION REFUSE AND SEWAGE DISPOSAL SYSTEM

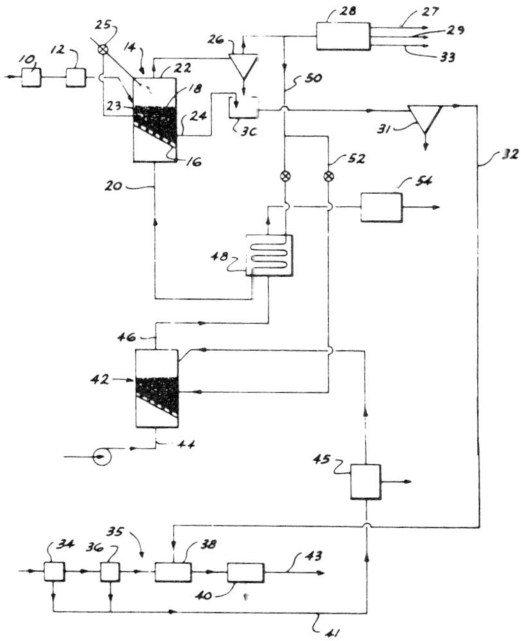

Source: U.S. Patent 3,472,186

The liquid fuel is discharged through line **29** and the water through line **33**. The two fuel fractions can then be transferred to a suitable point of use, for example, to a steam generating power plant. The solids portion which is not carried out of the fluidized bed chamber **14** by the volatiles can be stripped off through line **24**. This char is conveyed to slurry chamber **30**, where it is combined with the char discharged from cyclone separator **26**.

The particles **18** making up the fluidized bed are supported on a perforated plate **16**. The fluid for the fluidized bed is introduced through line **20** into the bottom of the chamber. Since a reducing atmosphere containing little or no oxygen is desired in the chamber, a good fluid to use is the combustible volatiles being discharged from the upper portion of the chamber. These volatiles, at a point downstream of cyclone separator **26**, flow through line **50** to heat exchanger **48**, where they are heated prior to their introduction into the fluidized bed.

In order to insure that no combustion takes place in the fluidized bed chamber in the event that oxygen somehow enters the chamber, and the temperature climbs above the ignition temperature of the refuse, a temperature sensing probe **23** is positioned in the fluidized bed. When this probe senses a temperature above a predetermined maximum, for example, 1300°F, it actuates a valve **25** positioned in a water line, to spray the chamber and reduce the temperature.

The solids slurry, or char slurry contained in mixing chamber **30** flows to a liquid cyclone **31**. Here any heavy inerts or metallic members can be removed. The char slurry then flows through line **32** to the sewage treatment apparatus **35**. The sewage treatment apparatus **35** comprises a primary treatment chamber **34**, where gross solids are screened out of the entering raw sewage. The sewage next flows to a clarification stage, which may be in the form of a settling tank **36**, where additional solids settle out of the effluent.

After removal of the settleable solids the sewage is thereafter transferred to a mixing chamber **38**, into which the char slurry is also introduced. The char or carbon reacts with the sewage in such a manner as to adsorb the fine particles of sewage solids and dissolved matter remaining after the primary treatment of screening and settling.

The treated sewage then flows to a secondary clarification stage **40**, which may be another settling tank where the solids settle out and are separated from the effluent. The clear effluent is discharged through line **43**. The separated solids from the various stages of the sewage treatment apparatus are conveyed through line **41**.

This sewage sludge may have a liquid content running higher than 95%. Thus it is advisable to pass this sludge through a dewatering device **45**, which may be a filter, centrifuge or other suitable piece of equipment for removing a sufficient amount of liquid so that the remaining concentrated sludge is capable of being burned. This sludge is then conveyed to a combustion chamber, shown in the figure in the form of a fluidized bed chamber **42**, where it is burned.

Air to support combustion of the sewage sludge and to maintain the solids in a fluidized state is supplied through lower inlet line **44**. In the event that the sewage sludge contains too much moisture and thus is not capable of maintaining ignition by itself, some of the combustible volatiles may be transported through line **52** to the combustion chamber **42**. The hot combustion gases leave combustion chamber **42** through exhaust duct **46**. After these gases give up heat to the combustible volatiles in heat exchanger **48**, they pass through gas cleanup apparatus **54**, and are then exhausted to atmosphere.

From the above, it can be seen that a system has been provided which will economically and efficiently dispose of both refuse and sewage. By subjecting the refuse to pyrolysis, gaseous and liquid fuels are derived which will be of economic value as fuels which can be burned in a steam generating plant. The char from the pyrolysis chamber can be used in place of other costly chemicals for treating sewage, so as to adsorb and coagulate the solids therein.

The sewage sludge is burned in a combustion chamber, with the heat therefrom being used for pyrolysis of the refuse. Thus it can be seen that a number of components have been arranged in a combination where the end products of each can be utilized to advantage in another portion of the overall system.

B.B. Reilly; U.S. Patent 3,060,869; October 30, 1962; assigned to Drano Corp. describes an incinerator for the disposition of municipal rubbish. It is characterized by the provision of a combined storage hopper and conveyor designed to receive rubbish from dump trucks and convey it to the incinerator. It is capable of accepting material at a much higher rate than the material is consumed, so that while collection of rubbish may continue over an eight hour day, for example, the incinerator may operate perhaps twenty-four hours, or for such longer period of time as is required to dispose of the day's accumulation.

More specifically, the plant has an elongated pit in which is an elongated bin, the bottom of which is formed by one or more vibratory or shaking conveyors, for advancing material from the receiving end of the pit to the incinerator at the opposite end of the pit. There are high stationary sides for the bin, so that the material can be stored to a considerable depth, thereby enabling the conveyor to receive material at a much higher rate than it is consumed, and in slack periods work off the accumulation. This avoids the objectionable practice of dumping the refuse into a receiving pit and transferring it with a grab bucket to a furnace feeding conveyor, and facilitates the rapid and safe dumping of the trucks.

The rubbish is mostly combustible and the shaking conveyor that feeds the rubbish shifts and agitates the rubbish as it is advanced from the storage bin to the incinerator. It has been found that if the sludge cake is dumped onto the rubbish as the rubbish is moved along the bin, it is gradually broken up and diffused through the rubbish, and when it is introduced into the incinerator chamber it is distributed through the mass of rubbish.

Some of the residual water in the sludge is absorbed into the rubbish so that the water content of the sludge is lowered and the sludge spread over, and through the rubbish which is more readily incinerated than in the conventional sludge incinerating plant. At the same time, municipal plant investment is substantially reduced, since one incinerator and one crew dispose of the two waste materials. Complete details of the apparatus and process are provided.

Liquid Phase Coking Process

H.V. Hess, R.F. Wilson and E.L. Cole; U.S. Patent 3,652,405; March 28, 1972; assigned to Texaco Inc. have found that it is possible to purify both raw sewage and municipal refuse with the following advantageous results: the resulting process lends itself to automation and has low investment and operating costs; both sewage sludge and solid refuse can be converted to commercially valuable products; the water effluent which is obtained is low in chemical and biological oxygen demand as well as in solids content; and valuable metals in the refuse can be easily salvaged.

The process comprises the steps of: shredding solid refuse from which hard to reduce or large particulates have previously been removed; slurrying the shredded material with a recycle stream of water resulting from a subsequent coking step; removing noncombustible materials from the resulting slurry; mixing the slurry with sewage sludge; and coking the resulting mixture under a pressure sufficient to prevent the vaporization of water thereby producing grease, clean water and clean coke.

Referring to Figure 1.13, it will be seen that the raw sewage from a domestic and/or industrial sewage system is discharged from conduit **1** to a bar screen device **3** and passed through conduits **5** and **9**, respectively, into and through comminutor **7** and grit chamber **11** to remove large objects and grit therefrom. The sewage then passes through conduit **13** into primary settler **15** where the primary sludge is removed through conduit **33**

FIGURE 1.13: REFUSE LIQUID PHASE COKING PROCESS

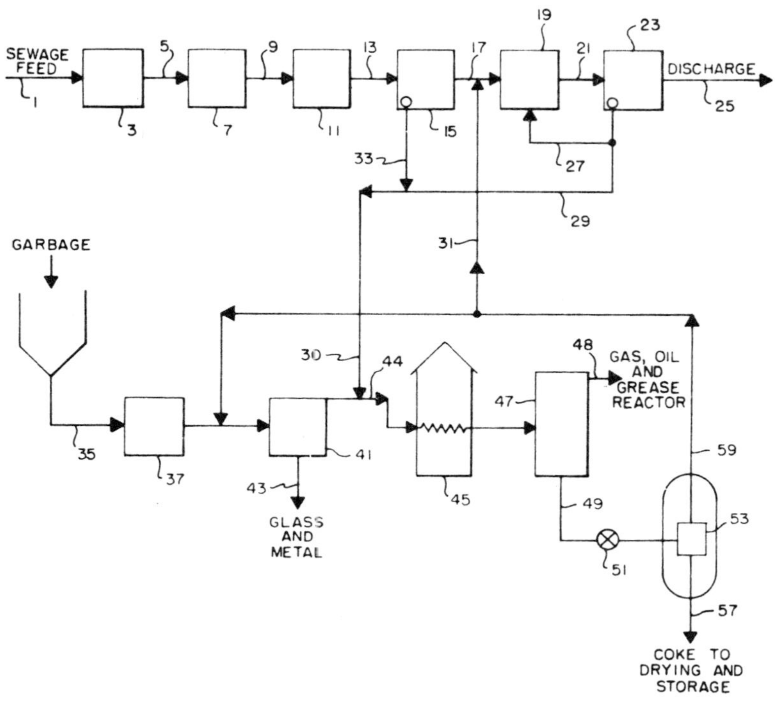

Source: U.S. Patent 3,652,405

The wastewater now substantially free of solid sewage is sent to an activated sludge treating chamber **19** via conduit **17**. Further removal of sludge takes place in final settler **23** to which the water sludge accedes through conduit **21**. Settler **23** serves to remove the activated sludge.

The water having a low BOD is discharged through conduit **25** to a receiving body of water. Conduit **27** is a recycle line for activated sludge. It is important that a large volume of vigorous activated sludge be maintained in vessel **19**. This is partially achieved by recycling a portion of the activated sludges. The excess activated sludge from vessel **23** representing that portion of the activated sludge not recycled to vessel **19** is combined with the sludge from primary settler **15** and the two sludges are mixed in line **30** and pumped to the tubular coking reactor **45**.

The incoming municipal waste is delivered by trucks and placed on conveyor line **35** to be shredded in grinder **37**. The shredded garbage is advantageously mixed with a recycle water stream from conduit **59** sufficient to provide a low viscosity slurry that can be processed in the gravity settler **41**. The slurry enters the settler near the midpoint and the heavier materials such as silt, glass, and metal particles sink to the bottom to be withdrawn through outlet **43** while the lighter combustible particulate materials float to the top. The combustible municipal refuse is withdrawn from the top of the settler through conduit **44**, mixed with the sewage sludge from conduit **30** and sent to the tube reactor **45**. In the tube reactor coking occurs at a high temperature and a pressure sufficient to prevent vaporization of the water.

The effluent from the coil is passed to the settler reactor **47** where the reactions are completed and the gas, oil and grease are separated from the water-coke slurry through conduit **48**. The slurry is sent via conduit **49** and pressure reduced to a separator **53** where the coke is separated from the water. The water from the separator, having a reduced BOD relative to the charge is discharged through conduit **59**.

A portion of this water is recycled to provide a workable slurry in vessel **41** while the excess water is directed to the activated sludge treating vessel **19** via lines **31** and **17** to achieve a further reduction in BOD. The following example is given to illustrate the process. The example shows how the process can be used to treat the sewage and municipal refuse for a city with a population of 30,000.

Example: The sewage from the septic sewers has typically 99.95% water and 0.05% solids, a BOD of 150 to 200, and a COD of generally greater than 200. A typical analysis of the sludge from the primary settler is as follows: solids, 5.0%; COD, 60,000 mg oxygen per liter; residue on evaporation, 50,000 mg/l; total volatile, 33,300 mg/l; and final residue, 16,700 mg/l. For this city, the total sewage sludge averages 30.0 tons per day. The municipal refuse for this city had the following typical analysis on a dry basis.

	Weight Percent Dry Constituent Basis*	
Constituent	With Metal and Glass	Without Metal and Glass
Paper	41.0	48.4
Yard Trimmings	11.1	13.1
Wood	7.1	8.4
Rags	2.8	3.3
Rubber	4.1	4.8
Plastic	0.3	0.4
Garbage	1.2	1.4
Metal	6.8	0
Glass	8.5	0
Moisture	17.1	20.2
Total	100.0	100.0

*All water accounted for as % moisture.

The total daily refuse production averages 75 tons per day. The incoming refuse is dumped by truck into a hopper from which the refuse is continuously removed onto a conveyor belt. While on the belt large inorganic particulates such as iron rods, automobile radiators, large stones, gallon bottles, etc., are removed and segregated by hand before being shredded in a hammer mill type comminutor. Recycle water amounting to 1,014 tons per day is added to the shredded refuse and this mixture is pumped to the settler where 11.5 tons of glass and metal are removed daily. The glass and metal-free refuse slurry (totaling 1,077.5 tons per day) is pumped and mixed in line with the 30.0 tons of sewage sludge. This mixture is then coked in dual reaction coils (each 600' long x 2" diameter) at 550°F for 1 minute total time with a maximum pressure of 1,100 psig. The coked mixture is allowed to settle and react in the settler-reactor (40' high x 2.6' diameter) which has a residence time of slightly less than 30 minutes.

The products obtained from the settler-reactor are gas, grease and coke-water slurry. The gas represents 1.1 weight percent of the charged solids on a dry basis (0.6 ton). The grease is a viscous semisolid when cooled representing 2.6 weight percent of the charged solids (1.4 tons). The grease fraction is a complex mixture of many organic compounds and can be used as a source of fuel. The coke-water slurry is sent to a filter where a dense nonputrescible coke is removed from the clean water. The dense coke represents 96.3 weight percent of the incoming solids on a dry basis (50.2 tons). This coke can be used as a fuel or as a mulch-fertilizer.

The resulting clean water stream represents 95.5 weight percent of the total incoming mixture (1,055.3 tons). A recycle stream of 1,014 tons is separated and pumped to the garbage shredder leaving a clean water effluent stream amounting to 41.3 tons. All operations take place in a closed system. No pollutants are exhausted to the air, as when refuse is incinerated. Water which is finally discharged is relatively low in COD and can be disposed of in bodies of water without polluting them.

Unfired Afterburner Apparatus

R.B. Bruns, D.J. Frey, E.J. Huyghebaert and A.B. Mindler; U.S. Patent 3,604,375; September 14, 1971; assigned to International Hydronics Corporation and Combustion Engineering, Inc. describe a process for achieving complete and relatively inexpensive combustion of solid, semiliquid, and liquid combustible waste materials.

The process involves incompletely burning a combination of such materials in a first furnace to produce a combustible gaseous exhaust stream, achieving complete combustion in a second furnace of combustible high heat value waste materials to produce a noncombustible gaseous stream, intimately mixing the two streams in an unfired afterburner, without the addition of heat from an auxiliary heat source, while maintaining a minimum admixture temperature above the temperature required for autoignition of the combustibles in the first stream, whereby the heat produced by combustion of the high heat value waste materials is utilized for the spontaneous ignition and complete combustion of the incompletely burned materials.

Liquid wastes of high heat value and of low heat value are separately stored, and are variously blended for feed to the first furnace and the second furnace in accordance with the heat requirements of the system to achieve complete combustion of all wastes being burned. The apparatus particularly adapted to perform the process comprises a pair of furnaces, a refractory-lined afterburner chamber having no auxiliary heat source, duct means interconnecting the afterburner chamber and each of the furnaces, and exhaust means defined by the afterburner for permitting exhaust of the burned gas stream after combustion is completed. The apparatus preferably includes appropriate draft handling, gas cooling, and particle removal means to render the gaseous atmospheric discharge relatively pollution-free.

Incineration and Calcination in Single Heating Unit

A process described by *P.J. Cardinal, Jr. and R.J. Sherwood; U.S. Patent 3,623,975; November 30, 1971; assigned to Envirotech Systems, Inc.* involves the heat treating and recalcining of lime sludge and sewage sludge obtained from a tertiary treating procedure in which gaseous combustion products, such as carbon dioxide, are selectively maintained at a low level adjacent the discharge end of the heating apparatus to eliminate or obviate the possibility of recarbonation of the recalcined lime produced in and discharged from the heating apparatus. Such selective limitation of the carbon dioxide level is achieved by omitting heat input sources adjacent the discharge end of the heating apparatus.

An important feature of this process resides in the utilization of only a single-heating unit, such as a multiple-hearth furnace, in which lime sludge and sewage sludge are simultaneously heat treated to recalcine the lime sludge and to incinerate the sewage sludge waste to the fullest extent possible. Advantages of such a simplified system are its substantial savings in cost due to the elimination of one furnace from the treating plant, thus reducing substantially the initial capital cost requirements. Additional operating cost savings are realized due to reduced fuel consumption by operating a single furnace to attain the same results previously produced in two separate furnaces.

Simultaneous Incineration of Combustible Waste and Sludge

D.M. Larger; U.S. Patent 3,815,522; June 11, 1974; assigned to Eau Et Assainissement Socea, France describes a process for simultaneously incinerating combustible waste and sludge. The process comprises first heating the sludge by heat exchange directly with the fumes of combustion from the incineration of the combustible waste, and then burning the sludge wherein, between the heating and the incineration, the sludge is allowed to bake and is separated from its water, only the residue of this separation being burnt while the water separated out is evacuated.

This process and apparatus are advantageous in that the heat of the fumes of incineration of the combustible waste is put to use and the cooling and a separation are effected outside the heat exchanger whereby it is possible to pass the sludge rapidly through the exchanger, which considerably reduces accumulation of the sludge and corrosion of the exchanger surfaces.

The apparatus shown in Figure 1.14 effects the simultaneous incineration of combustible waste and sludge in an incinerating furnace **1**. The stack **2** or circuit for evacuating the fumes from the incinerating furnace comprises, starting from the furnace proper in which the combustible waste is burnt, a vertical conduit **3** for the outlet of the fumes, a horizontal chamber **4** for incinerating the sludge and a vertical conduit **5** for evacuating the fumes.

Placed in the conduit **5** is a heat exchanger **6** having a nest of tubes which is connected by piping **7** to a sludge supply pump **8** connected to a sludge purifying station **9**, and by piping **10** to a level **11** intermediate the base and upper part of a baking or cooking and decantation vessel **12** which is heat insulated and under pressure. The base **13** of this vessel communicates by way of piping **14** and a reducing valve **15** with the upper part **16** of the sludge combustion chamber **4**. The upper part **17** of the vessel is connected to a steam or vapor condenser **18** by way of piping **19** and a reducing valve **20**, the vapor thus condensed being returned to the sludge purifying station **9**. The water decanted in the vessel **12** above the sludge is also returned to the sludge purifying station by way of piping **21** which is connected to the vessel above the level **11** and includes a reducing valve **22**.

This apparatus operates in the following manner. The sludge supplied by the purifying station is fed under pressure by the supply pump **8** to the heat exchanger **6** which is heated by the fumes which come from the furnace **1** for incinerating the waste and have passed through the sludge incinerating chamber **4**. This heat exchanger heats the sludge to its baking temperature.

The sludge is then fed to the baking and decantation vessel **12** under pressure in which it stays such time that, while it is maintained at its temperature by the heat insulation, it bakes and then decants. The thus decanted sludge is expanded in the valve **15** and then burnt in the chamber **4**. The vapor given off in the vessel **12** is expanded in the valve **20** then condensed in the condenser **18** and the water decanted from the vessel is also returned to the purifying station by way of the piping **21**. In this way the heat given off by the combustion of the combustible waste is employed for heating the sludge to its baking temperature while the sludge does not stay in the heat exchanger during its baking, which reduces soiling and corrosion of the heat exchanger.

Moreover, the baking of the sludge is employed for improving its decantation. Indeed the baking of the sludge brings about a modification in its structure and a coagulating effect which enhances the decantation.

FIGURE 1.14: SIMULTANEOUS INCINERATION OF COMBUSTIBLE WASTE AND SLUDGE

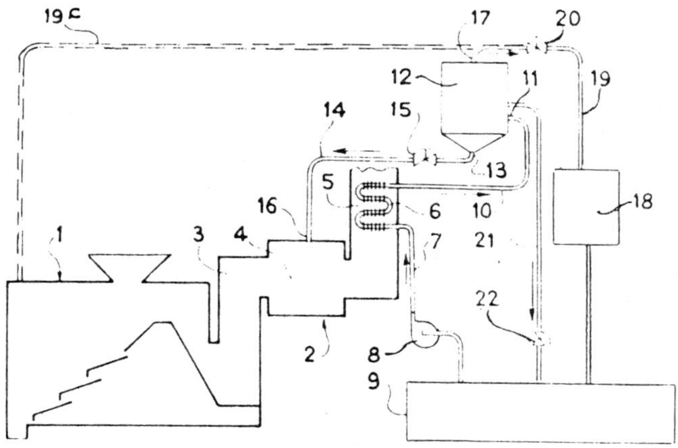

Source: U.S. Patent 3,815,522

This process permits reducing by two-thirds the amount of sludge to be burnt. Indeed, for example, for 1 kg of sludge to be treated it is possible to separate two-thirds thereof constituted by water, that is 660 grams, by employing about 190 kcal, whereas to evaporate 660 grams of the water of the sludge, more than 400 kcal would be required. This process is therefore of high interest from the point of view of energy consumption and permits starting with sludge having 96% water instead of 92% and thus simplify the sludge purifying station.

Furthermore, it is possible to obtain fumes coming from the incineration of the dried sludge and the combustible waste at a temperature higher than 700°C, at which temperature the odorous vapors are destroyed by oxidation which consequently results in no discharge of odors to the atmosphere. Another advantage of the combustion of the sludge in the fumes of combustion of the combustible waste resides in the fact that the oxygen of these fumes serves to oxidize this sludge and thus improve its combustion in the furnace and even permits burning it in the fumes, this oxygen acting as a comburent.

An advantage of the baking and decantation of the sludge in the same vessel under pressure is that it is possible to have a rather low baking temperature, namely 160° to 190°C. Thus, the decanted water which is returned to the purifying station is less charged, since the lower the baking temperature the lower the amount of impurities dissolved in the water.

Combined Processing of Solid and Comminuted Waste

F.G. McMullen; U.S. Patent 3,769,921; November 6, 1973 describes a waste materials processing system for combusting solid waste material on grates in combination with comminuted waste material being combusted in suspension within a combustion zone. Conduits are positioned external to the combustion zone around which exhaust gases from the combustion zone pass.

Comminuted waste material is driven through the conduits and by pyrolytic action emits gases which are collected and passed to a recovery system. Additionally, predried sludge and sludge vapors derived from sanitary sewage are incorporated into the combustion zone for burning. The combined waste materials processing system may be incorporated within a steam generating system in order to increase the overall efficiency of the boiler.

The combined combustion methods of burning trash in trash form and pulverized waste in suspension along with sludge derived from sanitary sewage results in the maintenance of a substantially self-sustaining burning operation. Further, the combustion of these materials is accelerated over that found when only one or the other type of burning is used which results in an overall increased efficiency of a steam generating unit.

Concentration of Sludge Using Incinerator Hot Gas Outflow

According to a process described by *J.L. Winther; U.S. Patent 3,702,596; November 14, 1972* garbage incineration and raw sewage sludge concentration are combined. The hot gas outflow of an incinerator is used to concentrate by dehydration raw sludge received from a sewage clarifier. After simultaneously reducing the moisture content of the raw sludge and scrubbing the exhaust gas, the gases from the incinerator are again scrubbed, typically with the outflow from the clarifier or alternately with raw sewage to remove noxious gases of combustion and particularly to prevent undesirable high vapor content. Referring to Figure 1.15a, a garbage incinerator **A** is shown having a fuel and air inlet **B**, and an ash outlet **C**. The gases of combustion are exhausted from the top of the incinerator to downward exhaust flume **D** and thence to upward exhaust flume **E**. The incinerator and its associated parts are generally of the type described in U.S. Patent 3,570,422.

Sewage is typically received at a clarifier **F**. The clarifier separates raw sludge, a mixture containing approximately 3% solids, from the overflow. This raw sludge is piped to baffles **G** and downflowing flume **E**. The liquid portion or overflow from the clarifier is piped to upwardly flowing flume **E** at baffles **H**.

Briefly, fuel inlet **B** conveys garbage into which concentrated sludge has been formed to a point at the uppermost portion of incinerator **A**. Thereafter, the garbage is deposited interior of the incinerator and combusted with a resultant ash flow outwardly at **C** and a gas flow outwardly to flumes **D** and **E**. At flume **D** raw sludge of a 3% solid content is introduced into parallel downflow with the gases of combustion.

The heat from the gases of combustion are transferred to the sludge. The moisture from the sludge is transferred to the downflowing gases. The resultant gas at the bottom of flume **D** has an increased moisture content while the outflowing sludge has a reduced moisture content. When the gases of combustion pass upwardly through flume **E**, sewage overflow is counterflowed through them. This counterflowing sewage overflow reduces the temperature of the exhaust gases to a point where moisture initially acquired in flume **D** condenses and falls out of the air.

The result is a fine condensation of the moisture in the air around solid particles entrained in the air with extremely efficient solid particle scrubbing. Moreover, the overflow when passed from the stack absorbs most of the noxious gases of combustion. The result is that a relatively low emission of moisture particles, and noxious gases occurs.

Fuel inlet **B** to incinerator **A** includes an enclosed tunnel **12** having an endless conveyor **14** placed therein. The lower end of the upper flight of the endless conveyor defines the surface onto which garbage is placed. Thereafter the garbage mixture is transported upwardly to the end of the conveyor and interior of the incinerator.

Sludge concentrate is typically sprayed onto the upper surface of the conveyed garbage from a header **20**. This sludge concentrate travels with and is combined with garbage on the conveyor. It will be therefore understood that the fuel for the incinerator consists of a garbage mixture sprayed with concentrated sludge.

It should be noted that with the sludge concentrator of this process in the exhaust gas outflow from the incinerator, that the treated sludge can be introduced directly into the primary combustion area of the incinerator. With the low moisture content of the sludge attained, no interference with the combustion of gases released from the primary incineration will occur. The decision whether to introduce the sludge onto the garbage conveyor in tunnel **B** or to introduce it directly into the incinerator will depend upon several factors.

FIGURE 1.15: INCINERATOR-SLUDGE CONCENTRATOR COMBINATION

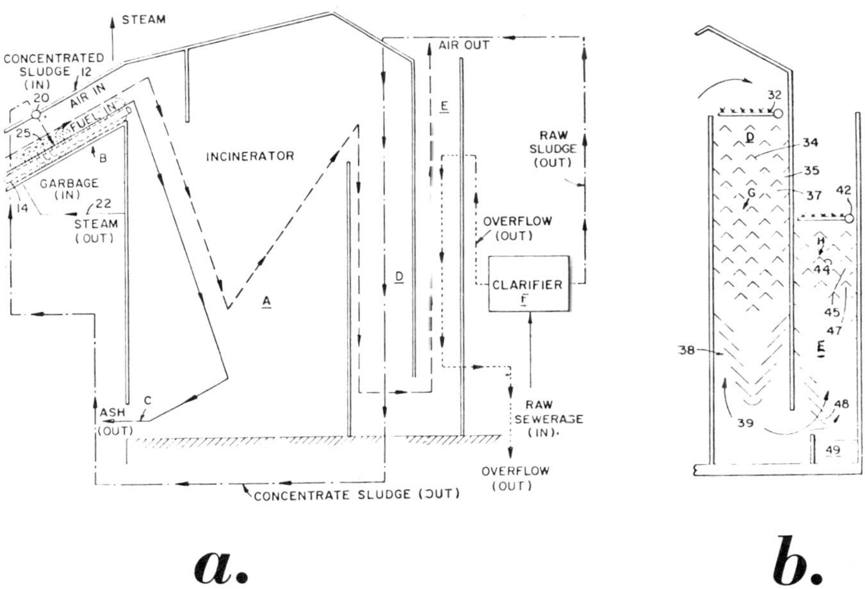

Source: U.S. Patent 3,702,596

These factors will include the nature of the conveyance equipment, the availability of steam for preheating, and the relative quantities of garbage and sludge being processed through the combined unit. A specific example will assist the understanding of this process. Typically, a commercial unit of the type described will be designed for the disposal of 3,000 tons of garbage per day. Consequently, 4,167 pounds per minute of the garbage must be introduced interior of incinerator A.

The heat balance is outlined in the following portion of the process. This balance assumes what are believed to be reasonable approximations for the treatment system. Abbreviations in the balance have been made including considering only the enthalpy of the air and moisture. Chemical heat, enthalpy of solids and heat losses are for the most part ignored.

It is an estimate that the average heat value of municipal garbage will be approximately 5,000 Btu/lb at a moisture content of approximately 30% by weight. Consequently, 7.5 lb of dry air will be required for the combustion of each pound of garbage for a total of 31,253 pounds of air per minute flow.

Assuming that the ambient air is at 70°F and 80% relative humidity the total inflow enthalpy of the air-fuel mixture will be approximately 21,788,317 Btu/min. Referring to the abovementioned U.S. Patent 3,570,422 it will be noted that the incinerator is designed to have a steam by-product. At least a portion of this steam by-product is channeled through piping **22** to a coil **25** interior of tunnel **12**.

Coil **25** keeps the inflowing fuel air mixture at a temperature in the range of 200°F. Incinerator **A** has inlets other than the inlet at tunnel **12**. Approximately 10% of the air will enter through the ash outflow and the sidewalls of the incinerator construction described in the above referenced patent.

After combustion has occurred interior of incinerator **A**, it is desired that air entering into the downflowing flume **D** be at an exhaust temperature of 1600°F. To achieve this outflow approximately 3,150 Btu/lb heat energy of the original 5,000 Btu/lb of fuel energy must be transferred to the exhaust gases. The remaining 1,850 Btu/lb will be consumed in thermal losses through the incinerator sidewalls, combustion inefficiencies, and the extraction of the heat for the steam.

Raw sludge will be added at the top of downflowing flume **B**. Typically, this sludge will pass through a series of baffles **G**. While Figure 1.15b describes baffles **G** in some detail, it should be understood that any number of baffles may be sufficient for the downflow. Typically, the scrubber should provide an agitated flow path on the wetted surface of the baffle for the downflowing raw sludge. The provided fluid flow path should be in fluid interference with but parallel to and in the same direction as the outflowing gases from the furnace.

The flow from the top of downflowing flume **D** to the bottom of downflowing flume **D** is designed to produce a solid concentration of at least 35% and preferably in the order of 40%. Accordingly, approximately 92% of the entering volume of 3% solid raw sludge will be vaporized leaving the remaining 8% as concentrated sludge.

It should be noted that the raw sludge flow rate will be relatively high. Typically, and assuming the 4,167 lb/min input of fuel, a raw sludge flow rate of 11,305 pounds per minute or approximately 2.71 lb of sludge at 3% solids per pound of garbage can be concentrated. It is significant to point out that this is a higher sludge to garbage ratio than is normally produced by a given population.

Referring to Figure 1.15b, downflowing flume **D** and a construction of baffles **G** can be illustrated. Typically, sludge is introduced into flume **D** at manifold **32**. The sludge is mixed with and falls downwardly with the exhaust gases. The sludge gas mixture falls onto a series of roof shaped baffles **34** aligned in alternate baffle rows **35** and **37**. Baffle rows **35** are aligned to intercept the flow path of the gas sludge mixture not intercepted by baffle rows **37**. Similarly baffle rows **37** are aligned to intercept the flow path of the gas sludge mixture not intercepted by baffle rows **35**.

Upon falling through alternate rows **35** and **37** of roof shaped baffles **34** the sludge passes to a series of collector baffles **38** and to a collector trough **39** at the bottom of downflowing exhaust flume **D**. At this point, the sludge is collected and piped for further treatment as a by-product of the system or preferably to the incoming fuel mixture to the incinerator. This has already been indicated in the description of Figure 1.15a.

At the bottom of downflowing flume **D** the temperatures of the exhaust gases will be reduced to a range approximating 167°F. Typically, 0.39 lb of moisture will be suspended in each pound of dry air for a total of 12,189 lb/min. From the bottom of downflowing flume **D** the exhausted gases will pass upwardly through upflowing flume **E**. The gases will be detoured upwardly and around baffles **H**. At the same time, sewage overflow from clarifier **F** will be passed downwardly from the top of upflowing flume **E** to the bottom of the upflowing flume **E**. The results will be a parallel and counterflowing agitated flow path of the sewage and air on the wetted surface provided by the baffles. Fluid flow in interference with the outflowing gases of the furnace will occur.

Again referring to Figure 1.15b, the scrubbing of the exhaust gases and upflowing flume **D** by baffles **H** can be conveniently illustrated. Typically, the overflow is introduced into the flume at a manifold **42**. The overflow is mixed with but falls downwardly and in opposition to the upflowing exhaust gases passing out of the incinerator.

The overflow gas mixture flow proceeds in the following manner. Typically, the overflow falls onto a series of roof shaped baffles **44** aligned in alternate baffle rows **45** and **47**. Similar to the baffle rows **35** and **37** in flume D, baffles **45** and **47** of flume **E** are alternately aligned so that overflow missing one of the baffles in one row falls upon an alternate baffle in an alternate row.

At the bottom of flume **E** the overflow is deflected by a series of collection baffles **48** and into a collection trough **49** where it is collected at a drain and moved for downstream processing. It will be noted that the distance between manifolds **42** and collection trough **49** is shorter than the distance between manifold **32** and collection trough **39**. This leaves the upper portion of the stack in the schematic of Figure 1.15b not occupied. In this area additional attachments to the incinerator may be placed. These attachments can include induced draft fans for pulling gases through the incinerator, air preheating and chemical deodorizing attachments if desired.

It should be realized that whereas overflow from a clarifier has been used as the scrubbing liquid, raw sewage with its relatively low solid content (on the order of 0.01%) can just as efficiently be used for scrubbing. Using the assumed heat flow consideration, approximately 761,025 lb of sewage overflow at 70°F will be entered into the upward portion of baffles **H** at the upper end of upflowing flume **E**. The resultant air temperature of the outflowing exhaust will be 80°F and the resultant temperature of the overflow will be 88°F.

In a process described by *T. Jacobivici; U.S. Patent 3,333,556; August 1, 1967; assigned to Von Roll AG, Switzerland* the preliminarily dewatered sludge is partially predried in a first stage in a predrying compartment at least approximately separated in a gas tight manner from the combustion compartment, such predrying action being carried out by means of a heating medium delivered to the predrying compartment.

This partially predried sludge is completely dried in a drying zone arranged in the combustion compartment, the drying being carried out under the action of radiant heat emanating from the combustion compartment, with the thus formed vapor clouds acting as moderator for the drying operation before their mixture with the combustion gases developed in the combustion compartment. The material now completely dry then arrives in a primary combustion zone and then in a subsequent so-called burn out or final combustion zone and a supplementary fuel is fired in the combustion compartment, while the vapors resulting from predrying are being mixed with the combustion gases, at a location which is disposed at least downstream of the combustion compartment, when referring to the direction of flow of the combustion gases. In so doing, predrying of the sludge can take place by means of flue gases diverted from behind the combustion compartment and which are then recycled, with the predrying vapors from the predrying compartment being guided into a furnace flue downstream of the combustion compartment.

Progressive Incineration

According to a process described by *H.W. Olsen; U.S. Patent 3,733,271; May 15, 1973* solid waste in garbage, trash and sewage is subjected to progressive incineration and reduced to inert residue capable of reclamation. Combustion gases are utilized for generation of steam and electricity, and are freed of sulfur and other atmospheric pollutants. Liquid sewage effluent is made potable by chemical treatment.

Referring to Figure 1.16, solid waste such as trash, garbage, and the like is delivered into a surge bin or tunnel **10** which is maintained at negative air pressure to prevent escape of odors into the atmosphere, as for example, by a suction line **11** of a pump or blower **12**. From the bin the waste material is conveyed to a chopper **13** where it is chopped into suitably sized fragments for delivery into a drying kiln **14**. A suitable air lock is provided between the bin and the chopper, through which air lock the material passes without materially affecting the negative air pressure in the surge bin. Similar such air locks are also provided at other points in the apparatus where their use is desirable or necessary.

Incoming sewage flows by gravity into an odor-tight comminutor **15** where it is converted into a thin slurry which is pumped to a multiple stage filter station **16**. The latter may be immediately adjacent to or at some considerable distance from the comminutor **15**. Solids filtered from the slurry at the filter station are conveyed to drying kiln **14**, while the liquid effluent is delivered by a pump **17** to a vacuum still **18** and also to a bubble tower **19**.

The drying kiln is indirectly heated by hot gases conveyed through a line **20**, and after passing through the kiln, these hot gases flow through a line **21** to heat the vacuum still. The kiln has an internal thread with lugs to distribute and agitate the waste solids for a maximum surface exposure, while vapors and noxious gases are withdrawn through the suction line **22** of a pump **23**, the outlet of which discharges through a line **24** below the water line of a water containing char tank **25**.

The dried solid waste is conveyed from the kiln to a first char kiln **26** which is equipped with a gas or oil burner **27** for use in starting up or when the solid waste input is so erratic that continuous flames in the kiln **26** cannot be maintained without such a burner. Like the kiln **14**, the kiln **26** also has an internal thread for agitating and exposing a maximum surface of the waste material to incineration.

During the incineration process in the kiln **26**, some of the combustible solids are burned completely while some of them merely become charred and are picked up by the combustion gases. These charred solids become deposited on a traveling screen **28** which takes them out of the flow path of the combustion gases which leave the kiln through a line **29**, and the charred solids are delivered as at **30** into the char tank **25** where they float on the water and may be skimmed off.

FIGURE 1.16: PROGRESSIVE INCINERATION

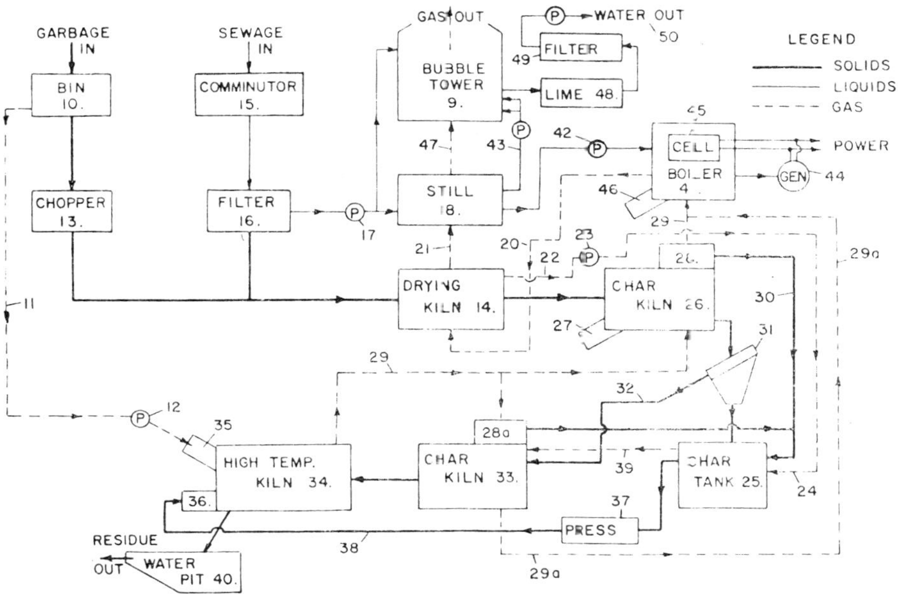

Source: U.S. Patent 3,733,271

Burning solids and incombustibles are discharged through an air lock from the char kiln **26** onto a vibrating screen **31** which passes most of small abrasive components to the bottom of the char tank **25** and diverts combustibles and larger incombustible components as at **32** into a second char kiln **33**. The latter is similar to the first char kiln **26** and also has a traveling screen **28a** for extracting charred solids from the flow path of combustion gases passing through the kiln **33**, such charred solids also being delivered into the char tank **25** as already explained in connection with the kiln **26**.

The burning waste and incombustibles in the second char kiln are discharged through an air lock into a high temperature, refractory-lined kiln **34**. Such materials may include heavy blocks of wood, rubber tires, metal pieces and glass, and the temperature in the kiln **34** is just under the melting point of glass, being provided mainly by a gas or oil burner **35** and supplemented by briquette fuel burning on a traveling grate **36** of the kiln.

Alternatively, the briquette fuel may be the main source of heat, supplemented by the burner **35**, such briquette fuel being obtained by skimming the aforementioned charred solids off the water in the char tank **25**, drying the same, compacting them into a briquette form in a press **37**, and then delivering them as at **38** to the grate **36** of the kiln **34**. Conveniently, odorous gases withdrawn from the surge bin **10** through line **11** may be delivered by pump **12** to be burned in the kiln **34**, as indicated. The noxious gases passing from the drying kiln **14** through line **24** into the char tank **25** rise through the water therein and are then passed as at **39** to be burned in the char kiln **33**. Ashes, metals, glass and other noncombustibles are dropped from the kiln **34** through an air lock into a water pit **40**. This shatters much of the glass or ceramics and cools the metals so that an inert residue may be recovered from the pit **40** for use as landfill or for metal reclamation if so desired.

The hot flue gases from the high temperature kiln **34** are conducted as at **29** through the first char kiln **26** and also as at **29a** through the second char kiln **33** to the furnace of a water boiler or steam generator **41**, from which the gases then pass as at **20** to heat the drying kiln **14** and as at **21** to vaporize water in the still **18**. The distilled water there produced is fed as at **42** to the boiler **41** and also as at **43** to the first and second stages of the bubble tower **19** as hereinafter described.

The steam generated in the boiler is delivered to the turbine of an electric generator **44** which produces current for auxiliary equipment of the disposal plant and also for electric arcs in labyrinth type sparking cells **45** provided in the furnace of the boiler. As the flue gases pass through these cells, their temperature is locally raised by the electric arcs to a point where nitrogen fixation occurs, and this allows sulfur dioxide in the gases to oxidize to trioxide.

Since these chemical reactions are very unstable at the high temperature at which they occur, the flue gases are cooled as rapidly as possible by exposure to the heat absorbing surfaces of the boiler. The boiler may be equipped with an auxiliary oil or gas burner **46** to assure its operation in the event of an interruption in the supply of hot gases from the kilns **34, 33** and **26**.

After leaving the furnace of the boiler as at **20** and passing through the drying kiln **14** and still **18**, the flue gases are cooled further and are delivered as at **47** to the bubble tower. This tower has several stages between which liquid does not communicate, the first two stages being charged with distilled water delivered from the still **18** as at **43**. The first stage traps fly ash and inert dust particles present in the flue gases and also absorbs some of the sulfur trioxide to produce sulfuric acid.

The second stage is designed to absorb practically all of the remaining sulfur trioxide. The third stage is a multilevel or tray stage into which sewage effluent from the filter **16** is delivered by the pump **17**. The effluent flows downwardly over the successive trays and is treated with acids produced by this process to oxidize any biotics in the water. After the treated water leaves the bubble tower, it is given a lime treatment as at **48** to precipitate the phosphates therein and is then filtered as at **49** to remove any solids, so that potable water is delivered as at **50**.

The apparatus and process facilitate disposal of solid material in both garbage and sewage by progressive incineration with the result that waste is reduced to ashes and incombustibles remain as an inert residue for use as landfill or for reclamation of valuable metals. Combustion gases are freed of fly ash, dust and sulfur, sewage effluent is made potable by chemical treatment, and both chemical and thermal pollution of the ecology is kept at an acceptable, benign level.

Vertical Elongated Tubular Combustion Chamber

P.U. Reusser; U.S. Patent 3,369,505; February 20, 1968 describes a method of burning rubbish and biological sludge in a vertically elongated combustion space. The process is comprised of the steps of charging a combustible rubbish having a heating value of between 1,400 to 1,800 kcal/kg and biological sludge having a water content of about 80% into the combustion space, supplying air to the body of material to be burned including selectively directing air into the body material at a number of vertical spaced locations throughout the body of the material, supporting the charged material within the combustion space, burning the material, moving the body of material as it is burned, and discharging the resultant slag from the burned body of material to the bottom of the combustion space.

A furnace for burning biological sludge and combustible rubbish mixed in a ratio of 1:1 is comprised of a vertically elongated tubular shaft having a combustion zone in its lower portion and a removable closure at its upper end for charging the material to be burned into the shaft. An outlet flue is provided from the shaft near its upper end. A movable grate is located near the lower end of the shaft for supporting a body of the material to be burned. Air inlet means are located in the sides of the shaft for directing air laterally inward into the body of material to be burned.

Additionally, vertically arranged lance means having air passageways therethrough are vertically mounted in the shaft and extend downward into the bed of material; the lance means are vertically positionable within the body of the material for selectively discharging air throughout the body of the material to be burned.

Further, air may be directed into the body of material to be burned from below by feeding air in through the grate. The body of the shaft may be circular or oval in configuration and the means for supplying air laterally into the body of material may be accomplished by rods extending radially inward into or above the body of material to be burned.

Moving Grate

T.R. Komline and W.H. Sanderson; U.S. Patent 3,322,079; May 30, 1967; assigned to Komline-Sanderson Engineering Corp. describe a method and apparatus for incinerating sewage sludge and municipal waste in which an ignited bed of the municipal waste is supported for continuous movement on a grate.

An upward flow of air through the fuel is produced by suitable means and the sludge is sprayed in finely divided particles into the upward rising air and combustion gases above the solid fuel bed, the upward rising gas and air currents impeding the downward movement of the particles to increase their exposure to the heating and drying action, some of the particles being entrained and subjected to combustion in the currents and other heavier particles descending onto the solid fuel for combustion after being subjected to such drying and heating action.

The incinerating apparatus illustrated by Figure 1.17 comprises an incinerator housing **10** supported on a suitable foundation **11**, and defining an enclosed horizontally elongated combustion chamber **12** within which is operatively disposed a generally conventional traveling or conveying grate structure **14**, as diagrammatically illustrated in Figure 1.17a. The grating **14** may be similar to that described in U.S. Patent 3,152,562, and is adapted to support for combustion on its upper surface a bed of solid fuel **B** which will normally comprise combustible municipal waste such as trash or trash and garbage intermixed with each other. Such municipal waste is delivered onto the forward or charging end of the grating by means of a charging hopper **16** communicating with the combustion chamber through an opening **18** in the front wall of the housing at a location above the forward or charging end of the grating. Material is delivered into the hopper from a charging floor **17**.

The grating is somewhat inclined downward from front to rear and is arranged in conventional manner to progressively convey the burning bed of the solid fuel rearward at a rate such that the material comprising the bed will be completely burned or oxidized upon reaching the rear or discharge end of the grating. The ashes will be continuously discharged from the rear end of the grating into a suitable discharge or collection means, such as exemplified by the ash pit **20** in Figure 1.17a, to be disposed of as desired.

As is usual, the grating is constructed to permit the free passage of combustion air upward through it and through the burning material supported thereon. For supplying a constant flow of air under forced draft upward through the forward portion of the grating, there is provided beneath the grate a plenum **22**, the open upper side of which communicates with the grate, and means such as the centrifugal blower **24** communicating with the plenum through a passage **26** for supplying air from the atmosphere into the plenum at superatmospheric pressure. The air is thus caused to move upward through the burning solid fuel bed to maintain the combustion of the material at a sufficiently high operating temperature.

FIGURE 1.17: INCINERATION OF SLUDGE ON MOVING GRATE

a.

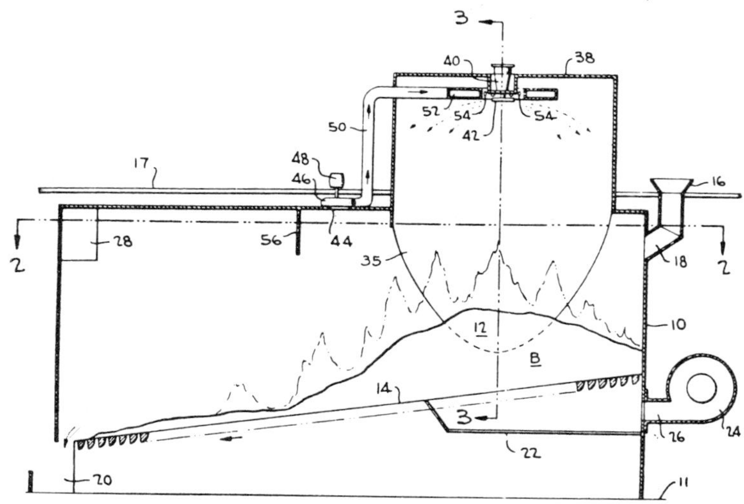

Vertical Section of Apparatus

(continued)

FIGURE 1.17: (continued)

b.

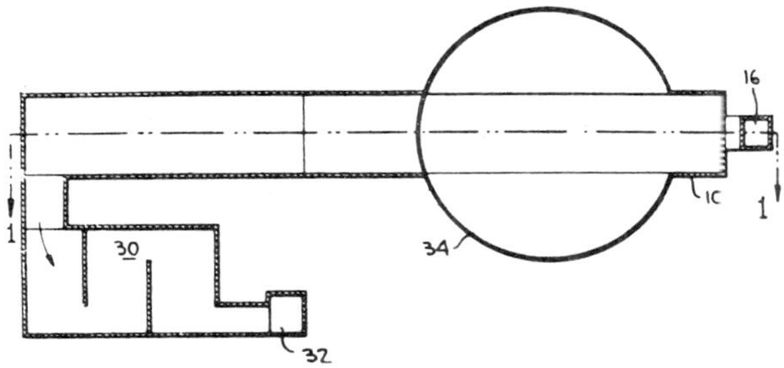

Plan Section Along Line **2-2** of Figure 1.17a

c.

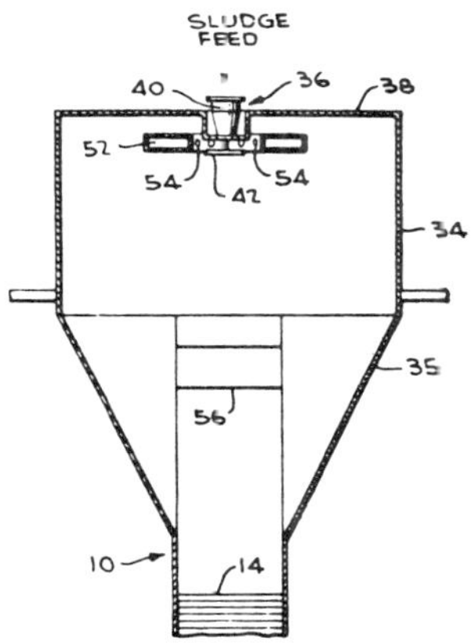

Vertical Section Along Line **3-3** of Figure 1.17a

Source: U.S. Patent 3,322,079

The upward rising air and/or combustion gases after passing from the plenum through the fuel bed **B** are guided or deflected rearward above grating **14**, by the inner configuration of the combustion chamber, so as to flow longitudinally toward the discharge port or flue **28**. From the flue, the exhaust gases pass through a conventional scrubber **30** for the elimination of fly ash or other entrained materials, then to and upward through a chimney or stack **32** for discharge into the atmosphere.

Located above the forward or charging end of the grate and plenum **22** is a cylindrical drying chamber **34**, having a conical lower end portion **35** merging and communicating with the incinerator housing **10**. The lower axial end of chamber **34** opens downward into the combustion chamber **12** in spaced relation above the grate so as to constitute in effect an upward extension of the combustion chamber. Wet or liquid sludge is delivered into the drying chamber by means of the centrifugal sprayer, or atomizer, generally designated **36**, supported through an opening in the closed upper end **38** of the spray drying chamber. The atomizer includes a motor **40** which is in driving relation with the rotary distributor **42** to rotate the latter at high speed about a vertical axis which is preferably coincident with the cylindrical axis of the chamber **34**.

A suitable form of centrifugal atomizer **36** is described in U.S. Patent 2,292,572, including means for supplying the liquefied sludge to the atomizer as well as air cooling means for enabling it to withstand better the high temperatures to which it is necessarily subjected in operation.

Located over approximately the mid-area of the grate in a fore and aft direction, and at the location of the maximum heat intensity of the ignited bed of solid fuel **B**, is the intake or inlet port **44** of a hot gas recirculating blower **46**.

The blower is illustrated here as being driven by a rotary electric motor **48** to deliver the heated gases through a delivery conduit **50** back into the upper portion of the spray drying chamber **34**, preferably through an annular duct or manifold **52** concentrically surrounding the centrifugal atomizer **36**, and having a plurality of discharge orifices, or nozzles **54** directed tangentially in the direction of rotation of the rotor **42** to contribute to the centrifugal dispersion of the particles by the distributing rotor and to produce a swirling and gradually downwardly spiraling movement of the incoming heated gases toward the open lower end of the spray drying chamber.

The primary function of these heated gases is to dry the particles projected outward from the atomizer. The swirling action of the air currents serves not only to increase the radially outward dispersion or spreading of the particles by centrifugal force, but also to create sufficient turbulence to slightly retard the downward settling of the particles and thus to increase the length of their stay in the drying chamber and the thoroughness of the drying action.

The heated gases recirculated through the blower may be at such temperatures that they will serve not only to dry the sludge particles, but also to wholly or partially oxidize some of them, namely the smaller or lighter particles, while the latter are suspended in such gases, so that when the swirling gases with such entrained oxidized particles emerge downward through the open lower end of the drying chamber, the particles thus entrained will be carried with the gases to and through the flue **28** without ever settling upon the solid fuel bed **B**.

Thus, to the extent that the particles are burned while in suspension, the fuel bed will be proportionately relieved of such tendency as they might otherwise have contributed toward clogging or smothering of the fire. The capacity of the incinerator to handle the sludge will be accordingly increased.

Since the drying chamber **34** constitutes in effect an extension of the combustion chamber, the sludge particles which settle upon the solid fuel bed move thereon directly from the drying chamber without any loss of heat, such as would occur were they separately dried and then transferred to the combustion chamber. Thus, the sludge particles require but a minimum amount of added heat from the solid fuel bed to bring them to ignition temperature. Also, it should be noted that, as the downwardly swirling heated gases emerge from the lower end of the drying chamber **34**, they must move beneath the intake **44** of the heat recirculating blower **46** which is interposed between the drying chamber and the flue **28**, and at the same time will move through and be subjected to heat from the upward moving zone or curtain of superheated gases which moves progressively upward from the intensely hot mid portion of the fire bed **B** and the intake **44**. Thus, any entrained particles of unburned or partially burned sludge will be subjected to complete oxidation upon passing through this zone erroute to the flue.

In order to avoid the movement of any significant or appreciable amount of such particles into the intake, the flow of air beneath such intake may be downwardly deflected by means of a baffle **56** depending from the top wall of the housing, rearward of the intake **44**. Such a baffle serves the further important function of deflecting the path of movement of the entrained particles downward toward the fuel bed to thus be exposed to an increased amount of radiated heat.

It will be apparent that, after passing through the flue **28**, the exhaust gases with any entrained particles of burned sludge will pass through the scrubber **30** which will serve to remove and collect the entrained burnt sludge particles and/or fly ash, the bases then being discharged to the atmosphere through the chimney or stack **32**.

Throughout their lateral movement from beneath the drying chamber to the discharge flue **28**, the exhaust gases and entrained particles flow immediately above and in proximity to the ignited fuel bed, so that the heat therefrom will complete combustion of all entrained particles and gases before their discharge.

It is desirable in the operation of the apparatus that the rate of charging of the solid fuel onto the grating **14**, the volume of air through the solid fuel bed **B**, and the rate at which the dried sprayed particles are delivered onto the fuel bed will be so related that the temperature of the exhaust gases discharged through the flue will be maintained at or above 1400°F so as to destroy all organic odors. In the operation of the apparatus, which is apparent from the foregoing, the burning bed of solid material **B** is supported on and moved continuously rearward by the grating for discharge of the ashes of the completely consumed material into the ash pit **20**.

The particles of sludge are delivered downward into the combustion chamber from the drying chamber at a location directly over the upward opening plenum **22** above the burning bed of the solid fuel, and are laterally spread or distributed across the combustion chamber by the centrifugal action of the atomizer and/or by the centrifugal forces created by the whirling streams or jets of air emerging from the nozzles **54**. By such distribution, the heavier particles when received from the solid fuel bed, will avoid smothering same. At the same time, the lighter particles which are entrained in the gases will be burned while suspended enroute to the flue **28**.

The air which is formed upward through the fire bed **B** by the blower **24** and plenum **22** not only accelerates the rate of oxidation and increases the temperature of the bed, but in addition is itself heated and utilized to provide an upward rising stream or flow of gases above the forward end portion of the bed and beneath the drying chamber, through which the heavier particles of material from the drying chamber must descend and be further dried and/or ignited enroute to the solid fuel bed.

In addition, by entraining some of the lighter particles and carrying them laterally over the fuel bed to the flue **28** while causing their combustion while thus entrained, the moving air stream reduces the load on the solid fuel bed and thus increases the overall capacity of the apparatus accordingly.

Dual Combustion Chambers with Mechanical Grates

R. Tanner, T. Jacobovici and F. Wuthrich; U.S. Patent 3,533,365; October 13, 1970; assigned to Von Roll AG, Switzerland describe a combined furnace for simultaneously incinerating refuse or garbage and sewage sludge, the furnace incorporating two distinct combustion chambers in direct and open connection having mechanical grates of adequate design and position to suit the specific requirements for both the refuse and sludge and to allow the final phase of incineration to be completed in common. The furnace further includes means for the common discharge of both the gaseous products and the residues of incineration.

Automated Sludge Batchwise Incinerator

In a process which is described by *R.P. Farrell, Jr. and J.F. Schulte; U.S. Patent 3,381,637; May 7, 1968; assigned to General Electric Company* the sewage sludge incinerator is operated in a batchwise manner, sewage sludge being charged and then after removal of water by filtration the sludge is heated and dried and finally burned. An unusual aspect centers in the combination of control means for regulating the weight of sludge deposited in the incinerator for disposal by combustion so that the volume of sludge charged into the incinerator on each successive cycle will be varied automatically according to the solids content of the sewage sludge feed.

Furnace Grates

According to a process described by *R. Knaak; U.S. Patent 3,808,989; May 7, 1974; assigned to Koppers-Wistra-Ofenbau GmbH, Germany* a mix of household refuse and sewage sludge is admitted onto a furnace grate over which it travels. During this travel the mix is dried, ignited and partially combusted. From a discharge end of the grate the partially combusted mix is discharged directly into a rotary furnace drum where combustion is completed. The combustion gases developing during partial combustion on the grate and during final combustion in the furnace drum are separately withdrawn and are thereupon united before being passed into a gas cooler.

W. Bastgen; U.S. Patent 3,871,285; March 18, 1975; assigned to Albert Klein KG, Germany describes an incinerator device for burning granulated or otherwise fractured combustible material. The incinerator comprises a grate to which is fed combustion air from below. The grate is in the form of a sieve or a sieve insert which is swingably mounted and to which swinging motions are imparted.

There is also described a method for processing residuals in wastewater, liquefied sewage and liquefied sludge. According to the method, part of the moisture content of flowable material to be processed is removed and the solids in the wastewater, sewage or sludge are fractured to desired sizes after partially drying the same. The thus obtained granulate is then dried by applying heat.

Finally, if desired, the now dried granulated material is burned preferably by using a glowing or hot bed continually coacting with an air cushion.

Movable Separation and Filtration Bucket

W.E. Dorn; U.S. Patent 3,714,914; February 6, 1973 describes a sewage disposal device operative to separate solids from liquids in raw sewage and completely incinerate the solids and treat the fluids to eliminate pollution at the source of the sewage.

A separation and filtration bucket movable between a settling chamber for containing fluids and a combustion chamber, receives intermittent charges of raw sewage for filtration and drainage into the fluid separation chamber and for transportation and depositing of the remaining solids into the combustion chamber for drying out and complete incineration of the solids for subsequent removal as inert ash. Means is provided for further treatment of the separated liquid sewage to render it substantially pollution-free effluent.

Referring to Figure 1.18, the process equipment comprises outer concrete shell **1**, which houses a primary settling chamber **2**, also a liquid treating chamber **3**, an access well **4** to allow for access to operating equipment, also equipment holding chamber **5** is located within the outer concrete shell.

FIGURE 1.18: SEWAGE DISPOSAL DEVICE

a.

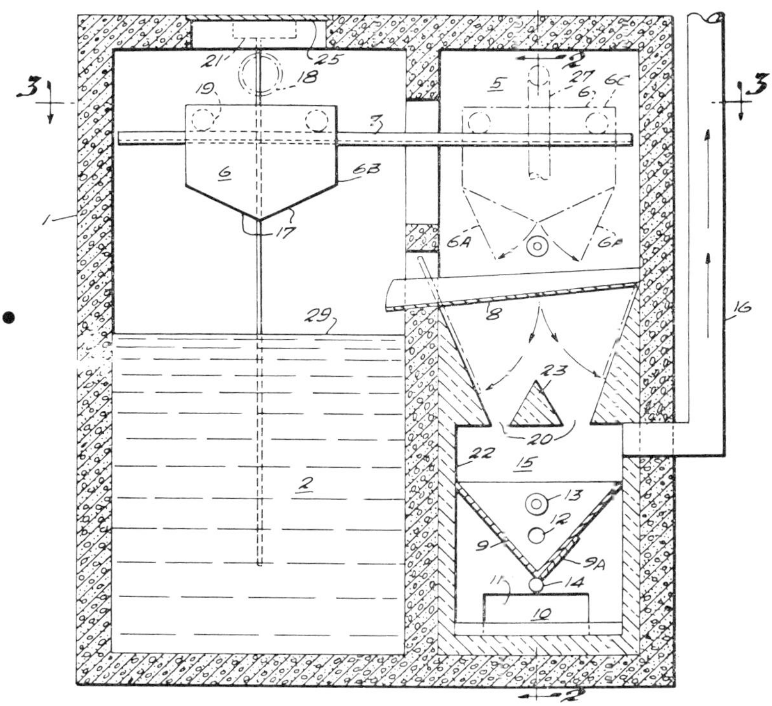

Sectional Side Elevational View

(continued)

FIGURE 1.18: (continued)

b.

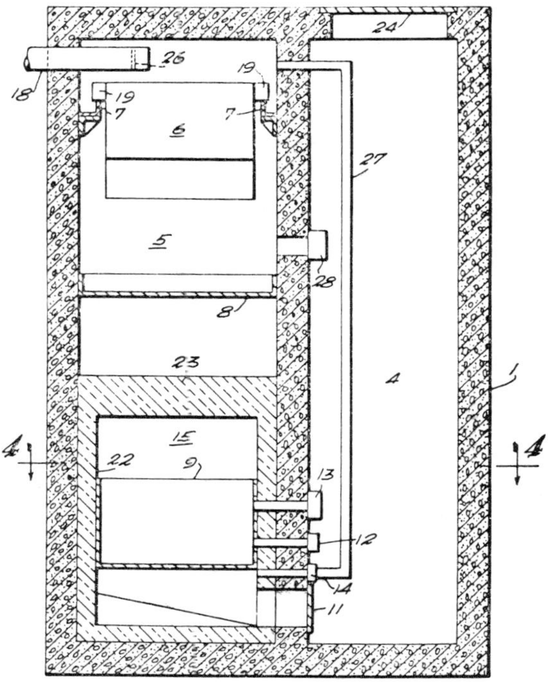

Vertical Detailed Section Taken on Line **2-2** of Figure 1.18a

c.

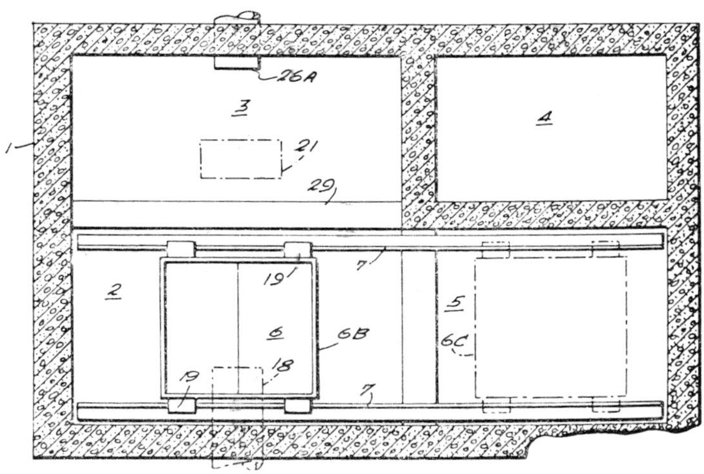

Horizontal Detailed Section Taken on Line **3-3** of Figure 1.18a

FIGURE 1.18: (continued)

d.

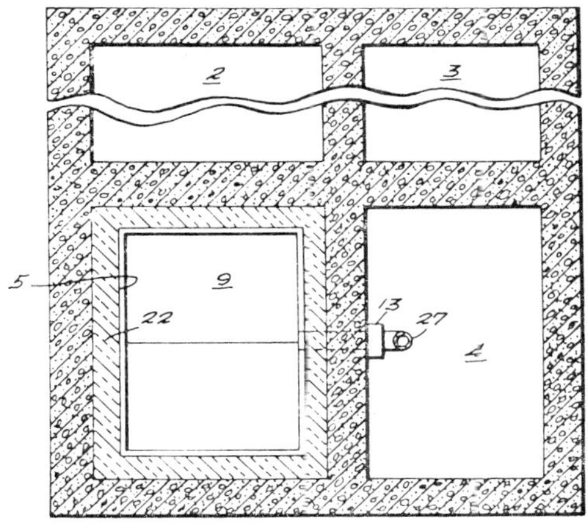

Horizontal Detailed Section Taken on Line **4-4** of Figure 1.18b

Source: U.S. Patent 3,714,914

Located within equipment holding chamber **5** is a separation and filtration bucket **6** which moves on rails **7**, between the primary settling chamber **2** and the equipment holding chamber. A drying and separation plate **8** is located directly below the separation and filtration bucket when it is in its operating position in the equipment holding chamber.

A cast iron grate **9** is located at the bottom of combustion chamber **15**. Both the cast iron grate and the combustion chamber are located directly below the drying and separation plate. Below the cast iron grate is ash pit **10**. The ash pit is made accessable by cleanout door **11** from access well **4**. A stoker and observation port **12** is located in the wall between the cast iron grate and the access well to allow for stoking and observation of the cast iron grate when required.

High pressure gas or oil burner **13** is located in a wall between the cast iron grate and the access well and fires directly into the cast iron grate to incinerate all dried sewage solids to a fine ash. The cast iron grate has a movable ash removal door **9A** for the removal of any crystallized ash that might accumulate within the iron grate.

Oxygen generating unit **14** is located in the wall between the ash pit and the access well and leading into space in the ash pit. The oxygen generating unit is required along with the gas or oil burner to raise the temperature of the combustion chamber up to 1800° to 2000°F which is required to burn all products of combustion being generated during the incineration period, especially hydrogen sulfide and sulfur dioxide gases. At the normal operation temperature of the device all of the gases given off will be burned to an odor-free gas that will be discharged into the atmosphere through outlet flue **16**, thus eliminating any contamination of the outside air. The separation and filtration bucket **6** is provided with two swing doors **6A** which open downward from the bottom which will allow for emptying contents of filtered solids onto drying and separation plate **8**. The separation and filtration bucket is also provided with a specially designed filter system **17** to allow for the separation of all solids from the liquids as received from the raw sewage feed line **18**.

All solids are retained in the separation and filtration bucket while the liquid is filtered into primary settling chamber **2**. The separation and filtration bucket is also equipped with wheels **19** which roll on rails **7**. The raw sewage feed line is located in such a position as to empty any raw sewage into the separation and filtration bucket when it is located in operating position **6B** in the primary settling chamber.

Heat from the combustion chamber **15** passes through adjustable vents **20** to regulate the amount of heat supplied to the drying and separation plate **8** and dry any solids that have been deposited on it by separation and filtration bucket **6** before passing the solids into the combustion chamber and finally into the cast iron grate **9** for final incineration.

The drying and separation plate is made up as two doors which open downward at the middle to allow for the emptying of the dried solids into the combustion chamber. Oxidizing and chlorinating unit **21** is located at the top of liquid treating chamber **3**, with a feed line running to the bottom of and inside side of liquid treating chamber **3**.

The combustion chamber and the ash pit **10** are lined with a super-high-temperaature refractory **22** which is both nonmoisture and nonspalling-proof. A heat separation block **23** is located at the top of the combustion chamber and is used for even distribution of heat at the bottom of the drying and separation plate.

The access door **24** is located in the top of the access well **4** to permit entry into same. The cleanout and access door **25** is located in the top of the primary settling chamber **2**. Outfall pipe **30** is located in the outside wall of the liquid treating chamber **3** for the purpose of draining treated effluent into drainage field or municipal sewer line.

The recirculating gas line **27** is located in the access well **4** and leading from the top section of the equipment holding chamber **5** to the bottom of the equipment holding chamber and then back into the combustion chamber.

In the combustion chamber the final burning of unburned gases collected from the top section of the equipment holding chamber **5** takes place before these gases are passed through the outlet flue **16** to the outside atmosphere.

The observation and stoking port **28** is located in the wall between the equipment holding chamber **5** and the access well **4** and leading into the space just above the drying and separation plate **8** to allow for observation and cleaning of excess material that has accumulated on the surface of the drying and separation plate when required.

Spray Technique in Reactor

According to a process described by *N. Rösler, W.G. Rudat and E. Albrecht; U.S. Patent 3,922,221; November 25, 1975; assigned to Rheinstahl AG, Germany*, sewage sludge is conditioned and subsequently dehydrated by directing sludge in the form of a spray into a sludge reactor tower after it has first been treated by floating surface aerators in a sludge storage tank. The sludge is first treated to control its pH factor so that it is below 4 and preferably between 2 and 3, and it is sprayed into droplets in the reaction tower at temperatures ranging from 40° to 100°C. Gases from a turbulent bed furnace into which a dehydrated sludge is fed are directed into the reaction tower at a level below a sludge bath in the bottom of the tower and it flows upwardly through the spray droplets to an outlet and to ash removing devices.

The sludge which is collected in the bottom of the tower is drained off to a dehydration device, for example a mechanical dehydrator such as a wire press or centrifuge. The sludge which is removed from the reaction tower is treated with an alkaline material to maintain its pH from between 6.5 to 7. The apparatus includes a sludge reactor tower having a lower end defining a sludge bath and with means for spraying a selected quantity of sludge into the tower. A hot exhaust gas outlet is located above the spray means and means for admitting flue gases into the tower are arranged above the sludge bath at the bottom of the reaction tower and below the spray means.

According to a process described by *J.E. Hanway, Jr.; U.S. Patent 3,738,289; June 12, 1973; assigned to Chicago Bridge & Iron Company*, waste-treatment sludge from which water has been largely removed is dispersed downwardly in an upwardly flowing atmosphere within a firebox where the temperature is maintained sufficiently high to dry and burn the sludge while it is in free falling condition, and ash and other products of combustion escaping upwardly to a gas scrubber and ash separator, with relatively clean gas being discharged to the atmosphere.

The firebox is preheated by a burner at the bottom. As necessary to maintain the required temperature, there is further burning of fuel at this burner or by addition to the sludge feed. Addition of fuel oil to the sludge before pumping the sludge facilitates pumping, and achieves mixing. Supplementary air is fed at the bottom and somewhat tangentially, to impart rotation to the gases within the firebox.

Feed System

According to a process that is described by *M. Ochi and F. Shibata; U.S. Patent 3,877,398; April 15, 1975; assigned to Ataka Construction Co., Ltd., Japan*, an incinerator for sludge or the like is provided with a gravity feed down an inclined surface. At least one site along the incline, where the material being fed is in contact with heated drying gases from the combustion chamber, the incline has a generally horizontal surface portion, swept by a multiple-, curved-blade rotary agitator which cuts and stirs the material to expose a greater part of the bulk thereof to the exterior for drying by contact with heated gases from the combustion chamber. The incineration of sludge means of the process, enables the fully automatic charging, stirring, drying and incinerating of that material as one continuous operation.

The breaking of the clods at the agitating station(s) promotes efficiency of the device by using heat thrown off the combustion process to prepare the sludge or the like to be more easily incinerated as dried, pulverized material.

Sludge Supply Manifold

L. Bakker; U.S. Patent 3,678,870; July 25, 1972; assigned to The Air Preheater Company, Inc. describes an incineration chamber that is especially adapted to burn liquid or semiliquid fuel in the form of a combustible sludge by increasing the residence time during which the fuel is in contact with the hot gases of combustion.

In an incinerator of the "starved air" type shown in Figure 1.19, trash or refuse to be burned is placed in an incinerator housing **10** enclosing a primary combustion chamber **12** that is preferably formed in an ellipsoidal configuration in the manner shown in U.S. Patent 3,419,700. The incinerator is formed with a loading or access door **14** at one end for the supply of solid material to be burned and an outlet port **16** at the other end for the exhaust of products of combustion. The inner wall of the incinerator is insulated by a layer of fire resistant material **18** capable of withstanding temperatures ranging upwards to approximately 2500°F.

The bottom of the vessel is adapted to receive a series of tubes **20** having openings **22** therein sized to provide adequate air for combustion of a charge placed therein. As an arrangement that protects the tubes from the corrosion and erosion of the burning waste the tubes may be embedded in the insulating material so only the exit openings **22** for the air are connected via valve **25** in line **27** to a suitable outside source of air for combustion whereby sufficient air may at all times be supplied to the primary combustion chamber.

In the same manner that tubes **20** are embedded in the insulation layer lining the floor of the incinerator, sludge supply tubes **26** are embedded in the insulation lining the upper portion of the incinerator. Thus liquid waste or sludge supplied to the upper ends of the tubes **26** may flow downwardly through the tubes so that it is heated while passing in heat exchange relation with the hot insulating material which lies in contact with the combustion gases. As the hot sludge is exhausted from the belled outlets **28** it usually has a low viscosity because of its high water content, but as it flows downwardly over the walls of the incinerator it quickly loses moisture by evaporation and is subsequently burned or subjected to pyrolysis according to a predetermined incineration process.

The tubes **26** are attached to a header or manifold **30** that is supplied with sludge from a source via an inlet pipe **32**. The sludge is preferably fed through pump **40** and valve **42** whereby the proper amount of sludge may be supplied to the tubes **26** and exhausted from the belled outlets. While some of the sludge may be blown from the belled ends **28** of the tubes into the incinerator chamber, a major portion of it will gravitate downwardly and follow the curvature of the wall as a slowly moving stream flowing to the bottom of the incinerator.

As the sludge flows down through the tubes **26** and out the belled ends thereof it is heated and the moisture therein is lost by evaporation leaving a dry residue or cake.

FIGURE 1.19: SLUDGE BURNER

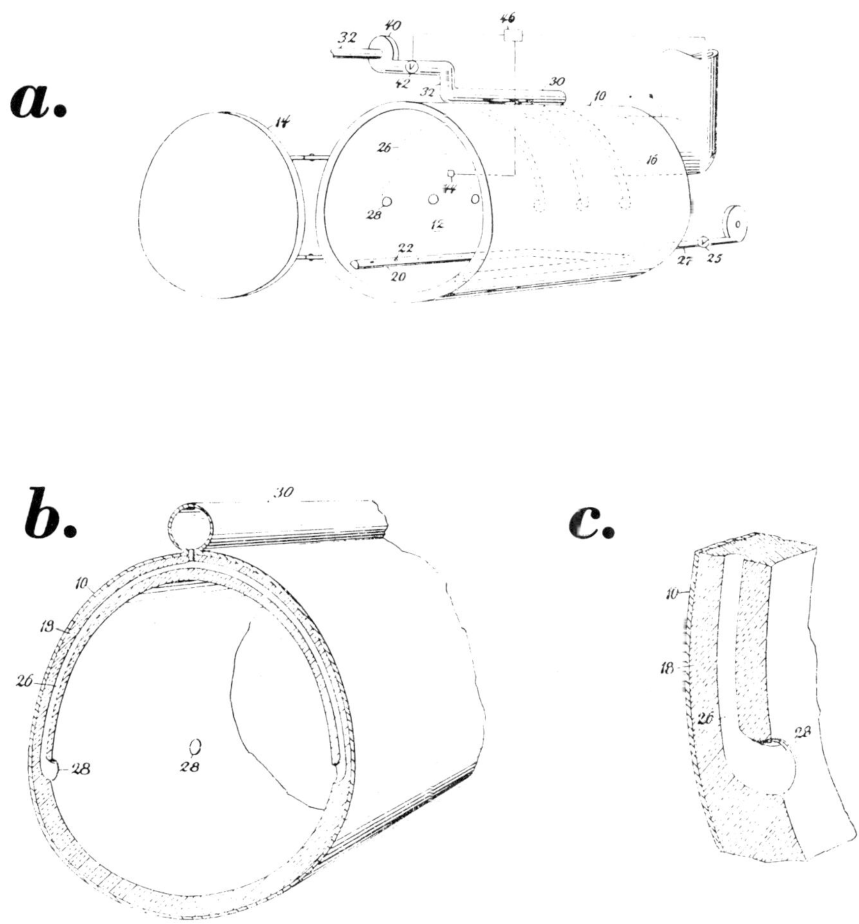

(a) Perspective View of an Incinerator Adapted to Handle Sludges

(b) Perspective View of an Incinerator Showing the Relationship of Sludge Supply Manifold and Supply Tubes

(c) Enlarged View of Sludge Supply Tube as Mounted in the Incinerator

Source: U.S. Patent 3,678,870

This cake is summarily subjected to a continued pyrolysis according to standard starved air incineration principles until only an inert ash remains.

To enable the effective burning of all sludge in spite of its varying calorific value, the flow of sludge to the incinerator chamber may be controlled in accordance with the temperature in the chamber obtained by the partial combustion of the waste material. Accordingly, a thermocouple **44** in the primary combustion chamber is adapted to sense the temperature of the heat generated therein and activate a controller **46** that regulates operation of the sludge supply valve **42** and the air supply valve **25**.

Thus the thermocouple **44** is adapted to sense the heat of the combustion chamber to activate a controller whereby a temperature in excess of a predetermined amount will move valves **25** and **42** toward a closed position. Similarly insufficient heat at thermocouple **44** will actuate the controller **46** to move the air and fuel valves toward an open position to increase the combustion rate in the chamber **14**.

Concentrated Sewage Injected into Hot Flue Gas

R. Mutke and P. Steller; U.S. Patent 3,587,488; June 28, 1971; assigned to Vereinigte Kesselwerke AG, Germany describe a process according to which rubbish and sewage are simultaneously disposed of while the amount of injected sewage is controlled so as to maintain a substantially overall effective burning operation of the disposal apparatus.

Shown in Figure 1.20 is a disposal apparatus for the combined disposal of rubbish and sewage and which is seen to comprise a separator unit **1** which, depending on the specific requirements of the apparatus, may be incorporated to separate or filter solid particles from the sewage disposed in this unit for such filtering treatment. This separator unit **1** may, for example, be a centrifuge. Following the filtering process of the sewage, the solids are conveyed to a combustion furnace **2** by means of a duct member **14** and are combusted or burned in this furnace. Also combusted in the furnace is the rubbish which is directly disposed there either manually or mechanically, as may be required.

The liquid sewage emanating from the separator unit **1** is subsequently conveyed to a heat operated evaporator or flashing installation, generally indicated at **15**, for concentration in the installation. In this installation the liquid sewage is initially received in a first evaporating stage in the form of evaporator chamber **5**, via a heat exchanger **3** and a feed pump **4**. From this chamber **5**, the partly evaporated liquid sewage is then conveyed to a second heat exchanger **3'** via a feed pump **6** and again conveyed back into the chamber via a pressure valve **7** for further evaporation.

This recirculation process of the liquid sewage to and from the chamber **5** via the feed pump **6**, the heat exchanger **3'** and the pressure valve **7** is repeated and the hot vapors resulting from this repeated evaporation process in the chamber are conveyed therefrom to the heat exchanger **3** via a duct member **8** for heating the liquid sewage emanating from the separator unit **1**. Following heating the liquid sewage in the heat exchanger **3**, the vapors are then disposed of in the form of a clean condensation product. The liquid sewage is constantly kept at a certain level by means of the feed pump **4**.

FIGURE 1:20: COMBUSTION PROCESS

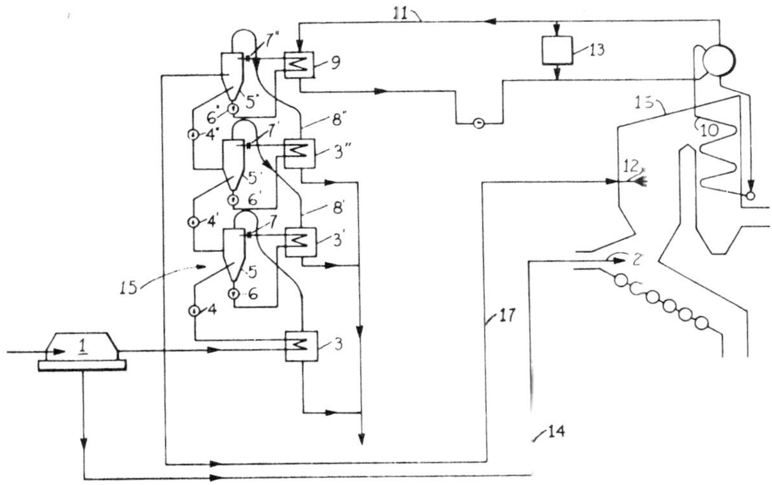

Source: U.S. Patent 3,587,488

The partly concentrated sewage resulting from this repeated recirculation and evaporation process in the first evaporator chamber **5** is then conducted to a second evaporating stage in the form of evaporator chamber **5'** the feed pump **6'**, the heat exchanger **3''** and the pressure valve **7'** via a feed pump **4'** for further recirculating and concentrating the sewage and, upon treatment of the sewage in the second evaporating stage, the sewage is then further conducted to a third evaporating stage in the form of evaporator chamber **5''**, the feed pump **6''**, the heat exchanger **9** and the pressure valve **7''** via a feed pump **4''**, for still further concentration treatment of the sewage

Likewise as regards the heating of the liquid sewage in the heat exchanger **3**, the hot vapors resulting from repeated evaporation of the sewage in the chambers **5'** and **5''**, are used to heat the sewage in the preceding stage, i.e., hot vapors are conveyed from the chamber **5'** to the heat exchanger **3'** via a duct member **8'** and hot vapors are conveyed from the chamber **5''** to the heat exchanger **3''** via a duct member **8''**. This arrangement is based on the countercurrent principle.

The heat necessary for heating the sewage in the third heat exchanger **9** is supplied by a heat exchanger **10** via a duct member **11**. The heat exchanger **10**, in this instance, is an air-water type heat exchanger operative for the production of saturated steam and is arranged in the path of a flue or smoke-duct **16** which is connected to the combustion furnace **2**. The flue **16** is operative to conduct the stream of hot flue gases resulting from burning the rubbish in the furnace, into engagement with the heat exchanger **10** to subject the same to heat so as to convert the water into saturated steam which then is supplied to the heat exchanger **9**.

As regards the three successive stages for concentrating the sewage, it is noted that, in this instance, the first stage incorporating the chamber **5** has a lower temperature than the second stage incorporating the chamber **5'**, and that the third stage incorporating the chamber **5''** has the highest temperature. In this manner the degree of evaporation and subsequent concentration of the sewage is gradually increased such that at the third stage the liquid sewage is completely concentrated.

The sewage thus fully concentrated in the chamber **5''** is then conveyed via a duct member **17** to the flue **16** and, by means of at least one atomizer **12**, is injected into the stream of hot flue gases passing through the flue. The concentrated sewage injected into the hot flue gases serves to cool these gases preferably to a temperature of 700° to 750°C at which temperature it has been experimentally found that the concentrated sewage is entirely and completely burned.

Further cooling of the flue gases to a temperature of about 200°C is effected in the heat exchanger **10** in which, as explained above, the hot flue gases convert the water in the heat exchanger into steam. Arranged intermediate the heat exchangers **9** and **10** is a thermoregulator **13** which serves to adjustably regulate the supply of saturated steam from the heat exchanger **10** to the heat exchanger **9**. By regulating this steam supply to the exchanger **9**, the degree of concentration of the sewage in the evaporator installation **15** is varied in that the rate of evaporation of the liquid sewage increases or decreases with the rise or fall of temperature.

In varying the heat and thereby the concentration of the sewage in the evaporator installation **15**, also the amount of concentrated sewage is adjusted so that when the temperature in the heat exchanger rises, the recirculation for evaporating the liquid sewage in the respective evaporating stages of the installation decreases which gives as a result that the sewage thus treated is supplied to the flue **16** and injected into the flue gases at a faster rate, while when the temperature in the heat exchanger decreases, recirculation for evaporating the liquid sewage increases with the result that the sewage thus treated is supplied to the flue and injected into the flue gases at a slower rate.

Increase in quantity or rate of concentrated sewage is then necessary when the combustion production in the combustion furnace **2** is increased, as a result of which also the temperature of the flue gases increases and since it is an object of the process to maintain the temperature of the flue gases at a preferred temperature of not less than 700° to 750°C, such increased rate or quantity of concentrated sewage injected into the hot flue gases serves to cool and maintain these gases at such temperatures.

Contrarily, decrease in quantity or rate of concentrated sewage is then necessary when the combustion production in the combustion furnace **2** is decreased, as a result of which also the temperature of the flue gases decreases, and such decreased quantity or rate of concentrated sewage injected in the less hot flue gases serves to cool these gases to a lesser degree thereby maintaining the same at the desired temperature.

In any event, the quantity or rate of concentrated sewage injected into the flue gases should be varied dependent upon the temperature of the flue gases and maintain the same at 700° to 750°C, at which temperatures it has been experimentally found that the concentrated sewage is burned in its entirety.

Alternatively, and in order to vary the concentration of the sewage in the evaporator installation **15**, the latter may consist of a great number of individual evaporating stages, and such variation of concentration of the sewage may be effected by either connecting or disconnecting a number of these evaporating stages, in which case the supply of steam supplied to the heat exchanger **9** may be constant.

For example, in concentrating sewage with a liquid content of about 98%, it is obvious that with the example, shown, more evaporating stages are to be utilized than with a sewage having a liquid content of about 95%.

In related work *R. Mutke and P. Steller; U.S. Patent 3,592,149; July 13, 1971; assigned to Vereinigte Kesselwerke AG, Germany* describe an apparatus for drying sewage having a high liquid content of 90% or more. The apparatus according to this process comprises a flue duct cooperating with a source of hot flue gases to conduct these gases along a predetermined path and in which at least one injecting member is operative to inject substantially liquid sewage into the hot flue gases passing through the flue duct. Mixing means are provided after the injecting member to mix the hot flue gases with the sewage injected therein to thereby dry the liquid sewage and simultaneously cool the hot flue gases. Complete details of the apparatus are provided.

Conveyor Unit

N.F. Pergola; U.S. Patent 3,655,048; April 11, 1972 describes a method and apparatus for treating and disposing of sewage which includes introducing raw sewage into a liquid separator, separating liquid sewage from the solid sewage and collecting the separated liquid in a collection reservoir. The solid sewage is passed onto a filtration conveyor where the carry over liquid is filtered by the conveyor into a filtration reservoir and the solid sewage is transported to an incinerator. The liquid in the filtration reservoir is recirculated to the collection reservoir.

An air supply and diffusion device introduces diffused air into the liquid in the collection reservoir to carry suspended solid sewage to the surface of the liquid in the collection reservoir. The liquid and solid sewage on the surface of the collection reservoir is passed over a weir onto the filtration conveyor where the filtration, transportation, recirculation and incineration steps are repeated.

The diffusion of air into the liquid, the passing of the liquid and the solid sewage onto the filtration conveyor and the recirculation of the liquid is continued for a set period of time after the raw sewage has stopped flowing into the separator. At the termination of the recirculation cycle, the liquid is passed from the collection chamber through baffles into an effluent reservoir where the liquid is treated to reduce the bacteria count.

Screw Conveyor

R.B. Burden, Jr.; U.S. Patent 3,472,185; October 14, 1969; assigned to General Incinerators of California, Inc. describes an apparatus which comprises a screw conveyor having a hollow shaft, one end of which terminates centrally within the base portion of a cylindrical burner tube. The casing of the screw conveyor terminates at a conically tapered nozzle end portion which encloses the open discharge end of the hollow conveyor shaft and projects into the burner tube centrally of its base.

As the shaft of the conveyor rotates, sludge fed into the input end of the conveyor casing from a dewatering means is conveyed to and deposited within the nozzle end thereof. At the same time air blown through the hollow shaft into the nozzle end of the casing forces a spray of sludge deposited within the nozzle out through nozzle apertures into the burner tube. The discharged sludge is whirled in a rotary path through the burner tube by blasts of primary combustion air introduced into the tube at its base as fuel is also injected into the tube and ignited to heat and effect combustion of the whirling sludge particles.

Referring to Figure 1.21, the apparatus includes a cylindrical burner tube **10** having an outer steel shell **12** and an inner refractory liner **14** defining a combustion chamber **16**. An insulating air space **18** between the liner **14** and shell **12** may be filled with noncombustible insulating material such as asbestos if desired. The burner tube has a cylindrical base constructed as a separate unit, including a concrete inner base portion **22** and an outer steel jacket **23** joined to the main shell of the burner tube by annular angle brackets **24**. The concrete base portion has a central opening **26** for receiving the projecting discharge end or nozzle portion of a screw conveyor **28**.

Positioned in the base radially outwardly from the screw conveyor **28** is a burner ring **30** having orifices **32** through which fuel such as gas or oil is injected into the combustion chamber. Fuel flows into the ring through a fuel line **34** from a supply source. Ignition means **35** on the ring ignites fuel injected into the chamber through the orifices **32**. The ring **30** is preferably constructed in segments of equal size, joined together by close nipple fittings, although a one piece burner could be used. The diameter of the burner ring is therefore adjustable by adding or removing segments as desired according to the requirements of a given installation and diameter of the burner tube for which the ring is intended.

A plurality of annularly arranged tuyeres **36** project tangentially into the combustion chamber **16** from the base portion **22** at a position radially outwardly of the burner ring **30**. The tuyeres are fixed at their inner ends to an annular metal manifold **38** fastened to the exterior jacket **23** of the base by the angle brackets **40**. A blower **42** connected to an intake opening **44** of the manifold blows primary combustion air into the manifold from which the air is discharged into the combustion chamber tangentially of such chamber by the tuyeres **36**.

Means are provided for feeding sludge into the burner tube. These means include the previously mentioned screw conveyor **28** which comprises a casing **50**, a hollow central shaft **52** within the casing and a helix, or blade, **54** extending about the shaft **52** between the casing and the shaft. The discharge end of the casing **50** projects into the combustion chamber **16** through the central opening **26** in the base **22** and is enclosed and conically tapered to define a nozzle **56** having discharge orifices **58** in its tapered flanks.

FIGURE 1.21: COMBUSTION PROCESS

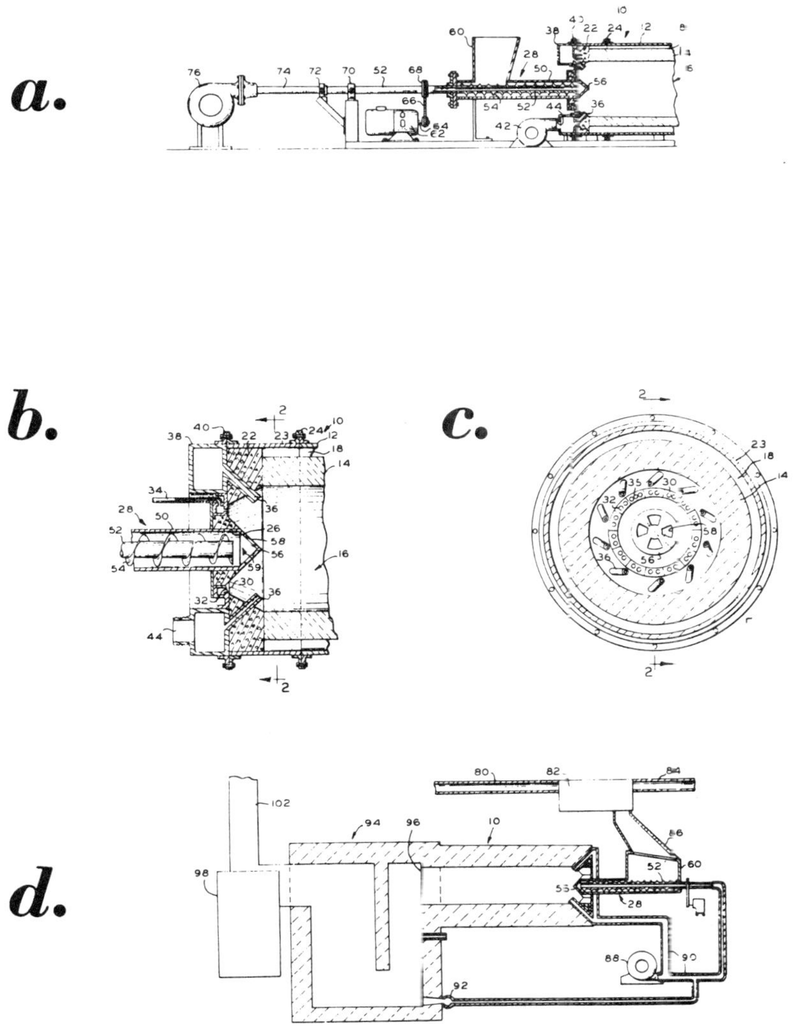

(a) Side View of Apparatus
(b) Discharge End of Conveyor
(c) View Along Line **2-2** of Figure 1.21b
(d) Complete Disposal System

Source: U.S. Patent 3,472,185

The shaft **52** and blade **54** of the conveyor terminate at about the position at which the tapered nozzle portion **56** of the casing commences so that an enclosed space **59** is provided within the nozzle for receiving sludge from the conveyor. A sludge-receiving means, including a hopper **60** adjacent the opposite input end of the casing **50** enables sludge to be fed into the conveyor.

The hollow conveyor shaft **52** extends horizontally beyond the input end of the casing **50** and blade **54**, and is connected to a power means for rotating the shaft in a direction to convey sludge into the nozzle **56**. The illustrated power means includes an electric motor **62**, the drive shaft **64** of which is operatively connected to the conveyor shaft **52** to drive the same by a chain **66** and sprocket **68**. The conveyor shaft **52** is rotatably mounted near its outer end in a bearing **70** and is also coupled at **72** to a stationary hollow shaft **74**, which is connected in turn to a second air blower **76** for blowing air through the hollow conveyor shaft **52** into the nozzle end **56** of the casing.

In operation, sludge that has been at least partially dehydrated, as, for example, by being passed through a centrifuge, is fed into the hopper **60** and conveyed through the casing **50** by the rotating screw conveyor into the interior of the nozzle **56**. Simultaneously air blown through the conveyor shaft **52** by the blower **76** forces the sludge collected within the nozzle out through the discharge orifices **58** in a spray into the combustion chamber **16**. Air entering the nozzle **56** through the shaft **52** is prevented from backing up between the casing and shaft by the continuous wall of sludge moved forwardly through the casing by the blade **54**.

Prior to the entry of sludge, the burner tube is preheated to combustion temperature by ignited fuel from the burner ring **30**. Within the combustion chamber, the sludge is whirled in a rotary path by a high velocity stream of combustion air introduced into the tube through the tuyeres **36**. The whirling particles of sludge are enveloped in a flame, heated and gradually combusted. By the time the particles have been moved to the discharge end, all but the unusually large particles will have been completely consumed. However, those particles that are not consumed can be subjected to further burning in secondary combustion means, such as, for example, the incinerator system illustrated in Figure 1.21d.

Figure 1.21d illustrates a complete system for destroying sludge and incorporates a burner tube **10** and screw conveyor **28**, similar to the ones previously described, as its primary elements. The system commences with a sludge feed line **80** through which sludge-bearing sewage is pumped into a centrifuge dryer or other dewatering means **82**, where the sludge is separated from the effluent and then deposited through a chute **86** into the hopper **60** and then into the screw conveyor **28**. The effluent is discharged from the centrifuge through the line **84**. A single blower **88** and its associated system ducts **90** supply air to the nozzle **56** of the conveyor through the hollow shaft **52**, and to an auxiliary, premix burner **92** at the base of a secondary combustion chamber, or incinerator **94**.

Sludge sprayed into the burner tube **10** from the nozzle **56** is burned and simultaneously whirled to the discharge end **96** of the tube and into the incinerator **94**, where incompletely burned particles are burned further and ultimately destroyed. For this purpose auxiliary burner **92** at the base of the incinerator injects premixed fuel and air into the incinerator and ignites the same. Gaseous products of combustion from the sludge are drawn with the aid of an induced draft fan through the incinerator **94** into a gas washer unit **98** where any ash is removed therefrom, before the gases are discharged into the atmosphere through the stack **102**. Summarizing the method by which sludge is conveyed and destroyed in the previously described apparatus, sludge is first at least partially dewatered and then is fed continuously by a screw conveyor into a partially enclosed space within the base of a burner tube, which space is defined by the tapered nozzle end **56** of the casing. A stream of forced air is then blown into the enclosed space behind the deposited sludge through the hollow conveyor shaft **52**, forcing the sludge out in a spray through orifices in the walls of the tapered nozzle **56**, into the burner tube.

Within the burner tube the sludge particles are whirled in a rotary path gradually to the discharge end of the burner tube while being destroyed by intense heat and flame at a temperature which may reach upwards of 2000°F. The gaseous products of combustion are then scrubbed and exhausted to atmosphere. Auxiliary fuel consumption will depend on the amount of water remaining as part of the sludge after the dewatering step The method has been found to work satisfactorily with sludge having a 15 to 20% solids content by weight, although optimum results are achieved with sludges having a solids content approaching 40% by weight.

Heat from Incinerator Used for Dewatering

A process described by *O. Salamon; U.S. Patent 3,552,333; January 5, 1971; assigned to Von Roll AG, Switzerland* involves the concomitant incineration of both solid refuse and sewage sludge. The thermal dewatering of the sewage sludge is carried out in suitable, convenient equipment in a separate manner from the incineration of the solid refuse together with the dried sludge.

The equipment generally comprises a vaporizer of known construction, e.g., a thin layer vaporizer and a drum vaporizer or volute vaporizer of different construction with its design and function adapted to deal with suspension of particulate solid matters. The heat required for the thermal dewatering process is provided by the incineration of the solid refuse and dried sludge in furnaces also of known design and function.

Hot flue gases from the incinerator, or alternatively, steam generated in a boiler heated by the flue gases, are used as an intermediate thermal agent for the transfer of heat from the incinerator to the vaporizer.

Figure 1.22 illustrates an operational layout for carrying out the process where the hot flue gases from an incinerator are used as an intermediate thermal agent to directly provide heat to a one-stage vaporizer for the liquid sewage sludge. Referring to the figure, an incinerator for solid refuse is designated by reference numeral **1** and is seen to comprise, in the usual fashion, a charging pit **2**, a combustion grate **3**, which generally is of mechanical type, and an uncooled combustion chamber **4** having refractory brickwork.

The refuse to be incinerated is charged at **0** and subsequently passes through the charging pit **2** to form a layer on the grate **3** at which point the refuse is ignited and burnt out in known manner, the noncombustible residues known as the clinker being discharged through the opening **5**. Hot combustion gases leave the combustion chamber as illustrated at reference numeral **6** and are led through a convenient duct into the primary or heating side **7** of a vaporizer where the hot combustion gases yield their heat which is eventually expelled by natural or induced air currents into the atmosphere at reference numeral **8**, the temperature of the combustion gases being not under some 570°F. Liquid sludge having a 92% water content is fed into the secondary side **10** of the vaporizer, while the dried sludge is withdrawn at reference number **11** and conveyed by convenient means such as the endless belt conveyor **12** into the charging pit **2** of the incinerator, there to be commingled with the solid refuse. The commingling of the dried sludge with the solid refuse may, of course, take place elsewhere than in the charging pit, such as for example, in the refuse bunker.

FIGURE 1.22: PROCESS FOR CONCOMITANT INCINERATION OF SOLID REFUSE AND SEWAGE SLUDGE

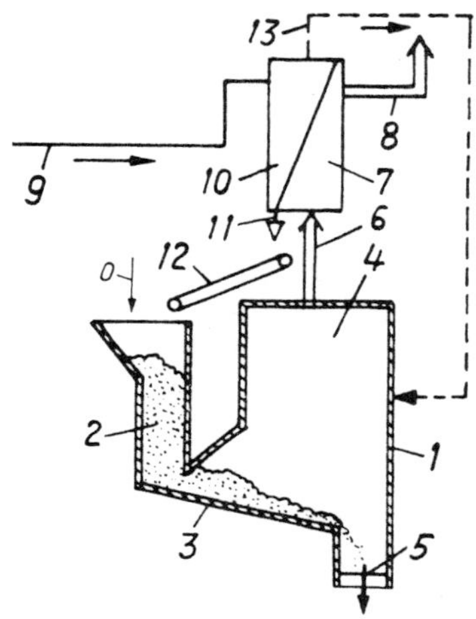

Source: U.S. Patent 3,552,333

The secondary water vapor which may be malodorous or otherwise affected, which results from the drying process in the secondary side **10** of the vaporizer is conducted at **13** into the combustion chamber **4** of the incinerator where the secondary water vapor is superheated so as to destroy any noxious substances which it may contain and thus render the incineration process and operation odorless. The general layout as depicted in the figure may be expanded, for example, by providing a two-stage vaporizer which may be advantageous and useful when the amount of sewage sludge is out of proportion to the amount of solid refuse.

In related work, *R. Tanner; U.S. Patent 3,529,588; September 22, 1970; assigned to Van Roll AG, Switzerland* describes a process for the concomitant incineration of solid refuse and aqueous sewage sludge. The solid refuse is burnt in a conventional incinerator whereas the aqueous sewage sludge undergoes a partial thermal dewatering by partially vaporizing the sludge water in a pressurized vaporizer. The vaporizer is heated by hot combustion gases effluent from the incinerator. The resulting sludge slurry is introduced in an atomized state into the combustion chamber of the incinerator or, alternatively, undergoes a further mechanical dewatering step to be brought into a semisolid consistency and is then added to and mixed with the solid refuse prior to its incineration to be burnt together.

Combustion of Low Solids Suspensions

In a process described by *Gustav Bojner and Gösta Bojner; U.S. Patent 3,413,937; December 3, 1968; assigned to AB Torkapparater, Sweden* the waste material from a purification plant, which is a suspension of a sludge in a liquid, is divided in at least one liquid phase and one semifluid phase, and the liquid phase is utilized as a washing liquid for hot gases obtained from a combustion of the waste material.

The liquid phase in the washing stage is heated and concentrated by the heat from the hot combustion gases and then conducted into the heating stage, while the semifluid phase is conducted directly into the heating stage, where the two phases again are combined and together heated for drying and combustion.

Referring to Figure 1.23, the equipment comprises substantially a rotary furnace or drum **1** which is surrounded by two rims or rings **2**, each rolling on two supporting rollers **3**. The driving means for imparting rotation to the drum may be of any known type. The drum is slightly inclined. The lower output end (to the left) which is the hot end and is somewhat enlarged, has a lining of refractory bricks or the like. In the higher input end of the drum, the washing stage **4** is formed, supplied with the liquid phase which is kept at a suitable level over the bottom portion of the drum.

Preferably a heat exchanger **5** is provided in the washing stage. The heating of the output end in which a combustion space **6** is formed, can be effected by any suitable burning means **7**, for instance an oil burner or a gas burner. Alternatively or additionally an inlet **8** can be provided for supplying hot flue gases from a furnace outside of the drum. The right hand portion of the heating stage **1**, immediately to the left of the washing stage **4**, constitutes a drying stage **1'**.

At the input end of the drum, there is provided an inlet **9** for the sewage waste, comprising a suspension of sludge and having a very low percentage dry or solid matter, and the inlet discharges into a concentrating apparatus **10** which has at least two outlets **11** and **13** and serves to divide the supplied sewage waste or suspension in at least one thin phase, viz the liquid phase, which has a still lower percentage of solid or dry matter, for instance, 0.5 to 2%, and is flowing out through the outlet **11**, and one thick phase, viz the semifluid or sludge phase, which has a corresponding higher content of solid matter and flows out through the outlet **13**.

The liquid phase is conducted through the funnel **12** into the washing stage **4** in order to absorb the foul-smelling matter and the dust in the hot gases obtained from the combustion, whereby the liquid phase is concentrated or thickened on account of the evaporation of water, while the hot gases flow through the washing stage (to the right in Figure 1.23). The semifluid or sludge phase is conducted into an especial feeding device, comprising a container **14a**, a screw conveyor **14b** extending through the same, and an associated driving motor **14c**. The screw conveyor **14b**, which is enclosed in a sleeve or tube, constitutes a feeding channel extending through the washing stage **4** into the drying stage **1'**, and does not deliver any material in the washing stage **4**. Only when delivered in the drying stage **1'**, the thicker sludge phase is recombined or mixed with the liquid phase arriving from the washing stage **4** but not being somewhat concentrated, the two recombined phases now being dried to the desired final percentage of moisture, and eventually combusted, if desired. During the rotation the charge of the drum is traveling through the same, to the left in the drawing, whereby the charge is successively dried or burned by the hot flue gases flowing to the right through the drum.

FIGURE 1.23: COMBUSTION OF LOW SOLIDS SUSPENSION

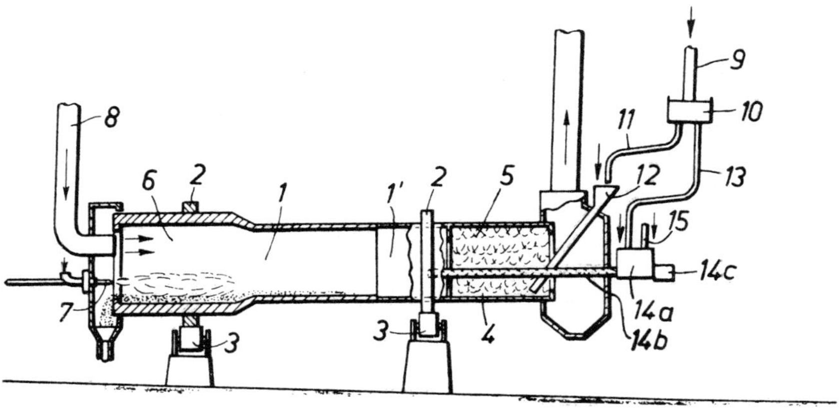

Source: U.S. Patent 3,413,937

It is possible to provide the container **14a** of the feeding device with an additional inlet **15** for other materials, such as garbage torn to pieces, or other combustible material. Thereby, the amount of combustible material in the combustion portion of the drum can be increased, if required, The provision of the additional inlet **15** has also a further advantage. In case the waste material to be treated would be so concentrated that it cannot be divided in the two different phases to the outlets **11** and **13**, respectively, the waste material can be supplied to the inlet **15** and be conveyed directly to the stages **1'** and **1** for drying and combustion, respectively.

In such cases the washing stage **4** should be supplied with water as washing liquid and should be provided with an outlet for discharging such water in any convenient well-known way. Thus, the arrangement can be utilized in many different ways and according to different practical conditions. There is provided both an oil burner **7** supplying the heat required, and an inlet **8** for hot gases from another combustion, for instance, flue gases having a temperature of about 800° to 1000°C. When the sludge material advanced through the drum, if desired together with garbage or the like from the inlet **15**, has become sufficiently dry, it is ignited in the output end of the drum, and then it is burnt perfectly during rotation of the drum. Therefore, in certain cases an especial burner can be omitted, but it is preferred to provide the oil burner **7** at least for igniting the dry material and the gases (methane) developed by the same during the heat treatment.

It should be noted that it is possible to so adjust the supply of combustible material through the inlet **15**, that the development of heat is sufficient for drying or concentrating the waste material, so that the total quantity is burned and only ashes are left in the combustion stage. It is also possible to provide for a balance between supply of heat and consumption of heat by regulating the supply of hot gases through the inlet **8** or by regulating the oil-burning at **7**.

The heat exchanger **5** in the washing stage can be a rolling system of chains and scooping sheet metal strips or the like, which serve also to keep the internal wall of the drum free from sticky coatings or depositions which otherwise may be formed on account of the sticky nature of the sewage wastes. For instance, this heat exchanger can be of the structure described in U.S. Patent 3,333,837, where the chains are suspended in the form of loops in a rolling cage. Now, during the rotation the chains are moving through the washing liquid while being cooled and wetted, and then they are lifted up into the higher free space, where the hot gases pass.

The foul-smelling matter and the dust carried by the gases are caught by the moisture on the chains and are washed away from the same during the rotation through the body of washing liquid on the bottom of the washing stage, whereby the concentration of solid matter in the washing liquids is increased on account of the evaporation of water. The scooping sheet metal strips or the like, when arranged, will continuously pour the washing liquid over the chains for effectively wetting the same.

The process makes it possible to dry, in a simple and effective way, sewage wastes with a very low percentage of solid matter, with or without adding garbage or other combustible material, to any desired final degree of moisture or to complete dryness or to combustion, by utilization of the heat value of the solid matter and supplied hot gases from an external source or, and according to the demand, heat from an oil burner, at the same time as the escaping gases are effectively washed from dust and foul-smelling matter, where all these operations can be performed in one single unit.

Mixing with Absorbent Carrier Material

H. Seiler; U.S. Patent 3,410,233; November 12, 1968; assigned to H. Seiler & Sohn AG, Switzerland describes a method for drying and burning of sewage sludge which involves mixing an absorbent carrier material to the sludge, drying the mixture by evaporation, separating the dried carrier material from the dried sludge, and recycling the dried carrier material to fresh charges of sludge.

The plant illustrated in Figure 1.24a comprises a combustion furnace **1**, which is preferably heated with residue oils and is also capable of burning dry sludge derived from the plant itself. The hot gases generated by the combustion are supplied to a rotary drum dryer **2**, which at its end adjacent the furnace is supplied by means of a feeder elevator **9** with a mixture of sewage sludge and an absorbent carrier material, e.g., sawdust or peat, which is derived from a charging device **8**. From the furnace heated drier drum **2** the dried mixture together with the hot gases and water vapor passes into a fan **3**, which conveys the same through a pipe **16** to a cyclone separator **4**. The latter discharges the separated water vapor and gases on top, while at the bottom the dried mixture composed of dried sludge dust and carrier material drops on a vibrator sieve **5**. The latter separates the dried sludge from the carrier material.

FIGURE 1.24: SEWAGE SLUDGE MIXED WITH ABSORBENT CARRIER MATERIAL

a.

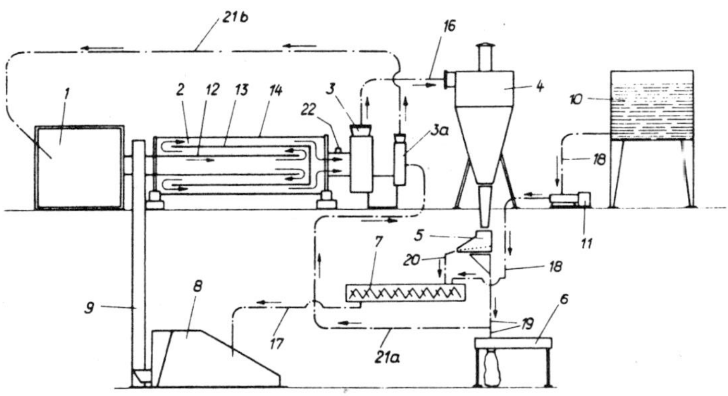

Diagram of Plant Apparatus

b.

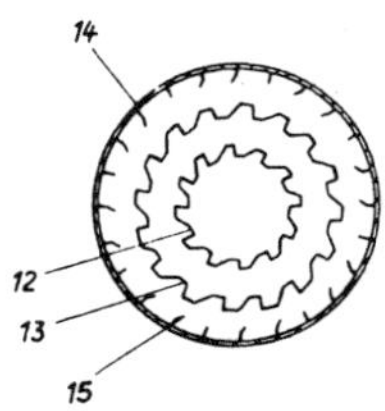

Drum Dryer, Transverse Section

Source: U.S. Patent 3,410,233

The dried sludge is passed through a pipe **19** into a bag filling device **6**, where it is filled in bags for being used, e.g., as fertilizer. It may alternatively be passed through a pipe **21a** to a fan **3a** and hence through a pipe **21b** as a fuel to the combustion furnace **1**. The wet sewage sludge is connected in a tank **10** and passed through a pipe **18** and by the aid of a sludge pump **11** to the mixer **7**, from where it is passed, mixed with the carrier material through a pipe **17** to the charging device **8**.

The carrier material is accordingly recycled and used all over again. The drier drum **2** comprises three concentrically telescoped cylinders **12, 13, 14**. As shown in Figure 1.24b, the outer cylinder **14** comprises inwardly directed curved vanes **15**. The walls of the two inner cylinders **12** and **13** are longitudinally fluted, having a sawtooth-like cross-sectional profile. The vanes **15** on the wall of the outermost cylinder **14** and those wall sections of the cylinders **12** and **13** which extend outwardly have substantially equal directions. The wet mixture of sludge and carrier material passes together with the hot combustion gases into the innermost cylinder, through which they flow axially in the direction of the arrows in Figure 1.24a.

At the right and end of the cylinder **12** the mixture of carrier material and sludge together with the combustion gases is deflected 180° and is passed into the space between the cylinders **12** and **13** through which space they pass in the direction of the arrows from right to left. At the left hand end of the space there takes place a renewed deflection of 180°, and the gases as well as the mixture, which is then completely dried, are passed by the fan **3** into the separator **4**.

The drying in the drum **2** accordingly takes place according to the principle of specific weight, i.e., the most rapidly dried and consequently weight relieved particles are continuously aspirated by the fan **3**. In the connecting pipe between the outlet of drum drier and the inlet of the fan, a regulator thermostat **22** is built in, which on the one hand automatically controls the fuel supply to the furnace **1**, and on the other hand automatically controls the dosage device of the charging means **8**, in order that the plant may always be supplied with hot gases of the right temperature and with the amount of mixture required.

Combustion in Presence of Carbonate Conditioning Agent

F.N. Kemmer, R.S. Robertson and R.D. Mattix; U.S. Patent 3,640,820; February 8, 1972; assigned to Nalco Chemical Company describe a method for improving the operational efficiency of sewage treatment plants whereby at least a portion of the sludge produced by such plants is converted into an active substance. At least a portion of this active substance is then used in these plants as a water-treating aid to improve the overall quality of the waste effluent waters produced by such plants mostly by removing soluble pollutants and the sludges may be efficiently disposed of without danger of pollution or other sanitation problems arising.

Referring to Figure 1.25, the plant is composed of a primary clarifier **10** into which raw sewage is fed through line **12**. The primary clarifier sludge is separated from the liquid phase of the sewage, with the liquid being passed to an aeration basin **14** through line **16**. After the liquid is aerated in the aeration basin, it is moved through line **18** to a secondary clarifier **20** where the additional solids are removed by sedimentation procedures. The supernatant effluent from clarifier **20** passes through line **22** to an optional tertiary treatment zone or stage **24**, where any remaining insoluble solids are removed to provide a final effluent which is discharged through line **26**.

In the primary clarifier, at least a portion of the soluble organic waste may be converted into a sludge. An optional step is to include an adsorption column **70** into which the final effluent is discharged through line **26** for removal of soluble solids. The effluent from column **70** is discharged through line **72**.

FIGURE 1.25: SEWAGE TREATMENT EMPLOYING CARBONATE CONDITIONER IN FURNACE

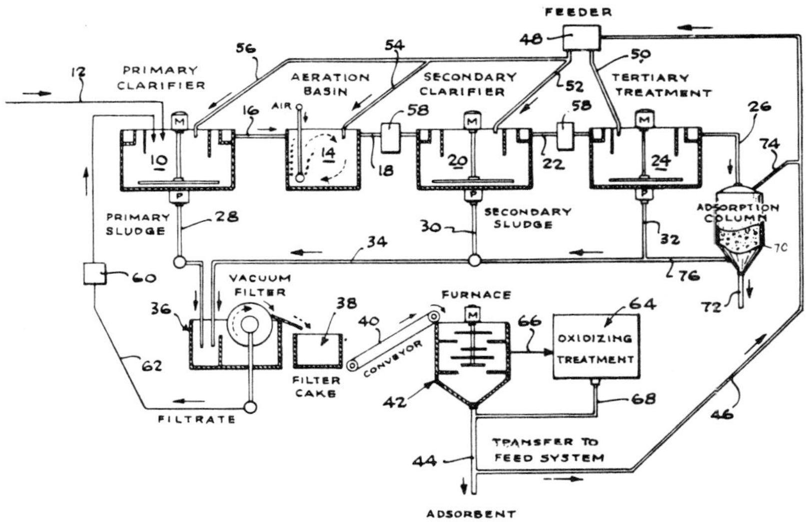

Source: U.S. Patent 3,640,820

Sludge removed in clarifiers **10** and **20** pass from these units through lines **28** and **30** respectively, and in some cases the fill from an exhausted column **70** from the tertiary treatment, could also be fed to a common feed line **34** where the sludges are combined and then deposited onto vacuum filter **36**. The combined sludges are dewatered in vacuum filter **36** and passed into a filter cake storage unit **38**, where they are removed by conveyor **40** to a charring and conditioning furnace **42**. A portion of the sludge from line **30** may be recirculated to line **16**.

Sludges entering furnace **42** are subjected to heat treatment in a controlled atmosphere with a conditioning agent present during which the residual moisture of the cake is removed and it is converted into the active substance of the process. The active substance is then removed from furnace **42** through line **44** and transferred through line **46** as a slurry to feeder **48**.

The active substance is mixed and fed through one or more of a series of supply lines **50, 52, 54** and **56** to a choice of lines **12, 16, 18, 22** and **26** where it is used to improve the clarity and quality of a supernatant effluent leaving any or each of these units. In some plants a single application point may produce the desired results, while in other plants a plurality of addition points may be necessary. In particular, the active substance reduces the soluble solids in the supernatant effluent. The active substance may additionally be used as a filter medium for line filters **58** and **60** which service line **18** and the filtrate feedline **62** on the vacuum filter **36**, and as the final treatment stage in column **70**. The active substance is made from a variety of sludges entering the vacuum filter **36** from the primary clarifier **10**, the secondary clarifer **20**, the tertiary treatment **24**, and the exhausted adsorption column **70**. The heating treatment in the presence of a conditioning agent produces the active substance of this process. Heating sewage in a furnace to high temperatures is known as incineration. Heating sewage in a furnace without a conditioning agent produces a waste ash which is disposed of relatively easily, but has no activity in terms of this process.

The furnace must be designed to withstand the fumes generated in the heating step. The temperature of the furnace can range from 300° to 900°C. Preferably, the temperature should be from 500° to 700°C. It has been found that at least 3 minutes are needed to achieve any sort of significant activation. The time can range from 3 to 200 minutes, depending upon the sludges used and the conditioning agents. The time and temperature can be varied to achieve the best working arrangement. The pressure should be at least one atmosphere absolute.

Conditioning agents can consist of any organic or inorganic compounds that are acidic and produce acidic by-products under the furnace conditions. By way of example, carbonate activating agents can be used. These compounds can be bicarbonate salts, carbonate salts or even carbon dioxide and carbon monoxide under appropriate conditions. The dosage of the conditioning agents should be from 0.1 to 50% by weight based on the dry weight of the sludges. Preferably, the furnace should contain from 1 to 25% by weight of the conditioning agents based on the dry weight of the sludges.

The active substance produced is a unique material in that while it usually contains at least 20% by weight inorganic material in the form of their oxides or in the form of inorganic salts, it may exceed 60% by weight inorganic material. In most instances the inorganic portion of the active substance produced in accordance with the process will range between 20 and 50% by weight.

Example: To illustrate the fact that the dried sewage sludges are capable of being converted into a useful active substance, the following is presented by way of example. Imhoff sludge, a partially digested raw sludge, was obtained from the Metropolitan Sanitary District of Greater Chicago and conditioned with an acrylamide coagulant type polymer, filtered and dried. Test results indicate that the average plant employing a properly running activated sludge process could be expected to produce sufficient active substance for its own needs. Any excess active substance may be used elsewhere in the plant. For example, the excess could be used to purify stack gases. The excess could be sold for industrial and agricultural uses. The level of recycle of active substance will vary.

Prior to being subjected to the heat treatment portion of the process it is beneficial that the sludge be sufficiently dewatered so that it contains not more than 85% by weight of moisture. The active substances may be used in any or all of the various clarification treatment operations normally performed in sewage treatment plants. Excess active substances may be sold as a commercial product. Preferably the active substance is added to the effluent from the aeration basin. The recycled sludge from the secondary clarifier returned to the aeration basin would then normally contain the active substance as well as the normal sludge. This may allow the size of the basin needed in the sewage system to be greatly reduced. The active substance can also be used in a tertiary treatment step as a final polish on the effluent.

Under some conditions it may be desirable to include a washing step after the active substance is discharged from the furnace. This washing step would remove soluble contaminants from the substance and allow them to be properly controlled in the sewage treatment system. The following tables illustrate some of the tests that have been run varying the conditioning agents, the amount of conditioning agent, time of conditioning and furnace condition.

Table 1 shows various carbonate compounds being used as conditioning agents. Carbon dioxide was effective when using a wet sludge. This is probably because an acidic carbon compound is formed. Sodium bicarbonate was effective when followed by an acid wash. In the following table, the temperature of the furnace was 600°C. The time in the furnace was about 10 minutes at temperature. The pressure was slightly over 1 atmosphere so as to maintain a positive pressure within the furnace. The moisture of the sludge was about 40% by weight. TOC is total organic carbon.

TABLE 1: $CO—CO_2—CO_3$

Activating Agent	% Removal TOC
CO	24
Dry CO_2-wet sludge	38
Wet CO_2-dry sludge	2
$NaHCO_3$ (25%)	1
$NaHCO_3$ (50%)	1
$NaHCO_3$ (50% with acid wash)	67
Na_2CO_3 (25%)	1
Na_2CO_3 (25% with acid wash)	18

TABLE 2

Sample No.	Temp. (°C.)	Yield	% Removal TOC
1	365°	49%	−42%
2	500°	36%	41%
3	600°	33%	41%
4	700°	20%	44%

TABLE 3

Sample No.	Temp. (°C.)	Time	Yield	% Removal TOC
1	600°	5	36%	53%
2	600°	10	33%	41%
3	600°	15	>33%	38%
4	500°	5	<44%	−155%
5	500°	10	44%	43%
6	500°	30	36%	41%

Table 2 shows the effect of temperature (time was 10 minutes). As can be seen from the table, increasing the temperature decreases the yield but increases the removal of total organic carbon. At 365°C, the removal of total organic carbon was a negative value. This indicates that the soluble organic species from the substance added contaminants to the system rather than removing soluble organics from the system. Table 3 shows the effect of time of activation. As shown in this table, increasing the time beyond a certain point decreases the yield and the removal of total organic carbon.

Compaction and Combustion with Coal

A process described by *H. Klesper; U.S. Patent 3,166,032; January 19, 1965; assigned to Passavant-Werke, Germany* is based on the concept of dehydrating sewage by compacting and of converting the compacted sewage into a thermally highly efficient and useful fuel. According to this process, the content of liquid of mud formed in sewage disposal systems is reduced by compacting the mud in the presence of powdered coal which is used to remove the liquid phase from the solid phase of the mud, and thereafter the content of coal of the mixture of coal and mud in cake form resulting from the process of compaction is increased to convert the mixture into a readily combustible furnace fuel. This fuel is subjected to useful combustion in a boiler furnace, preferably a boiler furnace forming part of a thermal power generating station.

An important step in carrying this process into effect is the use of powdered coal as a medium for removing the liquid phase from the solid phase of sewage which step may be carried out either by mixing powdered coal to the mud having a relatively large content of liquid and feeding this mixture of powdered coal and mud to conventional filtering machines, or by covering the filter surfaces of conventional filtering machines with a layer of powdered coal or by combining both ways of using powdered coal as a compacting medium.

H.L. Rekate and A.M. Hein; U.S. Patent 3,383,228; May 14, 1968 describe a process for disposal of garbage and sewage in the form of a solid, hard, cohesive, odorless and inert block of ashen material having a specific gravity greater than water. The process comprises the steps of destructive distillation of the waste material at high temperature and pressure, mixing with the residual mass a suitable binding agent; and compacting the mixture into a block.

OTHER PROCESSES

Shipboard Disposal System

A process described by *E.P. Foster, R.J. Kelly and W.L. Sage; U.S. Patent 3,875,357; April 1, 1975; assigned to The Babcock & Wilcox Company* involves a shipboard disposal system which eliminates the practice of overboard discharge of the shipboard sewage effluent and provides the hygienic alternative of chemically treating and disposing of the treated sewage in the combustion zone of the furnace associated with the ship's main propulsion boiler. The system also provides adequate capacity for storing the liquid-solids sewage mixture when it is not being injected into the boiler furnace due to low firing rate, or when the boiler is shut down.

Fresh water is delivered to the ship's galley, laundry and fresh water systems and according to the process, the effluent therefrom is conveyed for chemical treatment in a disinfectant contact tank with the exception of a portion which is filtered out and recycled through the ship's sanitary system before being separately conveyed to the disinfectant tank. There is a retention period in the tank to allow for complete chemical disinfection of the liquid-solids sewage mixture and use is made of rotary blade type agitators to promote the chemical treatment. The chemically treated sewage mixture is thereafter conveyed to a main holding tank where it is aerated and then discharged to a sump including a grinding apparatus for comminuting the solids contained in the sewage mixture to a particle size range which will readily burn when injected into the combustion zone of the ship's boiler furnace. The comminuted sewage leaving the grinding apparatus is screened through a centrifugal separator or hydroclone for the removal of any particles exceeding the predetermined size range, with these oversize particles being returned to the main holding tank for eventual regrinding.

The acceptably sized sewage is pumped to an injection nozzle which employs steam to atomize the sewage before it is injected into the combustion zone of the furnace associated with the ship's boiler. The combustion zone is maintained at a high temperature by the ship's fuel oil burners with the firing rate being regulated by the demands of the ship's propulsion and auxiliary equipment. The injection of sewage into the furnace combustion zone will only be allowed when the fuel input to the burner associated with the sewage nozzle exceeds a predetermined level. Whenever the fuel input is below this predetermined level, injection is discontinued and the sewage mixture is recirculated through the main holding tank with any excess being conveyed to one or more auxiliary holding tanks and recirculated therein.

Upon resumption of injection, the excess sewage is recycled into the main holding tank for eventual delivery to the furnace combustion zone. The sewage input to the injection nozzle is regulated to maintain a sewage to fuel input ratio resulting in substantially complete vaporization and incineration of the liquid-solids sewage mixture within the furnace chamber.

Sanitary Facility for Remote Locations

D.J. White and D.J. Dion; U.S. Patent 3,782,300; January 1, 1974; assigned to Mobile Systems International, Inc. describe a system for disposing of human waste and other sewage by incineration. The process of incineration is accomplished in a manner that breaks down the organic contaminants into unobjectionable and oxidized gaseous exhaust and does so in an efficient manner, that does not add significant quantities of other contaminants to the atmosphere.

The process utilizes an insulated tank that is capable of storing sufficient quantities of sewage to handle the particular installation. The sewage comprises waste matter suspended in an aqueous solution; thus, before any incineration can begin, the water must be driven off and large quantities of heat must be transferred to the water for this purpose. After the water is driven off, the heat must be effectively transferred to the sludge that remains until the sludge is completely incinerated and only a light ash remains. The ash is of such a small quantity that it need be removed only after long periods of use.

To accomplish the effective transfer of heat to the sewage during both phases of operation, the process employs a combustor. Air is supplied to the combustor by a blower that is capable of supplying air of a sufficient pressure to overcome the pressure head produced by the fluid in the tank and pressure drop through the piping to the tank. A unique colander design distributes the air from the blower evenly around an inner-core heat-chamber member by providing a solid area in the colander member immediately adjacent the input of blower air. This area deflects the air around the cylindrical colander member to provide substantially equal pressure at all points around an inner core cylindrical member that has a plurality of openings at one end to inject the combustion air immediately adjacent the fuel injection point and thereby provide for ignition of the fuel at one end of the chamber. Additional injector openings are spaced along the inner core cylindrical member, and the blower air is directed in such a manner that it tends to flow along the walls of the inner core member, to cool the walls and keep the extremely high temperature flame from contacting the walls directly. This design enables the combustor to be made of relatively unsophisticated and low-cost materials, while efficiently generating extremely high temperatures.

The high temperature gas generated in the combustor is carried by the injector means into the tank and injected into the sewage below the initial water level thereof. The gas bubbling from the end of the injector tube rises, by virtue of its low density, and is constrained to the vicinity immediately adjacent the injector tube by a collar surrounding the tube. This combination of a downwardly flowing gas which is constrained to upward movement in the immediate vicinity of the injector tube comprises a gas-lift device. The gas-lift device is effective both to provide a pumping action to circulate the sewage in a highly agitated manner, while it remains in suspension during the vaporizing process, and also serves the function of providing a long path of contact between the hot gases and the sewage, to transfer to the sewage substantially all of the heat contained in the high temperature gas.

After substantially all of the fluid has been boiled off, the process provides for the heating of the relatively dry material remaining, by direct contact with the gaseous stream. This stream is deflected from its downward path by a coaxial deflector. The effect of this deflector is to maintain scouring flow of gas across the bottom of the tank and up its walls in a swirling manner to continue to agitate the solid material until it is completely oxidized and burned.

The vapor and other exhaust gas exit the tank at the upper portion and enter a residence chamber. This chamber comprises a plenum chamber, where the air first flows upwardly and then downwardly to increase the residence time of the gases in the chamber. The chamber is maintained at an elevated temperature by direct injection of gas tapped from the gas delivered to the injector tube. The gas quantity is regulated to maintain the residence chamber at approximately 1600°F. This temperature is sufficiently elevated to oxidize any remaining contaminants, comprising organic material that has been carried with the vapor and other discharges from the tank.

Additional features of the process are incorporated to provide a system that is safe to use in proximity to people, without elaborate safety precautions. In particular, the temperature of the exterior of the device is maintained at a safe level through the use of an annulus surrounding an inner tank member to prevent the heat from raising the exterior temperature to an unacceptable degree and has the additional effect of acting as a preheater for the combustor. The air for the combustor is drawn, by the blower, through the annulus surrounding the tank and the air is heated by the high tank temperatures.

Similarly, the residence chamber is insulated by a dead air space between the interior residence chamber and the exterior of the exhaust system to provide insulation and finally, an exhaust pipe extends, vertically from the residence chamber, a sufficient distance so that the hot gases discharged cannot contact or cause damage to persons or property in the vicinity.

This process provides an improved method of disposing of sewage waste that would be applicable to use at temporary sites and remote facilities where standard sewage treatment facilities are not available.

Petroleum Waste from Oil Field Operations

W.J. Davies, B. Hatton, G.C. Boyd and H.E.W. Hanlan; U.S. Patent 3,704,676; December 5, 1972; assigned to Kenting Oilfield Services Ltd., Canada describe an incinerator capable of disposing of a mixture of sludge, water and debris, which normally collects in an open pit in an oilfield operation, without producing distasteful bellows of black smoke.

According to the process there is provided an incinerator having an open-topped chamber defining a combustion zone. Air supply means are arranged for feeding air into the chamber, the means being disposed for feeding a sheet of high velocity air over the combustion zone in a flow direction which is inclined downwardly from the top of one side of the chamber. Below the air supply means there is located a waste supply means for blowing an air-waste mixture into the combustion zone in a direction substantially parallel to the direction of travel of the sheet of air.

In the combustion zone a large ball of fire is established which is fed by the combustible constituents in the waste, and the ball of fire and the gases produced thereby are rolled at a very high speed by the high velocity air in the sheet which passes thereover. Because of the inclination of the sheet, the air, which forms the sheet, joins the tumbling ball of fire to provide a surplus of oxygen. The products of combustion cannot readily rise to escape the turbulent combustion zone since they are turned back by the sheet of moving air. Thus, the temperature within the combustion zone becomes extremely high, and clean and complete combustion is ensured.

Plasma Arc Process

A.W. Marr, Jr.; U.S. Patent 3,575,119; April 13, 1971 describes an apparatus for disintegrating and incinerating a concentrated slurry of solid organic material. The apparatus has an arcuate tunnel, a series of pairs of arc-forming electrodes spaced along and mounted within the tunnel and electrical source means connected to each pair of the arc-forming electrodes.

Referring to Figures 1.26a and 1.26b, there is a volute casing **10** which increases in cross-sectional area from a starting point **12** to an ultimate discharge point **14**, where gaseous products are discharged to a stack **16**. The volute casing **10** constitutes an arcuate tunnel and will be referred to as such. A pipeline **18** delivers the watered sludge to a vertical pipeline **20** which enters and proceeds through the gas discharge line **16** through a preferably vertical line **22**. This, of course, represents a heat exchanger transferring heat from the discharge gases in the stack **16** to the incoming sludge in the pipe **22**. The showing, of course, is simplified.

At the center of the volute casing **10** is a rotatable shaft **24** which, as viewed in Figure 1.26a and quite arbitrarily, rotates counterclockwise in the direction of the arrow **26**. A disc **28** surrounds the shaft **24** and rotates therewith. Peripherally arranged around the disc **28** but mounted in the volute casing **10**, are a plurality of heating stations, mutually substantially dentical, and generally designated in Figure 1.26a by **30**. These heating stations **30** may include a conventional electric arc but preferably are of the plasma jet type. A suitable discussion of this subject will be found in a bulletin entitled *Plasma Jet Technology,* available from the Superintendent of Documents, U.S. Goverrment Printing Office, under the identification number NASA-SP-033.

Figure 1.26c shows a cross section which includes part of the volute passageway **10** and cuts diametrically through one of the heating stations **30**. Since conventional arcs are old and well known, it s not deemed necessary to provide here a detailed disclosure of such an arrangement. Figure 1.26c therefore, describes an adaptation of the showing above referred to appearing in the publication *Plasma Jet Technology*.

In Figure 1.26c, the rotor or disc **28** is shown as having adjacent its periphery a series of ribs **32** which mate with grooves in the body portion **34** of a heating station **30**. The grooves are defined by ribs **36** formed in the body portion **34** which, with the ribs **32**, constitute a labyrinth seal against the egress of gases generated in the volute casing **10**. A cap **38** formed of refractory or at least insulating material is mounted in the body portion **34** and supports a preferably tungsten electrode **40** which has a conically tapered end **42**. A copper electrode **44** is also mounted in the body portion **34** and mates with the refractory cap **38**, as shown at **46** in Figure 1.26c.

FIGURE 1.26: ELECTRICAL ARC PROCESS

a.

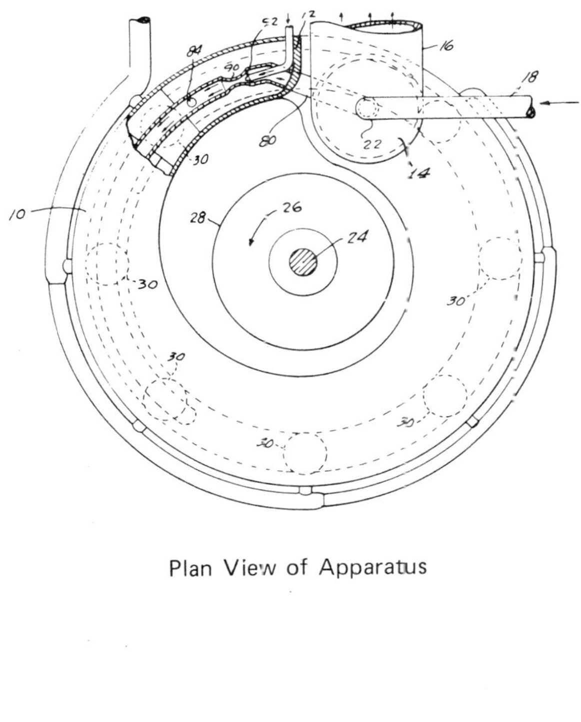

Plan View of Apparatus

(continued)

FIGURE 1.26: (continued)

c.

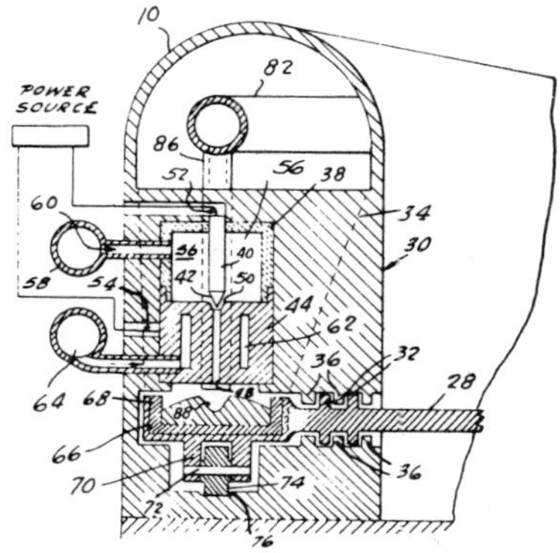

Section Along Line **3—3** of Figure 1.26b Showing Plasma Arc

b.

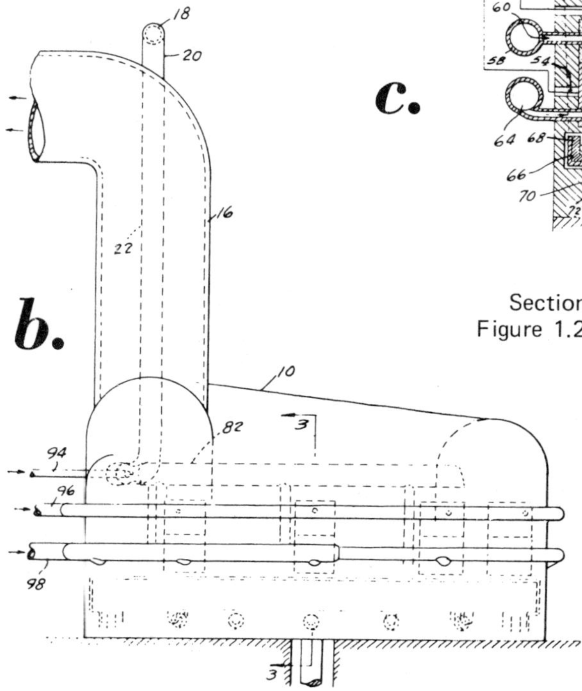

Side View of Apparatus

Source: U.S. Patent 3,575,119

The electrode **44** has a central bore **48** topped by a conical aperture **50** registering with but spaced from the conical end **42** of the tungsten electrode **40**. An electric arc forms in the gap between the conical end and the conical seat **50**. An arcing potential across the gap is supplied to the tungsten electrode **40** by an insulated connection **52** and to the copper electrode **44** by an insulated connection **54**.

In order to constitute the system of electrodes just described—a plasma jet arc rather than a simple arc—the cap **38** is designed to provide a cavity **56** around the electrode **40** and above the electrode **44**. The cavity **56** is supplied by a duct **58** with a constant stream of gas, usually air, which enters in the direction of the arrow **60** and is vented through the gap between the conical tip **42** of electrode **40** and the conical seat **50** of the electrode **44**, then through the passage **48**. This flow of gas in addition to creating the plasma jet effect has the incidental effect of acting as a coolant for the electrode **40**. The electrode **44** is cooled by the passage of a cooling medium through an intermediate duct **62** in the electrode **44** and supplied with coolant through a duct **64**.

Returning to the rotating disc **28**, it will be noted that the disc **28** has a circumferential trough **66** which is lined with a refractory **68**. At suitable intervals the trough **66** has formed thereon bosses **70** which support axles **72** on which are rotatably mounted a series of bearing wheels **74** which ride in a continuous track **76** provided in the body **34**.

As best shown in Figure 1.26a, the vertical sludge pipe **22** terminates near the bottom of the stack **16** in a horizontal line **80** which joins an arcuate section **82** centered on the axis of the shaft **24** and overlying the various heating stations **30**. The arcuate section **82** is provided at various points **84** with downward directed outlet pipes **86** which deposit sludge **88** into the trough **66** for which see Figure 1.26c.

In passing through the lines **18**, **22** and **80**, the sludge has become quite highly compressed and for this reason it is desirable to provide the section **82** at various points with a venturi section **90** into the throat of which is directed a jet **92**. The venturis **90** and jets **92** preferably are placed slightly in advance of each heating station **30** so that sludge delivered through an outlet **86** will have its density reduced, making it more amenable to the rapid evolution of gases and steam. The increased volume of gas and steam generated by the heating stations **30** is accommodated by the volute form of the casing **10**.

As best seen in Figure 1.26b, air is supplied to the line **82** by a header **94**. A second header **96** provides air to the cavities **56** and then to the arc. Coolant, however, is supplied to the copper electrode **44** through a header **98**. If additional air is needed for the purpose of consuming carbon formed as a residue of the decomposition of the organic material, it may be added as desired at various points around the casing **10**.

SLUDGE OXIDATION PROCESSES

AEROBIC PROCESSES

Integrated Oxygen System

J.R. McWhirter; U.S. Patent 3,670,887; June 20, 1972; assigned to Union Carbide Corporation describes a process which is an improvement in the treatment of BOD-containing water as widely practiced, where the latter is aerated in contact with activated sludge, the sludge is separated and at least part is recycled to the aeration zone as the aforementioned activated sludge.

In this improvement, first gas comprising at least 60% oxygen (by volume) is introduced and mixed as the aeration gas with the BOD-containing water and recycled sludge. One of these fluids is simultaneously continuously recirculated against the other fluids in the aeration zone. The liquid-solid contact time in the aeration zone is between 30 minutes and 24 hours, and the fluids are provided in sufficient quantity and rate to maintain the dissolved oxygen content (DO) of the mixed liquor above about 0.5 ppm.

The mixed liquor is thereafter separated into purified liquid and activated sludge, and the unconsumed oxygen-containing gas is discharged from the aeration zone. Second gas is provided having oxygen concentration at least 0.70 times the first gas oxygen concentration and at least in part comprising the unconsumed oxygen-containing gas from the aeration zone.

At least 85% by weight (dry basis) of the activated sludge separated from the mixed liquor is returned to the aeration as the recycled sludge. The fraction recycled refers to the average fraction recycled over a period of operation and allows for the fact that the recycle rate may vary from day to day or from hour to hour, and that the quantity of sludge stored in the separation facility may vary correspondingly. The second gas and the unreturned activated sludge are introduced to a digestion zone and mixed for liquid-solid contact time of at least 96 hours while simultaneously continuously recirculating one of the fluids in this zone against the other fluids. Waste sludge residue and third gas comprising at least 21% oxygen are thereafter discharged from the digestion zone. Figure 2.1a is a schematic view of process apparatus where a clarifier is located intermediate the aeration zone and the digestion zone. Figure 2.1b is a schematic view taken in cross-sectional elevation of apparatus arranged for series and cocurrent flow of oxygen gas and liquid-solid through three aeration tanks, a clarifier, and two digestion tanks in the manner of Figurer 2.1a.

FIGURE 2.1: INTEGRATED OXYGEN SYSTEM

a.

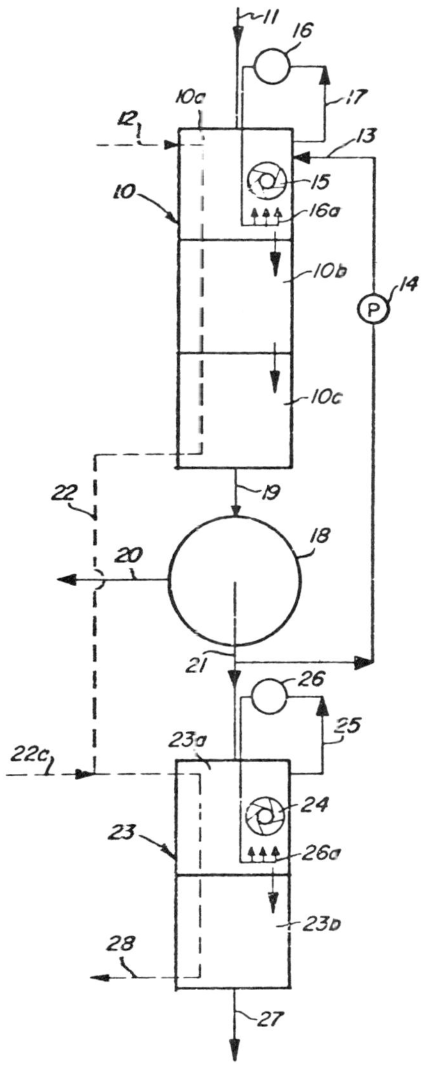

(continued)

FIGURE 2.1: (continued)

b.

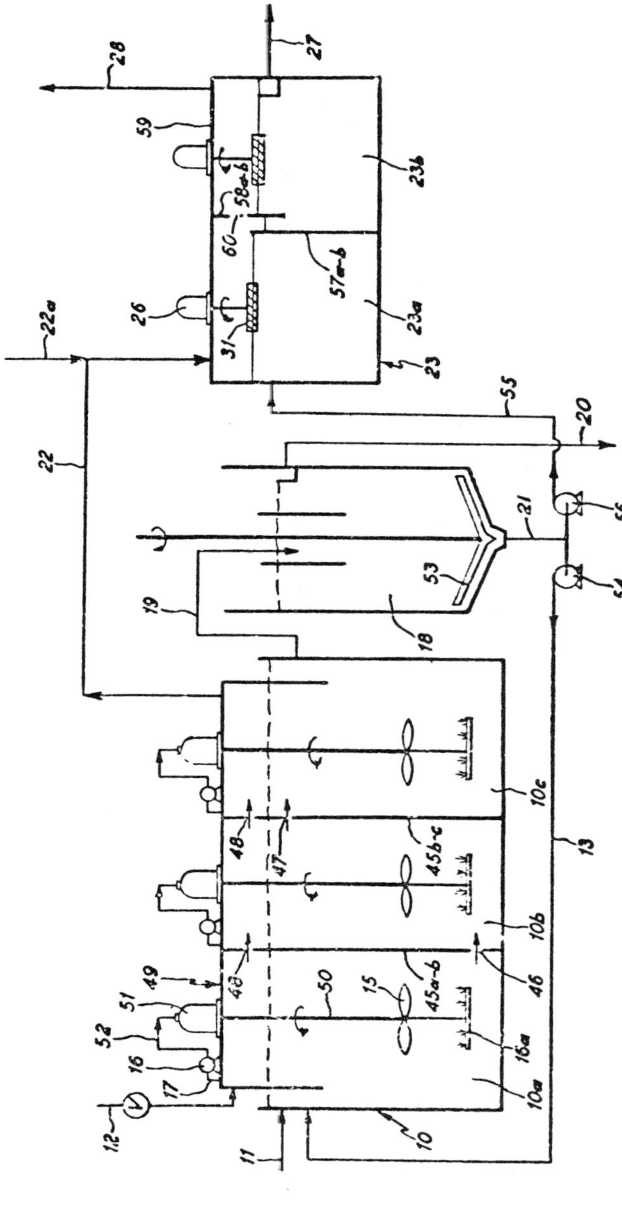

Source: U.S. Patent 3,670,887

Referring to Figure 2.1a, BOD-containing water, as for example sewage, enters aeration zone **10** through conduit **11**. An oxygen gas source comprising at least 50% oxygen enters zone **10** through conduit **12**. Activated sludge also enters zone **10** through conduit **13** having pump **14**.

The streams are intimately mixed in aeration zone **10** by mechanical agitation means **15**. The latter may comprise motor-driven impellers located near the liquid surface or submerged below the surface, and the oxygen gas may be introduced through conduit **12** either above or below the liquid. Such apparatus is known to those skilled in the art and should be selected to achieve high contact area between the fluids with minimal work expenditure. If the oxygen gas is sparged into the liquid, the bubbles should be small so that their total surface area is large and their buoyancy is low. Dissolution of oxygen is also aided by submerging the gas-dispensing means to a depth in the liquid where the hydrostatic effect is significant.

Means are provided for continuously recirculating one fluid against the other fluids in aeration zone **10**. For example pump **16** joined to the gas space by conduit **17** recirculates aeration gas to the lower portion of the zone for release as small gas bubbles through sparging. Aerating devices are commonly rated by the so-called air standard transfer efficiency which identifies the capability of the device to dissolve oxygen from air into zero DO tap water at 1 atmosphere pressure and 20°C. Suitable devices are those which have an air standard transfer efficiency of at least 1.5 lb O_2 per hp-hr. For these purposes the power used in rating the device is the total power consumed both for agitating the liquor and for gas-liquor contacting.

Although aeration zone **10** may comprise a single chamber it preferably comprises a multiplicity of stages **10a, 10b,** and **10c,** each having mechanical agitation means **15** and fluid recirculation means **16** (but for simplicity not illustrated in each stage). The unconsumed oxygen-containing gas from each aeration stage is preferably discharged and introduced to the next succeeding stage as the aeration gas (illustrated by a dotted line). Similarly, oxygenated liquor from each stage is discharged and preferably introduced to the next succeeding stage in cocurrent flow relation with the unconsumed oxygen-containing gas for mixing and recirculation of one fluid against the other fluids. Stage-to-stage oxygenated liquor flow is illustrated by solid arrows extending through the stage partitions.

Cocurrent gas-liquor flow through a multiplicity of aeration stages is preferred to satisfy the inherent variation in BOD of the water to be treated and of the succeeding partially oxygenated liquor. The feed gas representing the highest purity oxygen is contacted with the feed water in the first stage. Accordingly, the first or feed gas stage has the highest oxygen partial pressure and thus the highest oxygen transfer potential. Therefore, the high oxygen demand in this stage can be supplied without excessive power consumption.

The oxygen gas is introduced through conduit **12** to aeration zone **10** at sufficient rate to maintain the dissolved oxygen content of the mixed liquor above about 0.5 ppm. Oxygen is consumed at a rapid rate and with conventional air aeration the dissolved oxygen level in the aeration zone drops to near zero and cannot be maintained at levels suitable for aerobic microorganisms. This is overcome in part by using an aerating gas comprising at least 60% oxygen and preferably at least 90% oxygen. Also, the gas must be mixed with the returned activated sludge and BOD-containing water in an efficient manner to create a large interfacial area with minimal work on the fluids. Another feature of the process contributing to the dissolved oxygen level control is the aforementioned continuous recirculation of one fluid against the other fluids. The dissolved oxygen content of the aeration and digestion zones preferably should not exceed about 10 ppm because the work expended in achieving higher DO concentrations would tend to cause excessive attrition of the bacterial floc, reducing its agglomerate size and impairing its settling properties. Stated otherwise, the floc particles would be so small that the effluent becomes turbid and unsuitable for discharge into receiving waters.

In the activated sludge process, the recycled sludge consists essentially of flocculent agglomerates of aerobic organisms which have the ability, in the presence of sufficient dissolved oxygen, to first adsorb, then assimilate and oxidize the organic material of the feed water. This adsorption-assimilation sequence occurs in the aeration zone of the process and in Figure 2.1a the sequence is substantially completed upstream of separation zone **18**. The liquid-solid contact time in aeration zone **10** for organic food adsorption-assimilation is between 30 minutes and 24 hours. This time varies depending upon the strength (BOD content) of the waste water, the type of pollutant, solids level in aeration and temperature. A maximum retention period of 24 hours will usually provide adequate time to remove BOD from effluent, activate the sludge, and perform a reasonable degree of auto-oxidation if desired. Where several aeration stages are employed, the retention period in the aeration zone refers to the total time the biomass solids (the total bacteria present) together with the BOD of the feed water are held in all aeration stages.

In Figure 2.1a where the aeration zone **10** is entirely upstream of the clarifier **18**, the contact time is calculated as the total liquid volume of stages **10a, 10b** and **10c** divided by the volumetric flow rate of combined BOD-containing feed water and activated sludge recycle. If it is desired to extend oxygen treatment on the biomass beyond the period required to assimilate and oxidize the waste water's BOD, then the fraction of the organisms of the sludge which themselves are destroyed and consumed by biological oxidation can become significant. This auto-oxidation phenomenon (endogenous respiration) is described more completely in U.S. Patent 3,547,814. To the extent that auto-oxidation is practiced in the aeration zone the load on the digestion zone will be reduced.

However, carrying auto-oxidation to an extreme in the aeration zone should be avoided because it reduces the activity of the biomass to be recycled in the return sludge and impairs its settleability. Moreover, retaining the treated biomass under long-term aeration is expensive because the total quantity of biomass undergoing treatment is large relative to the excess sludge stream, and aeration basins become prohibitively large. If the sole objective is to assimilate and oxidize the BOD of the influent BOD-containing water, then a relatively short contact time will suffice.

Virtually complete assimilation of BOD has been experimentally achieved on municipal waste after two stages of aeration with 99.5% oxygen feed gas and cocurrent liquid-solid flow after a contact time of 37½ minutes in each stage or 75 minutes total. The effluent from the second stage was low in BOD (e.g., 20 to 35 ppm) and the sludge stabilized and settled well. In other tests where some auto-oxidation occurred, the total contact time was about 2.5 hours for three stages (50 minutes/stage) of cocurrent flow, followed by clarification after the third stage. In another test series with auto-oxidation and four stages of cocurrent flow and clarification after the fourth stage, the total contact time was 2.7 hours or 40.5 minutes per stage. BOD determinations made on the supernatant liquid from each stage indicated that treatment was essentially complete after two stages (total of 81 minutes' aeration) and furthermore, that the differences in BOD between the first and second stages was almost wholly accounted for by the completely mixed character (nonplug flow character) of the individual stages. It follows that had the first stage (40.5 minutes) been subdivided into two series stages of 20.25 minutes each, the purification of the water by adsorption and assimilation of BOD would have been largely complete after the first of the substages (20.25 minutes' aeration).

It also appears that a total aeration contact time for a six-stage system having these same characteristics may be 1.5 hours (15 minutes per stage). A total contact time of at least 30 minutes is necessary for oxygen absorption and BOD assimilation with the average municipal waste to reduce the BOD of the purified liquid to an acceptable level. A total contact time of less than 180 minutes usually provides substantial reduction of sludge production if auto-oxidation is to be practiced, permits avoidance of settling problems, and still permits a substantial reduction in aeration chamber capacity as compared with conventional practice. Present activated sludge plants for municipal wastes employing air aeration are believed to operate with total aeration contact time of 180 to 360 minutes.

The contact times are satisfactory for relatively low-strength municipal waste liquids having up to about 300 ppm BOD. For higher-strength wastes as for example those discharged from petrochemical plants, longer contact times on the order of 5 to 12 hours are necessary to yield an effluent of comparable purity. In the practice of this process, it is preferred to maintain high suspended solids concentration in the aeration zone. The BOD-containing water-activated sludge volatile suspended solids content is preferably at least 3,000 ppm and most desirably 5,000 to 9,000 ppm. One reason for this preference is that the solids concentration in the aeration zone affects the rates of the biochemical reactions occurring in the method.

In the treatment of municipal sewage, the suspended solids comprise: (1) biologically oxidizable organic material; (2) nonbiologically oxidizable organic material; and, (3) nonoxidizable nonorganic material. The nonorganic material such as sand and grit, and the nonbiologically oxidizable material such as polyethylene particles or paper are undesirable but unavoidable components of the BOD-containing water, e.g., sewage, entering the aeration zone. Normally relatively large particles, e.g., wood chips, are usually removed in a pretreatment step.

The major fraction of the total solids in the mixed liquor, e.g., 70% thereof, consists of bacterial floc (biomass) in the activated sludge recirculated from the clarifier to the aeration zone. The higher the concentration of bacteria, the more rapid will be the adsorption and assimilation of BOD, assuming other requirements are also met such as dissolved oxygen supply.

Returning to Figure 2.1a, the mixed oxygenated liquor is discharged from aeration zone **10** and passed through conduit **19** to the aforementioned clarifier **18** for separation into supernatant liquid and activated sludge. The former is drawn off through conduit **20** and the activated sludge is discharged from the lower end through conduit **21**. At least 85% by weight and preferably at least 95% of the activated sludge separated as stream **21** is returned through conduit **13** to aeration zone **10** as the source of active bacteria. Where several aeration stages are provided the activated sludge is recycled to the first stage **10a** where the oxygen gas and the BOD-containing feed water are also introduced. The activated sludge preferably has 12,000 to 50,000 ppm total suspended solids content (MLSS) and is preferably recycled to the aeration zone at flow rate relative to the BOD-containing feed water such that the recycling sludge/BOD-containing feed water volume ratio is 0.1 to 0.5.

The excess unreturned activated sludge not required to sustain the solids level in zone **10** is withdrawn from the process to the digestion zone **23**. The excess sludge is a small stream relative to the recycle sludge stream. The rate at which cells are wasted from the process cannot continuously exceed the rate of net cell production. If this occurs, the essential microorganisms will be lost and purification will cease. Only a fraction of the BOD feed to the system is used for synthesis of new cells and a fraction of cells already existent in the system is destroyed by auto-oxidation.

The excess cells often comprise on the order of 1% of the cells recirculated back to the aeration zone. It is reported that in high rate activated sludge processes characterized by low solids level in aeration and rapid cell synthesis, the excess cells may comprise a fraction on the order of 10% of the cells recycled. If cell wasting exceeds 15% (and less than 85% by weight of the activated sludge is returned to the aeration zone), the size and operating cost of the digestion equipment become unnecessarily burdensome.

The percentage of the sludge which is wasted to the digestion zone is an average quantity over a period of operation. In normal operation of an activated sludge system the schedule of sludge wasting may be variable and/or intermittent, and the rate at which sludge is recycled to aeration zone **10** may also be varied, e.g., in accordance with changes in the rate of feed water introduction. When sludge flow rates are varied the inventory of separated sludge held in clarifier **18** will vary correspondingly.

Unconsumed oxygen-containing gas is also discharged into conduit **22** from aeration zone **10**, and in particular from the final stage **10c** if multiple aeration stages are employed. Additional oxygen gas may be introduced to the system through conduit **22a** or directly to the digestion zone. The unconsumed oxygen-containing gas in conduit **22** and any fresh oxygen gas from conduit **22a** together comprise the second gas having oxygen concentration at least 0.70 times the oxygen concentration of the first gas introduced to the aeration zone through conduit **12**. Even though the oxygen consumed from the gas in the aeration zone may be equal to that consumed in the digestion zone, the oxygen purity change in the gas across the aeration zone will be far smaller than the change across the digestion zone.

This is due to the relatively heavy evolution of nitrogen and carbon dioxide from the small volume of liquid in the digestion zone. In a preferred case, the oxygen purity of gas fed to the digestion zone should be at least 80% of the oxygen feed gas purity to the aeration zone. An abrupt drop in oxygen purity occurs in the digestion zone, and without relatively high oxygen purity gas available to the digestion; prohibitively low oxygen partial pressure and prohibitively high power consumption will result. The unreturned activated sludge in conduit **21** and the second gas in conduit **22** are both introduced to digestion zone **23** for mixing therein for liquid-solid contact time of at least 96 hours. During this extended period, most of the excess biomass is destroyed by endogenous respiration and the sludge residue is rendered substantially inert. During this period, one of the fluids is simultaneously continuously recirculated against the other fluids. Mixing is facilitated by means **24** which can be identical to aeration zone mixer **15** or may be a different type preferred for moving the more dense excess activated sludge.

The illustrated fluid recirculation system comprising withdrawal-return conduit **25**, pump **26** and gas sparger **26a** is identical to the aeration zone fluid recirculation system but this is not essential. By way of illustrating alternative apparatus, rotating surface-type impellers might be used in the aeration zone **10** to serve both the mixing and the fluid recirculation functions by throwing relatively massive sheets or streams of liquid into the gas.

Multiple digestion stages **23a** and **23b** are preferred for the same reasons as multiple aeration stages **10a, 10b** and **10c**. Each digestion stage is provided with mixing and fluid recirculation means. The unconsumed oxygen-containing gas and the further oxygenated sludge preferably flow in cocurrent relation through the digestion stages. Waste sludge is discharged from the final digestion zone stage **23b** through conduit **27,** and third gas comprising at least 21% oxygen is also vented therefrom through conduit **28**.

Figure 2.1b illustrates apparatus which may be used to practice the Figure 2.1a series staged cocurrent gas-liquor flow system. The aeration zone **10** is in the form of a tank with vertical partitions **45a-b** and **45b-c** spaced to provide three compartments **10a, 10b** and **10c** as the aeration stages. These partitions extend substantially to the base of tank **10** and are joined thereto in fluid-tight relation. Oxygenated liquid-solid flow is provided by restricted flow opening **46** in the first-second compartment common partition **45a-b** preferably near or beneath the bottom thereof, and restricted opening **47** preferably in the upper portion of second-third compartment common partition **45b-c**. The unconsumed oxygen-containing gas flows from compartment to compartment through restricted openings **48** near the top of the partition.

The gas space above each compartment is enclosed by a common cover **49** to which the partition upper means are leak-tight joined. Accordingly, back mixing of oxygen gas from a succeeding compartment to a preceding compartment is avoided as long as a slight pressure differential is maintained. The mixing means for each compartment includes blades **15** submerged in the liquor and joined by a rotatable shaft **50** to suitable drive means such as motor **51**. Fluid recirculation means comprise withdrawal conduit **17** joined to cover **49**, pump **16**, return conduit **52** in flow communication with the inlet side of hollow shaft **50** driven by motor **51**, and sparger **16a** positioned at the lower end of shaft **50** beneath blades **15**.

The small oxygen gas bubbles discharged from the sparger **16a** by the pressure of pump **16** are distributed through each compartment in intimate contact with the liquor and rise through the liquor to the surface where the unconsumed portion disengages into the gas space along with the oxidation reaction product gases. The mixed liquor discharged from the third and final aeration compartment **10c** through flow-restricting conduit **19** is introduced to clarifier **18** for separation into supernatant liquid and activated sludge. Clarifier constructions are well-known to those skilled in the waste treatment art and may for example include rotatable scraper **53** at the lower end to prevent coning. The activated sludge is withdrawn through bottom conduit **21** and at least 85% by weight thereof is recycled through pump **54** in conduit **13** to first oxygenation compartment **10a** for mixing with the BOD-containing feed water and oxygen-containing feed gas. The purified liquid is discharged from clarifier **18** through conduit **20**.

The unreturned activated sludge is flowed through branch conduit **55** by pump **56** to digestion zone **23** in the form of a tank with lower vertical partition **57a-b** and spaced upper vertical partition **58a-b** dividing the tank into two compartments **23a** and **23b**. This sludge flow to and through digestion zone **23** may be continuous so that when multiple stages are employed, back-mixing of liquid across restricted openings is desirably inhibited. The liquid-solid contact time in digestion zone **23** is long compared to the corresponding time in aeration zone **10**, for example, 100-200 hours versus 1-2 hours.

Moreover the volumetric flow rate of liquid introduced to the digestion zone is very low compared to the liquid introduction rate to the aeration zone, e.g., the former is often less than 5% of the latter. Consequently, the restricted openings between stages of a multistaged digestion zone must be very small in order to develop sufficient pressure difference across the openings to prevent back-mixing. Small openings are, however, prone to become plugged, and provision should be made to clean such openings without interrupting operation.

The digestion zone **23** construction of Figure 2.1b avoids possible plugging of openings for interstage transfer of partially digested sludge. The sludge level in second stage **23b** is lower than the level in first stage **23a**, and lower vertical partition **57a-b** acts as a weir over which the sludge must fall in its passage from **23a** to **23b**. Gas staging is provided by upper vertical partition **58a-b** extending downwardly from the gas-tight cover **59** over the tank into the sludge. Restricted opening **60** in upper vertical partition **58a-b** permits flow of unconsumed oxygen-containing gas from first compartment **23a** to second compartment **23b** without back-mixing.

Instead of separate liquid-solid mixing means and fluid recirculation means as employed in the aeration tank, rotating surface type impellers **31** are provided for each compartment in the digestion tank **23**. Each impeller is joined by a shaft to power means as for example an electric motor **26**. The rotating impellers keep the solids in suspension and also throw massive sheets of liquid-solids against the oxygen gas in the overhead space beneath cover **59**. In this manner, the liquid-solids is continuously recirculated against the gas whereas in the previously described aeration tank the gas is continuously recirculated against the liquid-solid. It should be appreciated that two types of mixing and fluid recirculation apparatus are included in Figure 2.1b for purposes of illustration, and either type could be used in both the aeration and digestion tank.

The table below shows data which illustrate the advantage of an integrated oxygen system. The values in the table are computed for a system as described consisting of three oxygen gas stages in the aeration zone and two gas stages in the digestion zone. In all instances, the feed oxygen is 99.5%, and it is assumed that the DO level throughout the aeration zone is 6 ppm and throughout the digestion zone, 3 ppm. It is further assumed for simplicity that no extensive auto-oxidation is practiced on the mixed liquor upstream of the clarifier and that the total oxygen consumptions in the digestion zone and aeration zone are equal.

	Stage	O_2 absorption for separate O_2 supplies				O_2 absorption for integrated O_2 supplies			
		60%	70%	80%	92%	60%	70%	80%	92%
Percent O_2 in recirculating gas stream:									
Aeration	1	86.2	84.4	82.5	80.0	91.0	90.2	89.4	83.5
	2	82.3	78.7	74.3	67.0	89.6	83.6	87.5	86.2
	3	81.1	76.1	67.8	47.0	89.2	88.2	87.1	85.7
Digestion	1	69.8	65.1	60.3	54.8	75.0	69.4	63.0	54.1
	2	40.8	30.9	20.8	9.0	55.4	44.7	32.2	15.0
Relative power required for gas recirculation:									
Aeration	1	0.31	0.32	0.33	0.34	0.29	0.30	0.30	0.30
	2	0.27	0.28	0.30	0.34	0.24	0.24	0.25	0.25
	3	0.16	0.17	0.19	0.32	0.14	0.14	0.14	0.15
Total		0.73	0.77	0.82	1.00	0.67	0.68	0.69	0.70
Digestion	1	0.44	0.48	0.52	0.58	0.41	0.45	0.50	0.60
	2	0.83	1.17	2.02	14.00	0.58	0.74	1.11	3.50
Total		1.27	1.65	2.55	14.59	0.99	1.19	1.60	4.09
Aeration plus digestion		2.00	2.41	3.37	15.59	1.66	1.86	2.29	4.79

The table gives data for the integrated oxygen system and also for separate oxygen feeds to the aeration zone and to the digestion zone. Comparative values are shown for four different percentages of the total feed oxygen utilized in the plant: 60%, 70%, 80% and 92%. The data consist of values of oxygen concentration and values of relative aeration power requirement in each stage of the aeration zone and the digestion zone. The relative aeration power values are power requirements referred to a value of unity assigned to the total aeration power consumed in all three stages of the aeration zone when operated to utilize 92% of the oxygen fed thereto. Power required for holding solids in liquid suspension are not included in these ratios.

The table further shows total relative aeration power required in each zone, and total relative aeration power required in both zones combined for each set of parameters. An inspection of the aeration gas O_2 purities in the aeration zone shows a pronounced increase in oxygen purity when the gas system is integrated. For example, at 80% utilization, the gas purity in the third stage is 67.8% with separate oxygen feed streams and 87.1% with an integrated oxygen supply. Higher O_2 partial pressures are also obtained in the first and second stages of the aeration zone with the integrated system. The difference in oxygen purities becomes greater at higher oxygen utilization, and at 92% utilization, the O_2 purity in the third stage is nearly doubled when the gas system is integrated.

Considering the oxygen purities in the digestion zone, for separate 99.5% oxygen supply to the digestion zone, the O_2 purity in the first stage is far below the feed gas purity. For example, at 80% utilization, the first stage purity is only 60.3%. A further drop to 20.8% occurs in the second stage. By comparison, the O_2 purities in the aeration zone do not decrease nearly as much. This demonstrates the CO_2 equilibrium effect, and results from the incapacity of the small liquid volume in the digestion zone to retain the CO_2 as it is produced.

With the integrated gas system of this process, the same CO_2 equilibrium effect is noted but to significantly less degree. The O_2 purities in the aeration zone remain almost steady and at high values due to the high capacity of the liquid for CO_2 and due to the much larger volume of oxygen flowing through the aeration zone. The gas fed to the digestion zone is the same gas existent in the third stage of the aeration zone and its purity is relatively high. Again the oxygen purity drops abruptly in stages of the digestion zone but not to the low level observed with separate O_2 supply. For example, at 80% utilization, the second-stage gas purity in the digestion zone is 32.2% for the integrated system compared to 20.8% for the separate system, an improvement of over 50%.

The following conclusion is drawn for the 80% utilization case, and qualitatively is typical for all other cases considered in the above table; by integrating the gas system, the resultant 87.1% O_2 transfer gas to the digestion zone obtains higher O_2 gas purities in the digestion zone than when a separate 99.5% O_2 supply is provided therefor. Stated otherwise, integration produces higher oxygen partial pressures in the digestion zone despite a lower purity of oxygen introduced thereto.

It will be noted that the foregoing partial pressure advantage extends throughout the entire range of percentages of O_2 utilization from 60 to 92%. However, it is also seen that it is not advisable to attempt to achieve high utilizations on the order of 92% in the aerobic digestion zone or in an integrated system which includes the aerobic digestion zone. For 80% utilization, the O_2 purity in the second stage of the digestion zone drops to 20.8% or approximately that of air when separate oxygen supplies are provided, and to 32.2% when the O_2 supply is integrated. For 92% utilization, the O_2 purity at this point drops below that of air in either system, and therefore, it would be more advantageous to employ air for aeration of the second digestion stage.

However, if air were substituted for the oxygen-enriched gas in the second stage, the burden of obtaining high utilization of oxygen in the digestion zone would fall solely upon the first stage. The O_2 purity in the first stage will fall well below the value shown in the table above. In preferred practice, O_2 utilization greater than 80 to 85% should not be attempted when the staged gas system includes an excess sludge digestion zone according to the process. With respect to the relative power values of the table, it is seen that a value of 1.0 has been assigned the power requirement of a three-stage aeration zone achieving 92% O_2 utilization. Operating the aeration zone in this manner represents excellent economy as demonstrated in actual operation of the cocurrent, staged aeration system of the process using municipal wastewater as the BOD-containing feed.

Comparing the power consumed in the digestion zone for the same case, i.e., separate O_2 supplies and 92% O_2 utilization, the relative power consumed in the digestion zone is 14.6 times that in the aeration zone, demonstrating dramatically the adverse effect of the very low O_2 partial pressure (9% O_2) in the second stage. By comparison, the integrated oxygen system operated for 92% utilization requires only 0.7 relative power in the aeration zone and only 4.1 relative power in the digestion zone.

Again the economy of the integrated system is emphasized: with separate 99.5% O_2 feeds to the aeration zone and digestion zone, the total relative power value is 15.6; but with 99.5% O_2 fed to the aeration zone only, the total relative power is reduced 69% to 4.8 despite a gas feed purity of only 85.7% O_2 to the digestion zone. The second gas introduced to the digestion zone preferably comprises at least 80% oxygen. At least 70% of the oxygen in the first gas is preferably consumed in the aeration and digestion zones. The table below shows the power reductions obtained with the intergrated O_2 supply for all oxygen-utilization cases listed in the previous table.

% O_2 Absorption	Relative Power Separate O_2 Supply	Relative Power Integrated O_2 Supply	% Power Reduction
60	2.0	1.7	17
70	2.4	1.9	23
80	3.4	2.3	32
92	15.6	4.8	69

In summary, the first table illustrates that the third gas discharged from the digestion zone must comprise at least 21% oxygen. Otherwise the power requirement becomes inordinately high for continuous recirculation of one fluid against the other in this zone in order to achieve the required mass transfer of oxygen from the aeration gas. As previously stated, the feed (first) gas to the aeration zone should comprise at least 60% oxygen and preferably at least 90% oxygen (by volume). The first table is based on an oxygen feed gas purity of 99.5%. If the purity of the feed oxygen is reduced substantially, the effect on mass transfer of the diluent nitrogen introduced with the feed gas (assuming an O_2-N_2 mixture as this gas) begins to overshadow the effect of nitrogen and carbon dioxide evolved from the liquor during aeration.

Therefore, reducing the oxygen feed gas purity tends to diminish the relative advantage of the integrated oxygen system of this process over the separate oxygen supply systems to the aeration and digestion zones. Moreover, since the effect of reducing feed gas oxygen purity is to reduce oxygen partial pressures through the system, it is evident from the trend of data in the first table that the 21% oxygen vent gas lower limit composition will be reached at lower oxygen absorption efficiencies when the oxygen feed gas purity to the aeration zone is reduced.

For example, the first table suggests that with the 21% oxygen lower limit on the digestion zone vent gas, the O_2 absorption efficiency cannot exceed about 85% based on 99.5% oxygen feed gas to the aeration zone. However, if the feed gas is 80% oxygen, the O_2 absorption efficiency can only be about 75% consistent with the 21% lower limit on oxygen content of the vent gas. The digestion zone is much more sensitive to feed gas purity than the aeration zone because of the more pronounced effect of impurity evolution into the gas space. Accordingly, the oxygen feed gas purity to the aeration zone must at least be 60% to achieve reasonable O_2 absorption efficiency without excessive power consumption.

Warm Sludge Digestion with Oxygen

A process described by *N.P. Vahldieck; U.S. Patent 3,926,794; December 16, 1975; assigned to Union Carbide Corporation* relates to a method for integrating aeration of BOD-containing water and warm aerobic digestion of activated sludge with oxygen. It has been found that when oxygen gas is first introduced to the warm covered digestion zone and the O_2 partially depleted gas discharge therefrom is introduced to the covered wastewater aeration zone as at least the major part of the required oxygen, all at specified conditions, the oxygen purity of the aeration gas in the last-mentioned zone increases substantially as compared with the gas entering from the digestion zone. One of the essential conditions for achieving this oxygen purity jump is that the sludge temperature in the digestion zone be at least 10°C warmer than the liquor temperature in the aeration zone. The oxygen-purity of the wastewater aeration gas increases above that of the incoming off-gas from the digestion zone because CO_2 and water vapor are rapidly absorbed into the liquor. The larger flow of water in the wastewater aeration zone, as compared to the digestion zone, has a huge capacity for dissolving CO_2 and quickly reducing the CO_2 content of the incoming gas. The lower temperature of the liquor in the wastewater aeration zone results in a much lower water vapor pressure than that in the digestion zone and the water content of the incoming gas is rapidly reduced by condensation.

By way of example, in one instance where the entire O_2 supply to the wastewater reactor is furnished by the digester off-gas and contains 61% O_2, the aeration gas in the overhead space within the enclosed wastewater aeration zone contains 78% O_2. The substantial increase in oxygen purity in the wastewater aeration zone gas provides a relatively high oxygen partial pressure which in turn allows satisfaction of the high O_2 consumption rate required in this zone at very modest power requirements compared to prior art where either separate oxygen feed gas streams are fed to the two zones or the oxygen feed gas is first introduced to the wastewater aeration zone and the off-gas therefrom transferred to a warm digestion zone.

The process offers substantial power savings. By way of example, in one instance the total dissolution energy for this process, the wastewater aeration-to-digester gas flow system and the separate oxygen feeds system are respectively 163, 205 and 243 hp. More specifically, this process relates to a method for BOD removal from wastewater in a covered aeration zone and activated sludge digestion in a covered digestion zone by aeration with oxygen gas. First gas comprising at least 40% oxygen (by volume) is introduced into the covered aeration zone where it forms gas which is mixed with the wastewater and recycled sludge in the covered aeration zone and one of the fluids in such zone is simultaneously continuously recirculated against the other fluid in sufficient quantity and rate to maintain the dissolved oxygen content (DO) of the mixed liquor at least at 0.5 mg/l and the liquid temperature is maintained at least at 15°C.

The mixed liquor is separated into purified liquid and activated sludge, and unconsumed oxygen-containing gas is discharged from the aeration zone at a rate such that its oxygen content is no more than 40% of the total oxygen introduced to the digestion zone. Part of the activated sludge is then returned to the aeration zone as the aforementioned recycled sludge.

Second gas comprising at least 80% oxygen (by volume) and sufficient in quantity to provide part of the aforementioned first gas is introduced along with the unreturned activated sludge to the enclosed digestion zone. They are mixed therein and one of the fluids simultaneously recirculated against the other fluid in the digestion zone in sufficient quantity to maintain the dissolved oxygen content of sludge at least at 2 mg/l, and the total suspended solids content (MLSS) of the sludge at least at 15,000 mg/l. During this step, the sludge is maintained at temperature of at least 25°C and at least 10°C warmer than the wastewater aeration zone liquor temperature, but below 75°C. The digestion zone mixing and fluid recirculation are continued for sufficient duration to oxidize at least 60% of the biodegradable volatile suspended solids content of the sludge introduced to the digestion zone. Thereafter at least partially oxidized sludge and oxygen-partially depleted digestion gas of at least 40% oxygen purity are separately discharged at rate such that the oxygen content of the gas is at least 35% of the oxygen content of the second gas entering the digestion zone. The oxygen-depleted gas discharge from the digestion zone is then provided as at least the major part of the first gas introduced to the wastewater aeration zone.

Ambient Temperature Process

C.P. Gunson; U.S. Patent 3,151,063; September 29, 1964; assigned to Denver Equipment Company describes an aerobic process and system for stabilization of sludge for useful and easy disposition. In the process, the raw sewage or other waste products are clarified and the settled sludge removed. Thereafter, the sludge is broken up into small discrete particles. In doing this, the surface area of the particles is greatly increased, thereby lending greater surface activity and removing previously oxidized matter. During the particle disintegration, air is introduced under pressure substantially as the solids are being agitated, further enhancing the scrubbing of oxidized material that may have developed on the particle surfaces while at the same time forcing oxygen into the pores of and into close association with the sludge particles, thereby speeding up the desirable oxidation of the noxious odor-producing proteins and other compounds present in the sludge.

Relatively simultaneously with the agitation and aeration, a seeding with about 2% by volume of previously stabilized sludge is accomplished which functions as a catalyst for the reactions proceeding in the sludge mass. While this is a preferred percentage, lesser amounts may be used, although this may result in a slower initial reaction rate. After as much oxygen as possible is forced into the surrounding liquid and into close association with the sludge solids, agitation and aeration are stopped. The sludge mass is then allowed to remain quiescent for a period of time normally approximately one hour, the exact time being dependent upon the temperature.

During the quiescent period the oxidation process proceeds, the oxygen is exhausted, and the sludge solids become oxygen-hungry again. As the above oxidation proceeds, CO_2 gas is produced and is absorbed by the liquid. Aerobic bacteria thrive in a CO_2-saturated solution, and, as this type of bacteria develop, the oxidation processes proceed more rapidly. The above process of supplying oxygen to oxygen-hungry sludge while agitating is repeated until the protein matter and other compounds in the sludge are oxidized to a desired stable condition. It has been found that the agitation and aeration need not be simultaneous as intermittent operation accomplishes the desired results.

After the above has been accomplished, as indicated by bacteria count and desired oxidation of the proteins and other compounds, the resulting stabilized sludge is passed off to a conventional filtering, surge tank or other apparatus, and drying and/or sludge beds. It is important to note that no rigid control of temperature, alkalinity or hydrogen ion concentration is necessary. In fact, it has been found that if abnormal pH conditions are induced by new sludge masses having such characteristics the process is self-sustaining in that the reacting mass absorbs such condition and, of itself, returns to optimum operating conditions. In actual tests run in a large metropolitan area, sludge was accumulated for one day and seeded with four-day-old sludge. The agitator-aerator was run fifteen minutes an hour on the average. Samples were taken twice a week to to test for bacteria count and the desired stabilization of protein and other compounds by oxidation. If more air was needed, the agitation time was increased to approximately twenty minutes out of an hour. When satisfactory stabilization was obtained, the sludge solids were pumped to a sludge bed for drying. It is, therefore, apparent from the above that approximately four days are necessary for a given batch or mass of sludge to be stabilized.

The important features obtained by the process are: drastic reduction in the time required for desired stabilization of the sludge; reduction in the complexity and elaborateness of equipment necessary to accomplish the desired stabilization of sludge; obviation of the necessity of hydrogen ion concentration, and temperature control; the formation of a readily drainable product; and the overcoming of undesirable foaming.

Complete Mix System

J.D. Walker; U.S. Patent 3,649,529; March 14, 1972; assigned to Chicago Bridge & Iron Company has found that in an activated sludge sewage aeration system, aeration, mixing, and uniformity throughout a tank are improved by liberating air along lines positioned to generate in the contents of the tank rolls having their axes on a bias, that is, neither parallel to nor perpendicular to the tank walls or to the general direction of flow of the processed liquid. Preferably using a tank thus mixed, the known advantages of complete mixing are achieved, and may be enhanced, by feeding an intimate mixture of raw sewage and return activated sludge to the tank at feed points so scattered that no part of the tank is far from a feed point.

Figures 2.2a through 2.2i illustrate the use of this process in conjunction with a circular processing system common in package treatment plants. This system includes a circular tank system, generally **20**, having an outer wall **21** and an inner wall **22**. Radial partition walls **23** seal off and divide the outer portion of the circular system between outer wall **21** and inner wall **22** into discrete compartments or tanks. The general flow of sewage is from inflow line **29**, through premix section **30**, through aeration tank **24** to settling tank or clarifier **51**.

Clear effluent from that tank flows under skimmer ring **48** and overflows into trough **49** and is discharged to a stream, perhaps with further treatment. The sludge settling in tank **51** is drawn off through pipe **31**. Some flows through airlift unit **34** into aerobic digestion tank **33** where the solids are made relatively unobjectionable. The remainder of sludge from line **31** has conventionally been used to seed the raw sewage with useful bacteria and is called return activated sludge. Thus the process gets its name, the activated sludge process.

According to the process, return activated sludge from line **31** is vigorously mixed with the raw sewage in premix section **30** and the activated mixture flows through distribution flume **35** to enter tank **24** at a plurality of feed points **36**. These are located along a long side of tank **24** so that every point of the tank is only a short distance away from a feed point, although the location of air liberation prevents any concentrated flow to the outflow without extensive mixing within tank **24**. Thus, raw sewage inlet **29** empties into a premix section generally **30**. Sludge return line **31** permits introduction of a return portion of the activated sludge, settled in clarifier **51**, to be introduced into the premix section **30**. If desired, supernatant liquor drawn by decanting equipment **32** from digestion tank

33 may also be discharged into mixing section **30**. Also chemicals can be fed in at this point, as for phosphorus removal. Premixed sewage-sludge liquor, an activated sewage mixture, leaves the premix section **30** by way of trough or distribution flume **35** which extends substantially the entire length of tank **24**.

FIGURE 2.2: COMPLETE MIX SEWAGE TREATMENT SYSTEM

a.

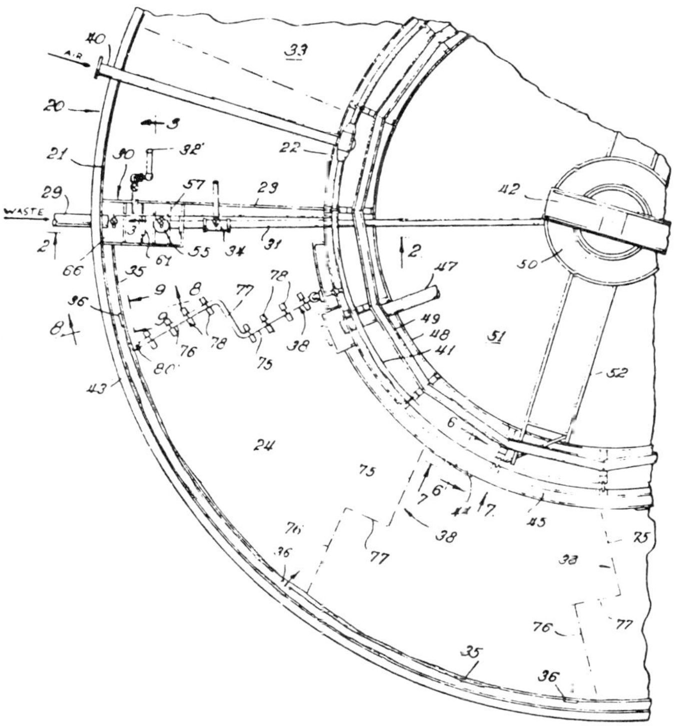

Portion of Circular Treatment System

b.

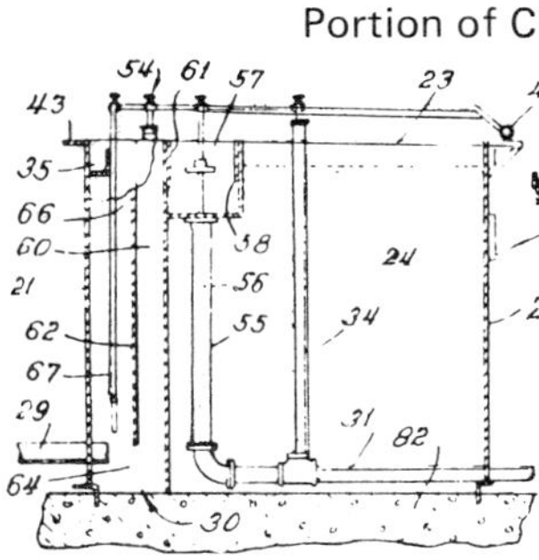

View Along Line **2–2** of Figure 2.2a

c.

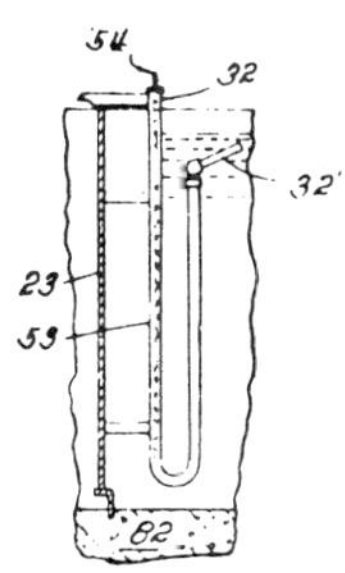

Cross-Sectional View Along Line **3–3** of Figure 2.2a

(continued)

FIGURE 2.2: (continued)

d.

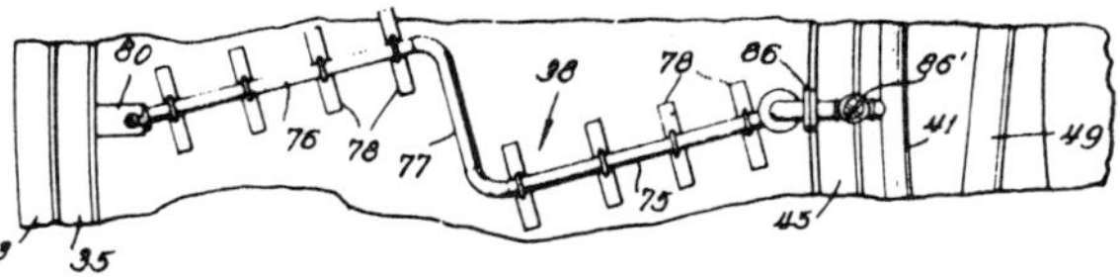

Enlarged View of Air Injection System

e.

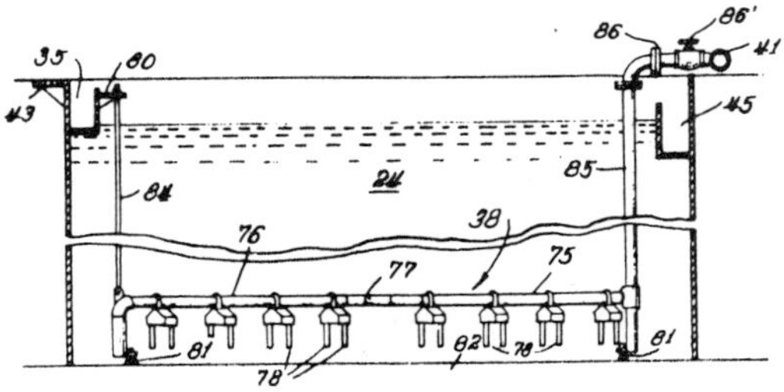

Vertical Cross-Sectional View of Air Injection System

f.

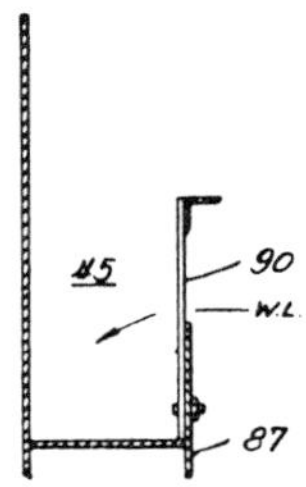

Outflow Weir, View Along Line **6–6** of Figure 2.2a

g.

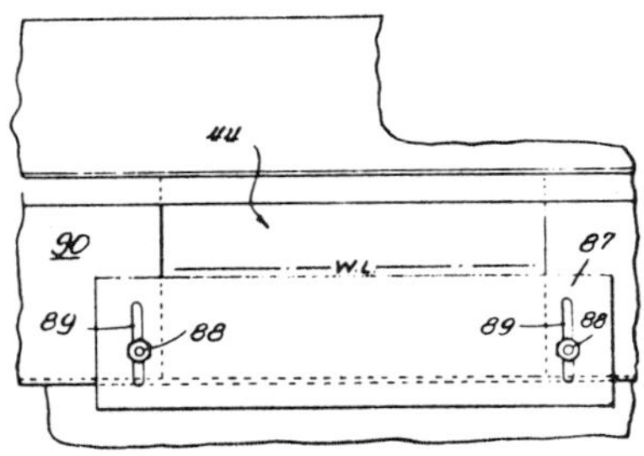

View Along Line **7–7** of Figure 2.2a

h.

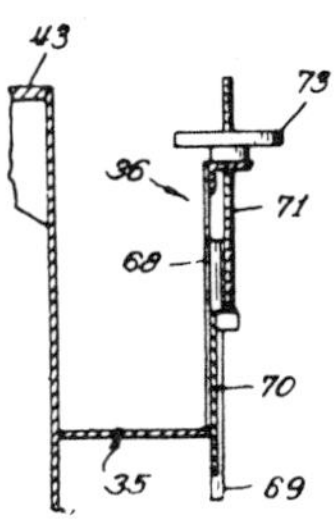

i.

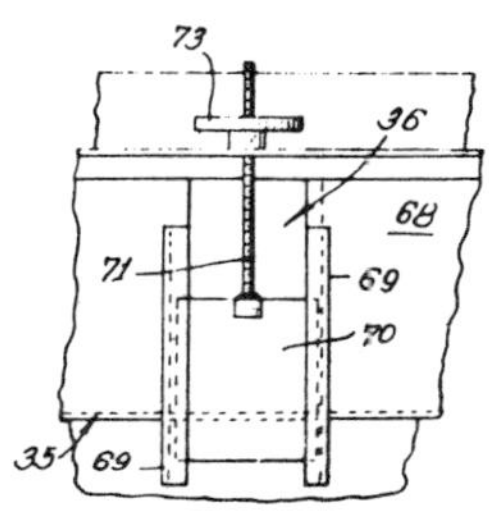

(h) (i) Adjustable Inflow Gate

Source: U.S. Patent 3,649,529

The activated mixture is discharged from trough **35** into tank **24** at uniform rates through a plurality of feed gateways **36**. The liquid in tank **24** is kept in motion as described below, with diverse flow directions. A large quantity of air is introduced into the liquid in tank **24** by means of a multiplicity of openings situated along air headers, generally **38**. These air headers are supplied through air supply pipe **40** and manifold section **41**. The configuration and construction of air headers **38** will be discussed in greater detail below. They are removed for servicing by conventional equipment, not shown, on rotating bridge **42**, which rides track **43**. Liquid leaving tank **24** is collected by way of discharge gateways **44** and trough or collection flume **45**, and led to inner tank **51** where solids are settled.

Thus, the overflow or effluent from tank **24** is introduced through conduit **47** to the centrally located inflow distribution structure **50** of inner tank **51**. Sludge is collected at the bottom of inner tank **51**, and this may be aided by rotating collector arms **52**. Details of some of the structures generally referred to above will now be discussed. The supernatant decanting means illustrated in Figure 2.2c includes an adjustable intake section **32'** and lift legs **53**. An air injection means **54** is used to lift the supernatant over radial wall **23** for discharge into premix section, generally **30**.

Vertical air-lift leg **55** for return sludge is fitted with an air pipe **56** having a sparger at its bottom, to raise return sludge which discharges into box **57** of premix section **30**. Return sludge discharged into box **57** can be diverted directly to tank **24**, if weir gate **58** is provided. Normally, however, the contents of box **57** are discharged into downward leg **60** of premix section **30**. Control may be provided by weir gate **61**. Baffle **62** directs the sludge (and supernatant decant liquor, if supplied) into the bottom region **64** of mix section **30** into which raw sewage is introduced through influent pipe **29**. The resulting mixture is simultaneously raised and vigorously stirred in upwardly extending chamber **66** of premix section **30** by air released from one or more spargers at the foot of air supply pipe **67**.

Enough air is released to achieve a high degree of intimate mixing with resulting particle-to-particle contact between sewage solids and the activated seed sludge. To this end violent agitation is desired, e.g., a turbulence equivalent to a velocity gradient of the order of 200 effective for 30 seconds or 700 for 5 seconds, or more. The velocity gradient here used is in terms of feet per second per foot. This is in contrast to mild mixing of sewage and return sludge in inflow channels sometimes provided heretofore, which was likely to be equivalent to a velocity gradient of only about 15.

To avoid using much space or energy, a very short detention period is preferred in premixing chamber **66**, not over ½ minute with rated flow rates, although up to two minutes would probably not be deemed prohibitively wasteful. The flow of the premixed process material from distribution flume **35** through weir gate **36** is regulated for uniformity by appropriate adjustments of the gates shown in greater detail in Figures 2.2h and 2.2i. Vertical wall **68** of flume **35** is fitted with gate-retaining elements **69** which retain vertically movable feed gates **70** against vertical wall **68**. The precise vertical position of each movable feed gate **70** is adjustable by means of the screws **71** turning handle **73**. The latter structures are perhaps best illustrated in the schematic presentation of Figures 2.2h and 2.2i. Air injection header assemblies, generally **38** include three distinct horizontal sections, namely, two elongated linear header sections **75** and **76** situated adjacent inner wall **22** and outer wall **21** respectively, and median connector section **77**, which connects sections **75** and **76**. A large number of gas injection outlets arranged in clusters **78**, are situated along headers **75** and **76** and optionally along section **77**. The header assemblies **38** are secured to the equipment at one end of each by brackets **80** which are fixed to the inner vertical wall **68**, and positioned by foot bolts **81** which rest on the floor **82** of tank **24**.

During installation, foot bolts **81** are adjusted to make the header dead level. Vertical rod **84**, and vertical leg **85** of the air supply system, provide means for lifting air header assembly **38** from tank **24** for inspection, and for cleaning if necessary. Either quick-opening couplers or flanges **86** are provided for convenience in separation of the header assembly **38**, and its attached components from air supply manifold **41**. Valves **86'** are preferably situated between flanges **86** and manifold **41**. Two-stage hoisting facilities are provided as described in U.S. Patent 3,339,901, in this instance using two double-winch hoists, one for rod **84** and one for pipe **85**.

It is noted that header assembly **38** is situated near but well above floor **82** of tank **24** to facilitate the movement of water in the tank to the region of header **38**. Header sections **75** and **76** are sufficiently elongated to generate a roll. Also, header assemblies **38** are spaced apart from each other a sufficient distance to permit development of a double roll extending away from each header **38**. The double roll is formed by the mass of liquor rising with the stream of injection air bubbles rising above header **38**. This causes a substantial flow of liquid in both directions away from the region of the headers at the top of the tank contents. When the outwardly moving liquid meets a wall or an on-coming flow, it moves downward. At the bottom, it is drawn toward the header to replace fluid moved upward from below header **38**, the overall pattern being referred to as a roll. This is distinguished from patterns in which the liquid is raised and drawn down in tight regions or columns.

It is particularly noteworthy that the gas injection orifices are arranged in a pattern along lines which are on the bias with respect to the general direction of displacement flow of the liquid being processed. The terms general direction of displacement flow of liquid being processed means the general direction to which liquid from the input tends to flow while no air is released in the tank **24**. Thus, the general direction of displacement flow of the liquid being processed in tank **24** is along the radii between outer wall **21** and inner wall **22**. Each of the header sections **75** and **76** are situated on the bias with respect to the radial lines. Shorter intermediate sections **77** are also on the bias with respect to the radial lines. Preferably the air injection orifices are situated only along sections **75** and **76**, with section **77** merely forming a connecting means to convey air into section **76**. In this form two separate sets of double rolls are generated by each of the headers, generally **38**. This causes a region of extreme turbulence between the rolls and at the surface immediately above section **77**.

The pairs of double rolls generated by headers **75** and **76** coact with the walls of tank **24** as well as with each other to bring about a degree of mixing previously unavailable in activated sludge processing equipment. It will be appreciated that a roll formed on the side of header **75** or **76** which is an acute angle formed by the wall and the header will continuously thrust liquid against the wall with varying degrees of vigor depending on the distance between the wall and the point above the portion of the header section involved. On the other hand, the roll formed in the region between the header and the wall which constitutes an oblique angle tends to move the majority of the water in the roll somewhat outwardly from the wall thus enhancing the formation of numerous eddies along the wall in that region. Moreover, the positioning of the headers **75** and **76** on the bias with respect to the radius of the circular tank results in the formation of rolls having axes and lines of thrust which are also on the bias with respect to the axes and lines of thrust of other rolls formed by the next adjacent headers.

The division of the flow of effluent liquor from tank **24** among its outlets is controllable by adjusting the height of movable gates or weirs **87**, after loosening nuts **88**. The slots **89** permit movement to a position higher or lower on vertical wall **90** of trough **45**. It is apparent that by feeding a unified activated sewage mixture at feed points so scattered along an activated sludge aeration tank, with high speed rolls, with substantially every zone of the tank being quickly reached by a portion of the inflow which has undergone turbulent mixing with the tank contents en route, the practical achieving of throughout-the-tank uniformity in large tanks is reached. A desired standard of uniformity, attainable when the process is fully used, is that if a tracer dye or the like is injected simultaneously at all feed points, there will after five minutes be no points in the tank except close to the feed points at which the concentration departs from the average concentration more than 10%.

By briefly subjecting the incoming raw sewage and return activated sludge to vigorous mixing, unified activated sewage inflow is provided so that volumetric equalizing of the flow among feed points will also achieve qualitative equalization offered. Additionally, very vigorous premixing yields a head start by quickly intermixing the raw sewage and the useful organisms in the return activated sludge, together with copious oxygen. Furthermore the vigor of the premixing accomplishes maximum tearing apart of dispersion of the sewage solids and return sludge flocs for maximum effective contact between them, and for maximum effective contact of this mixture with the entire body of liquor within the tank when mixing occurs within the tank, even though this latter mixing is not, at least for the most part, quite so vigorous.

By the disposition of the elongate air-liberation headers at a substantial angle of bias to the general direction of displacement flow and to the adjacent tank walls, a diverse flow to all parts of the tank results. With enough air liberation to produce high-speed rolls, and with offsetting of the ends of the rolls nearest each other, and preferably some axial spacing thereof, focal points of cyclonic or other greater than average turbulence are provided ensuring the repeated breaking up of any flocs and the penetration of oxygen and organisms to particle centers. Thus the maximum use of the space within the tank that is theoretically possible is more closely approached than previously in full-scale working tanks.

Aeration Technique

In a process described by *B.R. Brown and L.B. Wood; U.S. Patent 3,649,531; March 14, 1972; assigned to National Research Development Corporation* the dewatering time of a sludge obtained by anaerobic digestion of a primary sewage sludge is reduced substantially by subjecting the digested sludge to aeration with air or other suitable oxygen-containing gas. Typically, aeration is carried out for at least 4 hours, the sludge is allowed to settle, and then dewatered within two days. Aeration may be carried out by means of diffused air or, preferably, by mechanical means using a surface aerator.

Example 1: Anaerobically digested sludge is charged into a cylindrical concrete tank of dimensions 20 feet high and 6 feet diameter, which is fitted with an aerator consisting of ceramic tubular diffusers so arranged that they form 6 arms, each at an angle of 60° from the adjacent one, radiating from a central chamber into which air is passed. The porosity of the diffusers is such that very fine bubbles are produced. The sludge is aerated for periods varying from 6 to 24 hours, depending on the intensity of aeration, and then allowed to stand without aeration during which time it consolidates and sinks to the bottom, leaving a supernatant liquor substantially free from suspended solids. Typical results are given below.

	Period of aeration (hrs.)	Original sludge solids content, percent	Sludge solids content, percent after—		
			1 day	2 days	3 days
Run No.:					
1	24	2.3			3.4
2	24	3.4		4.3	
3	24	4.0	5.1		
4	24	1.9			4.0
5	24	2.1			3.8
6	24	2.4			5.5

Example 2: Digested sludge is charged into a cylindrical concrete tank of dimensions 40 feet diameter and 10 feet 6 inches working depth equipped with an aerator capable of producing coarse bubble aeration. The base of the tank is in the form of an inverted hollow cone. The diffuser is constructed from a series of 6-inch-diameter mild steel pipes, branching from both sides of a single pipe across the diameter of the tank and extending to the sides. On the underside of the pipe, along the whole length, a series of ⅛-inch-diameter holes are drilled and spaced regularly at one-inch intervals, totalling 2,056 holes. A connection for supplying air to the pipes is provided. The volume of digested sludge (approximately 400 tons) is aerated for 24 hours and then allowed to stand as in Example 1. Typical results are shown below.

	Period of aeration (hrs.)	Original sludge solids content, percent	Sludge solids content, percent after—		
			1 day	2 days	3 days
Run No.:					
1	24	2.7			5.1
2	24	2.8		3.9	
3	24	2.8			4.8
4	24	3.0			4.8
5	24	2.9			5.0

Modified Imhoff Tank

D.F. Peck and J.C. Troy; U.S. Patent 3,919,086; November 11, 1975; assigned to NUS Corporation describe a technique for modifying a conventional Imhoff tank, which utilizes an anaerobic wastewater treatment process, to obtain an activated sludge system which is an aerobic process. The modifications comprise: providing a plurality of air sparges in the lower portion of the digestion chamber of the Imhoff tank to aerate the wastewater; providing a pump means in the tank which has its suction end near the apex of the inverted pyramid-bottom of the tank and its discharge end disposed in the gas vent area of the tank; and, relocating the feed to the tank into the gas vent areas rather than into the center settling compartment which is conventional.

Spiral Motion Circular Aeration Chamber

P.H. Huisman; U.S. Patent 3,939,067; February 17, 1976; assigned to Gypsum Industries Ltd., South Africa describes a method of treating sewage using the activated sludge process. The process involves imparting to sewage introduced into a circular aeration chamber a spiral motion about the chamber center, using the motion imparted to the sewage to cause liquid-borne sludge from the chamber to flow along a canal around the chamber and returning at least part of the liquid-borne sludge in the canal back into the aeration chamber under the influence of gravity for recirculation. The apparatus is illustrated in Figure 2.3.

FIGURE 2.3: CIRCULAR AERATION CHAMBER

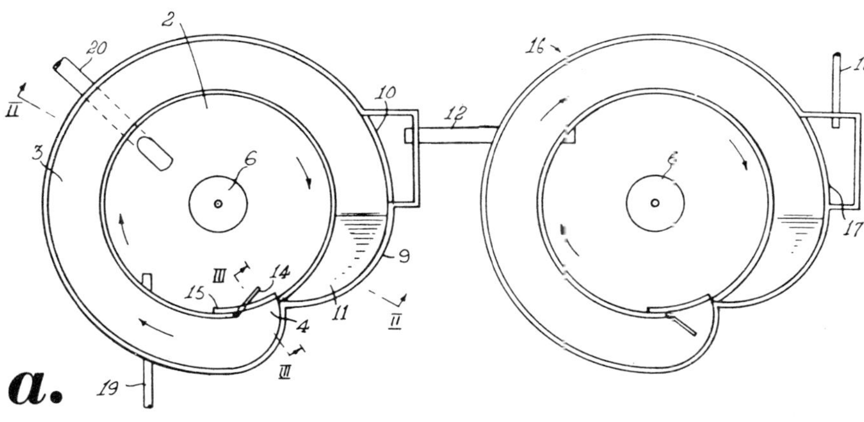

Plan View of an Aerating Installation

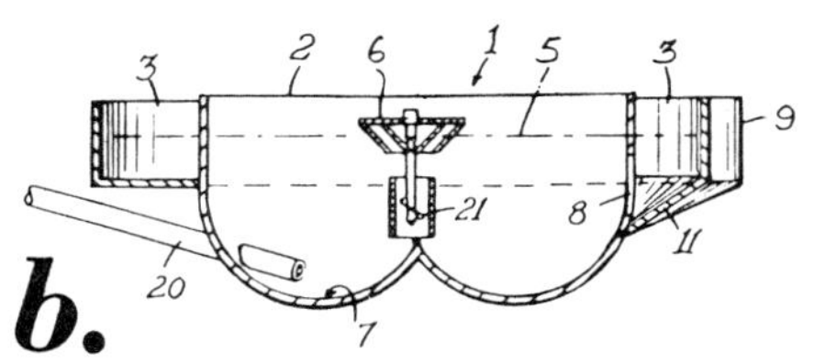

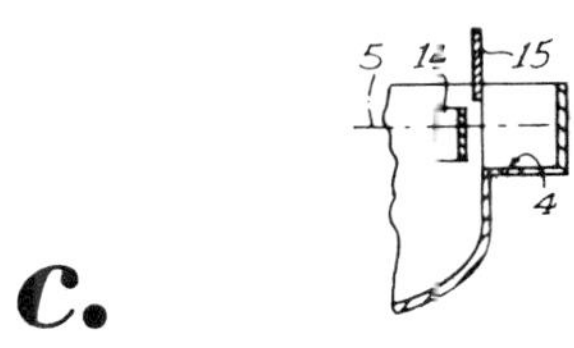

Sectional Elevation Along Lines II–II Sectional Elevation Along Lines III–III

Source: U.S. Patent 3,939,067

The aerating installation comprises a tank **1** having a central aeration chamber **2**, circular in plan and a canal **3** extending along the periphery of the aeration chamber. At one end the canal **3** has an overflow weir **4** from the aeration chamber into the canal. The weir **4** is located beneath the normal liquid-sludge level **5** in the chamber **2** and is so dimensioned as to enable liquid-borne sludge to be diverted into and driven along the canal under the action of a spiral motion about the chamber center of the sludge in the aeration chamber **2**.

This motion in the sludge may be created by an aerator **6** mounted centrally in the aeration chamber and preferably of the conoidal type. The aerator causes circular and upward motions of sludge in the chamber and these motions combine to give a spiral flow which pushes the sludge along the canal **3**. To facilitate this flow, the floor of the chamber **2** is preferably formed as an endless channel **7** of arcuate cross section. The aerator **6** preferably has a stirring mechanism **21** mounted thereon for rotation adjacent the floor of the chamber **2**.

The end of the canal opposite the end with the inlet weir **4** has an underflow weir **8** in the chamber wall for returning to the aeration chamber **2** at least some of the liquid-borne sludge in the canal **3**. This weir **8** is situated adjacent the inlet weir **4** and has a deflecting wall **9** of the canal located to cooperate therewith. Adjacent to the deflecting wall **9** and upstream is an outlet from the canal in the form of another overflow weir **10** preferably adjustable in height. The floor **11** of the canal slopes downwardly from the deflecting wall **9** to the weir **8** to enable sludge to be returned from the canal **3** to the chamber **2** under the action of gravity.

As shown, the deflecting wall **9** is displaced progressively towards the aeration chamber center over the length of that part of the canal having the inclined floor **11**. This is to enhance even distribution of sludge into the aeration chamber. Of the lightweight and heavy sludge passing along the canal **3** the former will discharge over the outlet weir **10** into a conduit **12** while the latter will pass through the underflow weir **8** to the chamber **2**. The weir **10** will be so adjusted as to provide an appropriate ratio of sludge returned to and removed from the tank **1**.

Means are provided for regulating the quantum of flow into the canal **3** and may conveniently be in the form of a swing gate **14** mounted at the inlet to the canal. A shield **15** is preferably mounted over the inlet weir **4** to prevent turbulence of liquid entering the canal **3**. The conduit **12** leads to a second tank **16** which is the same as the tank **1** described above. The outlet from the tank **16** is in the form of an overflow weir **17** which discharges into a conduit **18** leading to a sludge settling installation. In use, sewage is fed into aeration chamber **2** through conduit **19** after being passed through the usual screens. The sewage is aerated in the aeration chamber **2** together with activated sludge from the sludge settling installation. During aeration nitrification also takes place but the passage of the sludge along canal **3** is slow and nonturbulent, giving it an opportunity to denitrify.

At the underflow weir **8** at the end of the canal the heavier of the liquid-borne sludge in the canal is returned to the aeration chamber by sliding down the inclined floor **11** of the canal under the action of gravity. In this way the heavier sludge is recirculated and digestion is enhanced by virtue of the sludge being subjected to more aeration than is usual in these installations. The lighter sludge passes over the weir **10** to the second tank **16** where the process is repeated. It is considered that the microbial life digesting the sludge will be different for the heavier and lighter sludge and by separating these into two tanks overall digestion will be improved. It may be necessary to remove sludge periodically from the bottom of the aeration chamber through a conduit **20** or the like.

It is stressed that the use of the canals **3** in the tanks **1** and **16** allows denitrification to take place so that denitrified sludge is passed to the settling installation. The continuous recirculation of sludge prevents too much sludge being passed to the settling installation and hence prevents overloading of the latter. Also, recirculation increases the sludge concentration in the tanks and hence the bacteria concentration which improves digestion and conversion to gas.

Pressure Release Used to Break Up Solids Particles

According to a process described by *W.J. Bauer; U.S. Patent 3,939,066; Feb. 17, 1976* sewage is caused to flow through a hydraulic system which subjects the sewage to a sudden substantial drop in pressure, along with an attendant increase in velocity, and violent turbulence or agitation. By way of example, the pressure may be decreased on the order of 20 psi in 1/10 second or less, and the sewage may be subjected to turbulent flow while under a vacuum. The resulting material evidences a marked increase in the amount of suspended solids and a marked decrease in the amount of settleable solids, as compared to the input sewage material.

It is believed that the sudden decrease of pressure causes the larger organic solid particles to disintegrate or explode, perhaps due to expansion of absorbed gases or gases generated within the particles by anaerobic bacterial digestion. Such gas-producing phenomena may continue within particles of solid wastes, even though the exterior surface has been exposed to aerobic conditions. In any event, the sudden pressure drop and violent turbulence apparently cause comminution of the solid particles into many smaller particles and thorough mixing, resulting in apparent homogenization of the particles in the sewage.

The resultant sewage material then may be subjected to bacterial adsorption and ingestion, or biological treatment as it is called. The production of many smaller particles results in a greatly increased total surface area of solid particles for exposure to the action of natural bacteria which consume the organic sewage as food. If such bacteria are brought into intimate contact with such particles, as can be done during the turbulent flow and agitation, the bacteria will adsorb the particles of sewage on their surfaces for later consumption as food. This adsorption can take place in a matter of seconds or minutes.

The bacteria and associated adsorbed particles are then settled or otherwise separated from the remainder of the sewage mixture and removed as a sewage sludge. The remaining effluent has a greatly reduced quantity of organic waste materials in it, comparable in many instances to the character of effluent from an activated sludge treatment plant.

The sludge may be kept in an aerobic condition by addition of air or oxygen so that the metabolism of the bacteria and the associated consumption of the attached organic wastes by these bacteria will proceed at a rapid rate. A portion of this sludge, following a time of aeration or oxygenation, may be recycled back into the system in the pressure drop and/or agitation zones for adding and mixing bacteria with the small particles of waste materials produced by the passage of additional sewage through the system to enhance the adsorption and ingestion processes.

Pressure-Controlled Oxidation

According to a process described by *D.F. Othmer; U.S. Patent 3,772,187; November 13, 1973* domestic sewage and other polluted waters may be treated under a pressure of ½ to 5 atmospheres gauge or more with dissolved oxygen, or air, to supply the BOD. The oxygen-containing gas is drawn into the suction of a feed and recycle pump which agitates the liquid-gas mixture, and increases gas solubilization as it is being compressed. Simultaneously, any solids which are present are comminuted in being pumped to the pressure oxidation tank. The higher than atmospheric pressure increases oxygen solubility, concentration, and hence chemical or biochemical reactivity.

Thus, a much smaller residence time and vessel is required, the polluted water is withdrawn and depressurized to atmospheric pressure at which pressure some of the dissolved oxygen and other gases are released due to lower solubility, and vented. Most of the liquid after depressurization is recycled to the influent stream for repressurization and additional oxygen dissolution. The recycle may amount to 2 to 50 times, and some sludge obtained may also be recycled. A water turbine may recover some of the mechanical energy used by the pump. The process may be used by itself or as an adjunct to other processes for treating polluted waters, and also for the oxidation of sludge.

Pump Design for Use in Septic Tank

K.A. Digney and A.P. Gallauresi; U.S. Patent 3,896,027; July 22, 1975 describe a method for increasing the level of air or other gas entrained or dissolved in a liquid. The principal application is to enhance aerobic decomposition in sewage by recirculation through a pump where a high level of dissolved oxygen is maintained by mixing atmospheric air into the sewage as it passes through the pump.

The process is described in the context of a domestic-type sewage system, namely, an enclosed septic tank of conventional design. A centrifugal pump is arranged with a liquid intake opening below the sewage level within the tank, and an outlet which may be either above or below sewage level, but which in any case discharges the sewage back into the tank. The pump impeller blades are powered by a motor, preferably located outside the tank or in a separate chamber thereof, connected to the blades, or to intermediate support structure, by a rotating shaft.

The pump blades extend in a helical path from the central shaft to their outer ends and are enclosed in a chamber having inlet and outlet openings. The blades are essentially hollow, having open ends at their tips communicating with a central opening in the shaft. The shaft opening extends axially through the shaft from an air intake opening above sewage level to the communicating opening with the hollow interiors of the blades.

The sewage is propelled through the pumping chamber by the centrifugal action of the blades, passing from the inner to the outer blade ends. As the sewage passes over the outer tips of the blades at a very high velocity, a stream of air is injected directly into the sewage due to lower pressure being formed as it passes over the blade openings at the tips. The blade movement itself also tends to expel by centrifugal force the air inside the hollow interior, and a fresh supply of air will be drawn in through the shaft air intake. Thus, the steady flow of air through the hollow drive shaft and impeller blades is intimately mixed with the sewage flowing through the pump to maintain a high level of dissolved oxygen, thereby stimulating aerobic bacterial action and insuring proper sewage treatment.

Compartmented Unit

P. D'Aragon; U.S. Patent 3,888,767; June 10, 1975 describes a sewage treatment tank for the treatment of sewage from houses, hotels, garages, schools, small industries, and the like where normal sewage collector systems do not exist. The sewage tank contains no mechanically moving parts, and is composed of two adjacent compartments, a first compartment, with a centrally located sewage inlet having a V-shaped launder, open at the bottom. In this first compartment, most of the solids in the sewage are separated from the liquids, and are digested by anaerobes. The liquids are referred to as effluent.

The effluent passes from a first compartment (a) to a second compartment (b) where it is submitted to the action of aerobes in the presence of injected air. In each compartment there is a pipe immediately adjacent to the wall separating the compartments, and preferably centrally located. These pipes contain a plurality of holes in their upper portion, immediately below the roof of the compartments, which permits the continuous evacuation of gases produced in both the compartments. Both pipes also permit the removal of digested sediments from the compartments through the base of the pipes at predetermined intervals for cleaning.

The effluent of compartment (a) is fed towards the bottom of compartment (b) through a suitable slot in the wall separating the compartments, and a pipe connection with the gas exhaust pipe of compartment (b). The products of digestion of the sewage in compartment (a) by anaerobes, include: methane, carbon dioxide, ammonia and amines, hydrogen sulfide and organic sulfur compounds, phosphine and organic phosphorus compounds, and some insoluble humus.

The products of digestion of the effluent from compartment (a) in compartment (b), by aerobes in the presence of injected air include: carbonic gas and carbonates and bicarbonates, some humus, water, nitrites and nitrates, sulfates and phosphates. Sufficient air is injected into compartment (b), so that the effluent from the compartment contains dissolved air and a nil BOD. In compartment (b), compressed air is injected at the bottom end of a slightly inclined inverted tray, the tray having a series of small open compartments, similar to ice trays for refrigerators, so that the air travelling from the bottom end to the top end of the tray is in immediate contact with the effluent for a relatively long period of time, the air then passes under a series of similar trays, which are overlying and mounted above each other, so that the oxygen contained in the injected air has sufficient time to effect the biological digestion by aerobes.

In one example a tank having compartments of equal volume, to treat sewage from a dwelling occupied by eight persons would be about 5 feet wide, 6 feet high and 8 feet long. This size is determined from the assumption that a single person produces approximately 60 gallons of sewage per day, and considering the volume taken up by digested sediments, sufficient space must remain for easy removal of the gases. Each compartment is approximately 5 feet wide, 6 feet high and 4 feet long. The tanks may be made of many materials; wood rendered impervious, or plastics are both satisfactory materials.

For the easy removal of the accumulated sediments in both compartments (a) and (b) of the process, pipes for evacuating the gases have an internal diameter of about 4 inches, so that a hose or a pipe may be easily inserted in pump. Some sediment should always remain in both compartments, to allow the biological digestion of both anaerobes and aerobes in their respective compartments (a) and (b) to continue.

To have good biological digestion by aerobes in compartment (b), an excess of air is preferably injected. About 2 cubic feet of air per gallon of sewage is the preferred quantity. In this case, about 1,000 cubic feet of air are required. A relatively small compressor operating under a head of about 2 pounds gauge pressure is satisfactory and a motor using 12 to 15 watts is required. The motor should preferably be operated continuously.

The final effluent is practically free of coliforms, microorganisms and other bacteria. Soluble salts are contained in the effluent, and some of these salts such as nitrites, nitrates and phosphates may be objectionable, as they constitute excellent fertilizers, and in reaching either a lake or a river, they promote the growth of aquatic plants.

Therefore, in a preferred example, a third small compartment is added in which the effluent of compartment (b) comes in contact with lime, thereby producing nearly insoluble salts of nitrites, nitrates and phosphates. At other places where it is feasible, plants and/or trees could be made to grow along the path of the final effluent, where fertilizing salts contained in the effluent would be mostly absorbed by the plants and/or trees.

Package Plant

A process described by *D.A. King; U.S. Patent 3,917,532; November 4, 1975; assigned to Davis Water & Waste Industries, Inc.* relates to an improved tank arrangement in the type of system which employs a central sludge settling tank surrounded by an annular outer tank where aeration and digestion of the sewage occurs. The apparatus is shown in Figure 2.4.

A package sewage treatment plant adapted to be easily erected in the field is generally depicted at **10** in Figure 2.4a. The treatment plant includes an outer cylindrical tank wall **11** and a concentric settling tank wall **42**. The space between these two cylindrical walls is normally divided into contact and reaeration areas in an aeration section, a digester section and a chlorine contact section.

Sewage to be treated enters the plant **10** at a plant inlet **12**. An intermediate wall, shown generally at **20** and formed of arcuate sections, divides the contact and reaeration areas of the aeration section into inner and outer arcuate portions **13** and **18** and **14** and **19** respectively. The intermediate wall **20** extends only as far as vertical radial walls **40** and **41**, and does not extend into the digester section. The sewage enters the contact area **19** of the aeration section of the tank through opening **15**. The sewage activated sludge mixture flows in a counter-

FIGURE 2.4: PACKAGE SEWAGE TREATMENT PLANT

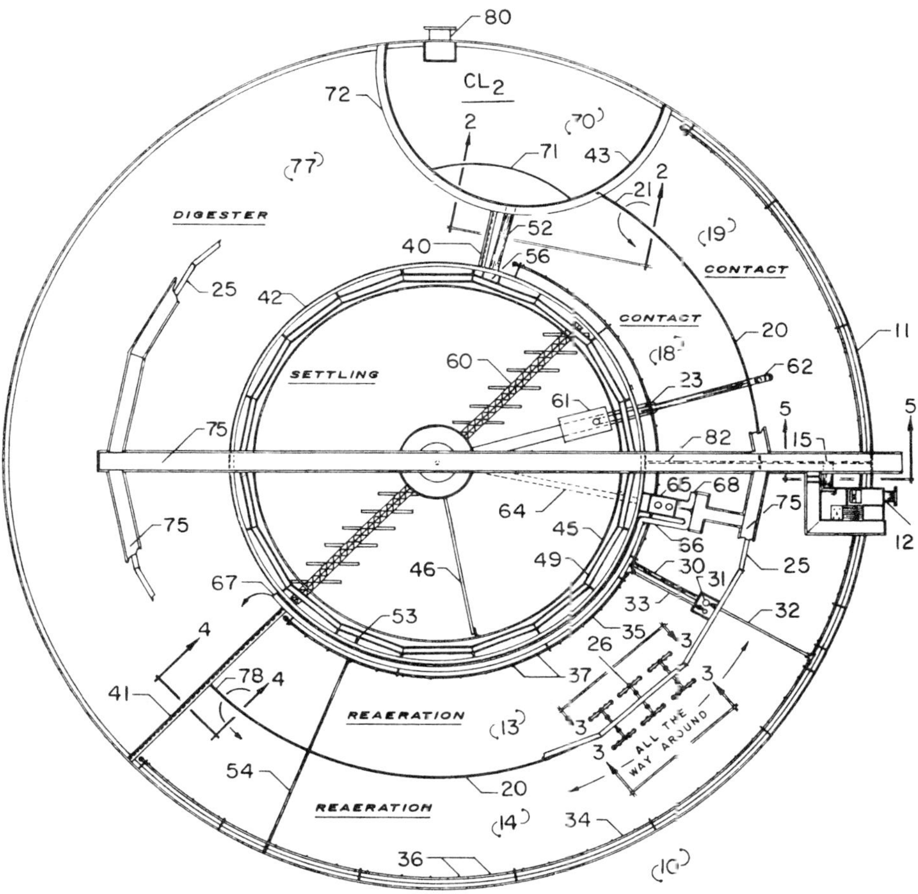

a.

Plant View of Treatment Plant

(continued)

FIGURE 2.4: (continued)

b.

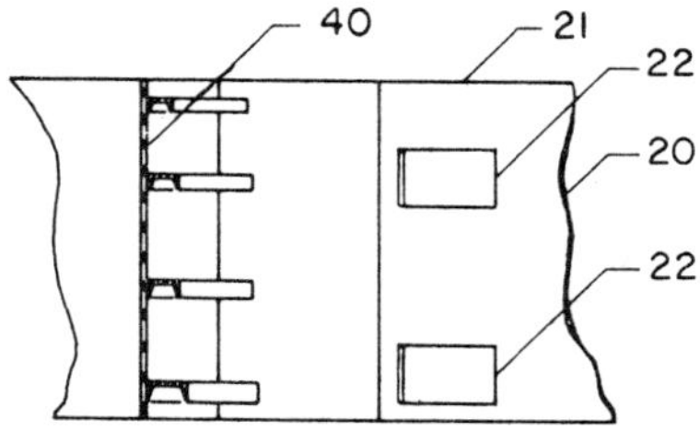

Sectional View Taken Along Lines **2–2** in Figure 2.4a

c.

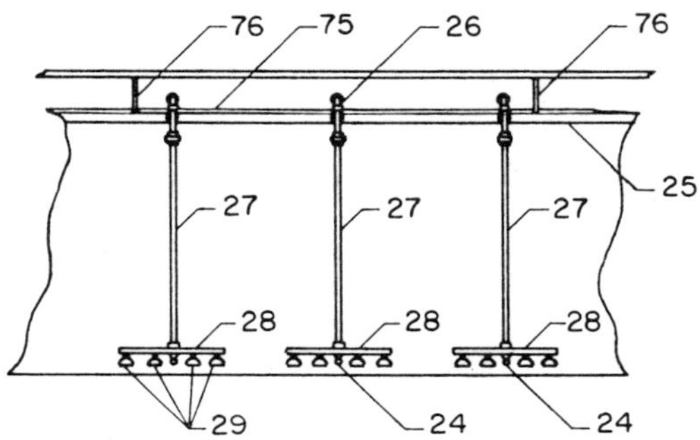

Sectional View Front Elevation View
Taken Along Lines **3–3** in Figure 2.4a

d.

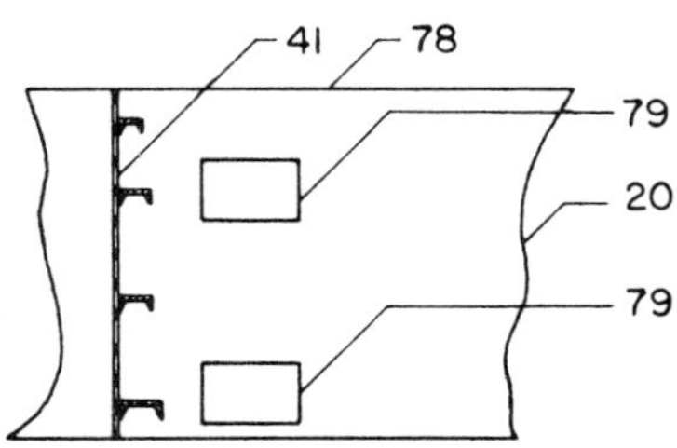

Sectional View Taken Along
Lines **4–4** in Figure 2.4a

e.

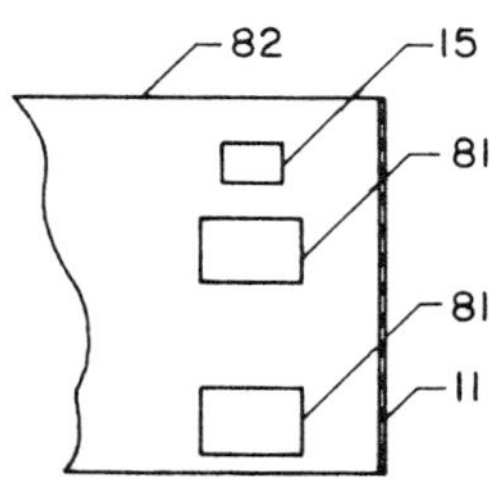

Sectional View Taken Along
Lines **5–5** in Figure 2.4a

Source: U.S. Patent 3,917,532

clockwise direction in outer arcuate portion **19** from the influent point **15**. Communication is provided between outer section **19** and inner section **18** through holes **22** formed in end portion **21** of the intermediate wall **20**, as shown in Figure 2.4b. Sewage from the outer arcuate portion **19** is thus allowed to pass through the end portion **21** of intermediate wall **20** near wall **40**, as shown by the arrows, to reach a settling tank inlet **23**.

As long as the sewage and activated sludge remain in the contact or reaeration areas, air is injected to provide oxygen for the microorganisms that break down the organic material in the sewage, and also to provide adequate mixing to bring the microorganisms, organic material and oxygen into intimate contact and to keep sludge from settling in the aeration areas. As shown in Figures 2.4a and 2.4c, air is supplied from blowers to an air header **25** mounted on the intermediate wall **20**. The air header **25** forms a full circle around the plant providing air to the digester **77** as well as the contact and reaeration areas. Air flows from the header **25** through descending pipes **27** to conventional air diffusers **26** having spaced injector portions **29**.

The air header **25** and air diffusers **26** can be mounted on intermediate wall **20** in any suitable manner. In the preferred embodiment brackets **24** mount diffusers **26** on wall **20** near, but spaced from, the bottom of the tank (usually about 2 feet). The injector portions are so located to prevent clogging thereof while allowing maximum mixing.

As the air bubbles injected through the injector portions **29** rise, they impart energy to the liquid in the tank causing it to roll from intermediate wall **20** to the outer wall **11**, and from intermediate wall **20** to the settling tank wall **42**. Since the size of the treatment plant, the distance from the settling tank wall **42** to the outer wall **11**, is limited by the distance that this rolling or mixing can be maintained, it is apparent that the provision of the intermediate wall **20** according to this process, makes possible the extension of the size of package treatment plants since rolling need be maintained only in the distances between the intermediate wall **20** and the inner and outer walls rather than in the larger distance from the inner wall to the outer wall.

In addition to providing this advantageous result, the intermediate wall **20** also provides support for the aeration piping **25, 27** and prevents short-circuiting of the sewage from the plant inlet **12** to the settling tank inlet **23**. In order to effectively dissipate the froth that is created in the aeration section, a pipe **30** brings liquid from the settling tank to a froth spray pump **31**, which supplies the liquid to froth spray nozzles **36, 37** through pipes **32, 34** and **33, 35**, respectively. The settling tank has a skimmer arm **46** cooperating with scum baffle **45** to help prevent scum build-up on the surface of the settling tank.

Scum box **61** communicates through scum return **62** with the contact area of the aeration section of the plant to eject scum from the settling tank. A rotating scraper mechanism **60** collects settled sludge from the bottom of the settling tank. Collected sludge is returned from the settling tank to the reaeration section of the plant via sludge return pipe **64**, box **65** and opening **68**. Waste sludge from the settling tank also is distributed to the digester portion of the plant through sludge return pipe **64**, box **65**, conduit **66** and opening **67**, as shown by the arrows. Sludge can be removed from the settling tank bottom by any conventional means. Sludge returned from the settling tank is discharged into the inner reaeration area

13 and flows in a clockwise direction from sludge discharge point **68** to the vertical radial division wall **41**. Communication is provided between outer section **14** and inner section **13** through holes **79** formed in end portion **78** of the intermediate wall **20**, as shown in Figure 2.4d. Sludge from the inner section **13** is allowed to pass through the end portion **78** of intermediate wall **20** near wall **41**, as shown by the arrows, into the outer section **14**. The sludge then flows in a counterclockwise direction in outer section **14** to the vertical radial division wall **82**.

Communication is provided between outer section **19** and outer section **14** through holes **81** formed in the vertical radial division wall **82**, as shown in Figure 2.4e. Sludge from the outer section **14** is allowed to pass through the vertical radial division wall **82** into the outer section **19** beginning a new treatment cycle.

Clear effluent from the settling tank flows under a scum baffle **45** and overflows into a trough **49** for eventual discharge into a chlorine contact tank **70** via a settling tank effluent pipe **52**. Chlorine is injected via a pipe **54** into a trough **49** at a point **53**, the point of the trough **49** furthest from a duct **56** leading to settling tank effluent pipe **52**, in order to provide the greatest time of contact between the chlorine and the effluent before the effluent is discharged.

The chlorine contact tank **70** is enclosed by an arcuate wall **43** and the outer plant wall **11**. Underflow and splash baffles, **71** and **72**, respectively, may be provided. The relatively unobjectionable liquid in tank **70** is finally discharged through plant effluent pipe **80**. A walkway **75** may be provided for manual inspection of all portions of the plant. If the walkway **75** is provided over intermediate wall **20**, it may be supported thereby by a bracket, such as bracket **76** in Figure 2.4c.

Cyclic Process

J.S. Earle, E.W. Schoeffel and F.J. Zimmermann; U.S. Patent 3,272,739; September 13, 1966; assigned to Sterling Drug, Inc. describe a process for the treatment of sewage sludge, utilizing heat and air in a manner which permits unattended operation.

According to the process, a preheated first portion of sewage sludge is directly heated at superatmospheric pressure in a heating zone adapted to pressurized operation with a hot gas to between 100° and 225°C, heating is discontinued, the sludge is maintained at at least 100°C while gaseous oxygen is passed there through for at least a portion of time at a rate in excess of which it is absorbed, thereby achieving a wet air oxidation, until the specific resistance of the sludge ($sec^2/g \times 10^7$) is reduced to less than 250, the pressure in the heating zone is partially reduced, at least a portion of the heat in the gaseous phase of the thus-treated sludge is transferred directly to a second portion of sludge to preheat the second portion, preferably to about 50° to 100°C and preferably at about atmospheric pressure, the thus-treated first portion of sludge is discharged from the pressurized zone and the thus-preheated second portion of sludge is also treated in the thus-described manner.

Sludge obtained from any stage of a sewage treatment process can be used in the process, including raw, primary, activated, concentrated primary, and digester sludges. Commercial wastes containing substantial amounts of solids with comparable levels of chemical oxygen demand (COD) can also be used, e.g., wastes from wool, scouring and desizing operations in textile operation, spent beer in industrial biological fermentations and effluent from paper and pulp making operations.

Limited Wet Oxidation Process

According to a process described by *G.H. Teletzke and E.M. Pogainis; U.S. Patent 3,256,179; June 14, 1966; assigned to Sterling Drug, Inc.* the digested sludge from the digester of a gas-producing sewage plant is subjected to a wet air oxidation at a temperature of at least about 120°C under pressure in the presence at all times of excess gaseous oxygen so as to reduce by less than 55% the COD of the sludge, the residual undissolved solids and resulting effluent are separated, the former then being concentrated and at least a portion of the latter cooled to about 75° to 108°C and the cooled portion returned to the digester, the amount of cooling being regulated so as to maintain the digester at a substantially constant temperature between about 30° and 60°C when the effluent is returned to the digester, the amount of ammoniacal nitrogen returned to the digester by the effluent being regulated so as to maintain the ammoniacal nitrogen content of the digester at all times below 1.0 g/l, preferably between 0.1 to 0.5 g/l.

In the process of U.S. Patent 3,060,118 the effluent from a wet air oxidation in which the COD of sewage sludge is reduced from 60 to 85% is returned as nutrient to a biological treatment. This system works well when the biological treatment is aerobic treatment. However, the effluent obtained from that process is a relatively poor nutrient medium for digester organisms. However, if the oxidation is conducted under conditions where less than 55% and more preferably 45% or less of the COD of the sludge is oxidized, the resulting effluent is more acceptable to the digester organisms. This can result in an increase in gas production. The digested sludge resulting from such recycling is lower in COD and thus more inert biologically.

This lower level of oxidation can be achieved by lower oxidation temperatures. For example, if the oxidation is conducted between 120° and 175°C, less than 10% oxidation is achieved in one hour at 160° and at 8 hours or longer at 120°C. Some degree of oxidation occurs at 120°C or higher but at the lower limit, it is often one percent or less. Therefore, the oxidation preferably is conducted at between 120° and 165°C, e.g., 120°C for 2 to 16 hours, 140°C for 1 to 8 hours, 150°C for up to 4 hours and 160°C for one hour or less.

From the above, it can be seen another way of limiting the degree of oxidation is to employ temperatures which are capable of producing higher oxidation levels but limiting the oxidation time so that an oxidation level of less than 55% is achieved. For example, about 10% oxidation is achieved at about 175°C and about 45% at 225°C in 60 minutes. Therefore limiting oxidation times to less than 10 minutes at 120° to 200°C, and less than 5 minutes at 200° to 225°C ordinarily will limit the oxidation to less than 55%.

Ordinarily, the oxidation should not be limited by reducing the oxygen gas input so that the oxygen is completely consumed as the resulting effluent is more toxic to digester organisms. Therefore, the oxidation should ordinarily be conducted in the presence at all times of excess gaseous oxygen, preferably an amount which leaves from 0.5 to 2.5%, e.g., 1 to 2%, gaseous oxygen in the gaseous phase of the oxidized mixture. The oxidation is conducted in the manner described in U.S. Patents 2,665,249, 2,824,058, 2,903,425, 2,932,613 and 3,060,118 but employing reaction conditions whereby oxidation is limited to less than 55%.

In the sewage plant shown in Figure 2.5, sewage enters the primary system by pipe **1**, after passing through the conventional screens, etc., to primary settling tank **10**. The settled solids from tank **10** are pumped through pipe **13** to the primary thickening tank **14**. The primary effluent is pumped through pipe **11** to the aerobic aeration system **20** and then through pipe **21** to the final settling tank **22** where the now innocuous final effluent is discharged from the plant through pipe **23** and a portion of the settled activated sludge is returned to the aeration system by pipes **25** and **27** and the remainder joins the settled solids from tank **10** by pipe **25**.

FIGURE 2.5: LIMITED WET OXIDATION PROCESS

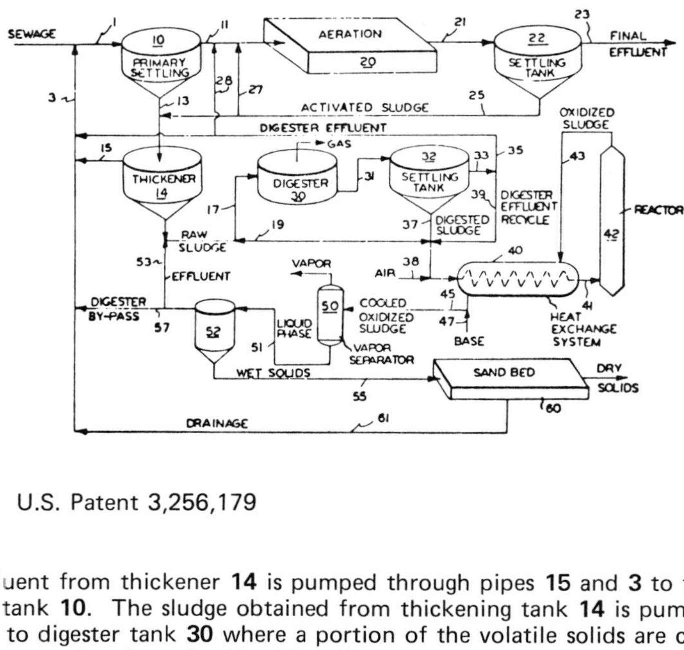

Source: U.S. Patent 3,256,179

The effluent from thickener **14** is pumped through pipes **15** and **3** to the primary settling tank **10**. The sludge obtained from thickening tank **14** is pumped through pipe **17** to digester tank **30** where a portion of the volatile solids are converted to methane and carbon dioxide. The digested solids are displaced from digester tank **30** through pipe **31** to settling tank **32** where the supernatant is separated from the digested solids.

The digester effluent is pumped thorugh pipes **33, 35** and **3** to primary settling tank **10** or through pipes **33, 35** and **28** to aeration system **20**. The digested sludge is pumped through pipe **37,** mixed with air from pipe **38** and the mixture pumped through the heat exchange system **40** where its temperature is raised to an oxidizing temperature. The hot mixture is pumped through pipe **41** to reactor **42** which provides the residence time required to achieve the desired degree of oxidation. The oxidized mixture is pumped through pipe **43** to the shell of the heat exchange system **40** where a portion of its heat is transferred to the incoming mixture. The partially cooled oxidized mixture is pumped through pipe **45**, where optionally its pH is raised a selected amount by a base supplied by pipe **47**, to vapor separator **50** where the pressure of the mixture is dropped to about atmospheric and the vapor phase discharged from the system. The energy dissipated in so doing can be used to run turbines for power and the heat energy captured by heat exchangers for use in the system

The liquid phase with suspended solids is displaced through pipe **51** to solids separator **52** where the oxidized sludge solids are separated from the effluent. The effluent is pumped through pipes **53** and **17** to digester **30**. The now innocuous separated solids are discharged through pipe **55** to sand bed **60** or other filtering system where, e.g., the final dried solids are separated from the residual liquid, which is returned to the primary settling tank **10** through pipes **61** and **3**. Primary sludge can be passed directly to the heat exchange system through by-pass pipe **19**. Digested sludge can be recycled to digester **30** through pipes **37, 19** and **17**. Effluent from the digester settling tank **32** is passed to the heat exchange system when desired by pipes **39** and **37**. Effluent from solids separator **52** is passed to primary settling tank **10** by by-pass pipe **57**.

In a typical operation, 10 million gallons per day (mgd) of raw sewage containing 10.4 tons of suspended solids having 8.35 tons of biological oxygen demand (BOD) is pumped to tank **10**, where effluent containing 4.13 tons of suspended solids with a BOD of 5.42 tons is separated from the remainder of the suspended solids. The aeration system **20** reduces to negligible values the BOD of this effluent, which is discharged from the plant as final effluent after separation from the resulting activated sludge in settling tank **22**.

The thickener tank **14** produces a primary sludge at a rate of 45,000 gallons per day (gpd) containing 9.5 tons of suspended solids of which 6.34 tons are volatile (organic) solids. This sludge, along with 19,950 gpd of effluent from solids separator **52** containing 0.7 ton volatile solids, is pumped to digester tank **30** where the mixture is converted to gas, leaving a digested sludge which is separated from 43,500 gpd of digested effluent in settling tank **32**, the effluent being returned to the primary system.

The 25,000 gpd of digested sludge containng 3.16 tons each of ash and volatile solids is mixed with sufficient air to reduce the COD of the sludge by about 50%. The sludge-air mixture is passed through the heat exchange system **40** where its temperature is raised to 140°C. The hot mixture is then passed to reactor **42** where the mixture is maintained at an average temperature of 205°C in the reactor **42** until half the COD of the mixture is eliminated by oxidation. The sludge-air ratio of the incoming mixture is continually adjusted so that the vapor phase of the oxidized mixture contains from 0.5 to 2.5% oxygen.

The hot oxidized mixture is passed thorugh the shell of the heat exchange system **40** where its temperature is lowered a regulated amount such that the effluent obtained from solids separator **52** has a temperature within the range of 75° to 105°C, depending upon whether the operation is summer or winter. The temperature of the hot effluent is regulated so that the digester is maintained at a constant 32°C. The hot mixture obtained from the heat exchange system **40** is reduced to about atmospheric pressure in vapor separator **50** by venting the vapor phase. The hot effluent obtained from solids separator **52**, at a temperature which when mixed with the contents of the digester **30** maintains the mixture at 32°C, is recycled to the digester **30**. The wet separated solids obtained from solids separator **52** containing 40% solids and 6.32 tons solids are spread on a conventional sand bed where they rapidly drain dry to a friable cake at a rate many times faster than conventional digester sludge.

In a variation of the above process, the separated effluent from the oxidized mixture, having a pH of 4.3, is mixed with 2 to 8 grams of $Ca(OH)_2$ per liter of mixture or equivalent amount of other base, and 10 to 20% of the water is then evaporated, thereby permitting from 85 to 98% of the ammoniacal nitrogen to escape with the vapors in vapor separator **50**, which increases gas production of the digester in addition to increasing percent COD reduction, the result regularly achieved by the process. In a continuous run, the ammoniacal nitrogen level of the contents of the digester is measured at daily intervals and the amount of base added to the hot oxidized mixture is varied in direct response to variations of the level.

When the ammoniacal nitrogen level approaches 1.0 g/l, enough base is added to raise the pH from 7 to 11 or even higher. When the level drops to below 0.8 g/l, the addition of the base is stopped. In another variation, the sludge-air mixture is oxidized at a temperature between 120° and 150°C with the heat necessary to maintain that temperature being provided by steam, added directly or indirectly to the hot sludge-air mixture just prior to entering the reactor. The separated effluent, containing about 5 g/l or less volatile solids when the oxidation temperature is 120°C and about 17 g/l or more when the oxidation temperature is 150°C, or above, contains from 0.2 to 0.5 g/l ammoniacal nitrogen and is a substantially superior nutrient for the digester than effluent resulting from oxidations at above 150°C.

In related work which is described by *W.B. Gitchel, C.A. Hoffman and E.W. Schoeffel; U.S. Patent 3,272,740; September 13, 1966; assigned to Sterling Drug, Inc.* an innocuous cured organic sewage sludge having optimum processing characteristics is produced by the wet air oxidation of a continuous stream of sewage sludge in the presence of a continuous stream of a substantially constant volume of gaseous oxygen under conditions of temperature and pressure which reduce by 5 to 45% the COD of the sludge while adjusting at least one of the temperatures to which the mixture is heated, the sludge flow and the COD of the incoming sludge in a manner directly proportional to variations in the gaseous oxygen content of the gaseous phase of the oxidized mixture so as to maintain the sludge at all times in the presence of gaseous oxygen during the wet air oxidation.

It was found that an innocuous organic cured sewage sludge having optimum processing characteristics and whose suspended solids are useful as a soil conditioner, composting ingredient and fill is obtained if the COD of the sewage sludge is reduced by wet air oxidation from 5 to 45%. An even superior product is obtained if the COD reduction is from 10 to 35%, desirably about 10 or 15% up to 30%. Its specific resistance to filtration and water holding capacity is lower than sewage sludge treated by the Porteous process (U.S. Patent 2,075,224) and its processability in solids separation systems is superior to the ash obtained from sewage sludge oxidized by the procedure of U.S. Patents 2,665,249 or 3,060,118. The solids obtained by filtration of the cured sludges are useful as a component of soil conditioners and fertilizers because of their fiber and insoluble humic acid content.

W.B. Gitchel, C.A. Hoffman and E.W. Schoeffel; U.S. Patent 3,359,200; December 19, 1967; assigned to Sterling Drug, Inc. describe a process for the oxidation of sewage sludge, more particularly to a continuous partial wet air oxidation of sewage sludge and other oxidizable aqueous suspensions.

In the process, a continuous stream of sewage sludge is oxidized by a nonautogenetic partial wet air oxidation with a continuous stream of an oxygen-containing gas. The stream of sludge is preheated to a temperature not exceeding about 150°C, which is below the desired reaction temperature, by indirect heat exchange. The preheated sludge is then heated to the desired reaction temperature by directly injecting steam into the heated sludge to the temperature at which the oxidation is to be conducted. The reaction temperature is maintained substantially constant by varying the heat energy supplied to the preheated sludge by the steam in response to variations in reaction temperature.

This process reduces and virtually eliminates fouling and scaling of heat exchange surfaces; provides a precise yet simple, inexpensive and reliable means for regulating reaction temperature; simplifies start-up procedure and thus lends itself to short, e.g., one shift, runs; and permits use of low COD sludges because reaction temperature can readily be reached.

Oxidation of Acidified Sludge to Improve Filterability

W. Schotte; U.S. Patent 3,649,534; March 14, 1972; assigned to E.I. du Pont de Nemours and Company describes a procedure for improving the filterability of the activated sludge by, in sequence, (1) adjusting the pH of the sludge to a value in the range of about 2.0 to 5.0 (preferably about 3.5), (2) heating the sludge to a temperature in the range of about 120° to 170°C (preferably about 150°C) while maintaining a preselected pressure below which the water content of the sludge remains primarily liquid and (3) oxidizing the sludge by contact with an oxygen-containing gas (preferably air) for a period of time (typically 2 hours) sufficient to reduce the chemical oxygen demand of the sludge by about 5 to 20%.

Specific filtration resistance is evaluated by a standard Buchner funnel filtration test described by P. Coackley in *Biological Treatment of Sewage and Industrial Wastes,* volume II, pp 270 to 291, Reinhold Co., N.Y. (1958). The procedure involves the use of a size 1A Coors porcelain Buchner funnel with an internal area of 38.5 cm². This is mounted on a graduate through the intermediary of an adapter connected to a manometer and a vacuum system. A Moore vacuum regulator maintains a constant vacuum of about ½ atmosphere. All tests are made with three pieces of filter paper, of which two pieces are No. 41 Whatman paper disposed on the bottom, with either a Whatman No. 42 (fine) or a Whatman No. 4 (coarse) on top. The time required to collect a given volume V, of the filtrate is expressed by the equation:

$$\theta = \frac{\mu r c}{2PA^2} \cdot V^2 + \frac{\mu R}{PA} \cdot V$$

θ = time, sec
μ = viscosity of the filtrate, poise
r = specific filtration resistance, sec^2/g
c = suspended solids concentration in initial slurry, g/ml
P = filtration pressure, g/cm^2
A = filter area, cm^2
V = volume of filtrate, ml
R = resistance of the filter medium sec^2/cm^2

Strictly speaking, the equation shown above should also contain the dimensional constant, g_c = 981 (mass-g) (cm) (force-G) $(sec)^2$, where r and R should be multiplied by g_c. However, very few investigators have included g_c and, for ease of comparing results, the constant has also been omitted here. Measurements were made by recording the volume of filtrate collected as a function of time. The stopwatch was started when about 10 to 20% of the filtrate was collected in order to allow some time for the formation of a thin layer of cake. The table below affords a comparison of filtration resistance for typical ordinary materials, [e.g., carbon, $CaCO_3$ and $Fe(OH)_3$] and for prior art sewage sludge processes with this process:

Comparison of Filtration Resistances

	Filtration Resistance $sec^2/g \times 10^7$
Powdered activated carbon	0.3
Precipitated $CaCO_3$	2.0
Gelatinous $Fe(OH)_3$	300.0
Activated sludge, untreated	(*)
Teletazke report supra-activated sludge, partial oxidation	14.0
Activated sludge, acidified heated	62.3
Activated sludge, heated	7,750.0
This process	1.0-2.0

*From 30.0 to 6,000.0

Process Conducted in Sunken Shaft

According to the process described by *W.J. Bauer; U.S. Patent 3,449,247; June 10, 1969* wet oxidation of a mixture of combustible refuse and fluid sewage is carried out in the lower portion of a shaft extending into the earth a sufficient distance to provide the desired pressure by the head of fluid material in the shaft.

In carrying out this process, a first vertical shaft is sunk in the earth to a depth on the order of one mile. This shaft will preferably extend into an impervious rock strata and the shaft may be lined as necessary to form a continuous outer conduit. A second conduit which may be within the outer conduit, communicates with the lower portion of the shaft. The depth of the shaft is sufficient to produce the pressures required for wet oxidation through the action of gravity.

An appropriate mixture of air, water and combustible material, which may comprise sewage sludge or sewage liquids enriched by combustible refuse to obtain the requisite proportions of combustible material, or other oxidizable wastes, is directed into the shaft. Air may be introduced into the mixture through the use of blowers at or near the surface entrance to the shaft, whereby the air is introduced under relatively low pressure. The relationship of conduit size and flow rates is such that the velocity of the water mixture downward in the shaft is of sufficient magnitude to carry this air, which may be in the form of bubbles, against the influence of gravity.

The entrained air is thus compressed continuously as the mixture of air, water and combustible materials flows downward in the shaft, until sufficient pressure is obtained at the requisite temperature in the lower portion of the shaft for oxidation of the combustible materials to take place. Additional air for the wet oxidation process is initially compressed by a compressor on the surface, at the top of the shaft and is directed downward through a second shaft to the oxidation zone. The weight of the vertical column of compressed air in the second shaft raises the pressure of this air to the pressure of the oxidation zone at the level of introduction of this additional air.

Powdered Activated Carbon

L.A. Pradt and J.A. Meidl; U.S. Patent 3,876,536; April 8, 1975; assigned to Sterling Drug Inc. describe a process of treating sewage sludge or night soil by wet air oxidation, followed by biological oxidation of the liquid phase by aeration in the presence of biomass and powdered activated carbon. In the system set forth in Figure 2.6, the sewage sludge or night soil is supplied to a storage tank **1** where it is transported in admixture with air through a heat exchanger **2** to a reactor **3** where the sludge or night soil is subjected to wet air oxidation. The wet air oxidation takes place between a temperature of 150° and 375°C and at a pressure sufficient to maintain most of the water present in the liquid phase, i.e., between 150 and 4,000 psig.

The extent of oxidation preferably ranges between 30 and 70% reduction in COD. The reactor contents after oxidation are returned through the heat exchanger **2** and then passed to a separator **4** where the gas and liquid phases are separated. The liquid phase contains suspended solids such as sand, described as ash. The ash is removed by settling and decanting the liquid, or by filtration, and the liquid sent on to an aeration contact tank **5** containing a biogrowth or biomass suitable to effect biological oxidation of the organic solutes present in the waste liquor. Powdered activated carbon in aqueous slurry is maintained in the aeration tank to the extent that the concentration of carbon is at least 500 ppm and preferably in the range of 10,000 to 20,000 ppm by weight.

After a period of 10 to 150 hours of aeration contacting, preferably about 50 hours, the liquid and biomass-carbon pass to a clarifier **6** where the mixture of powdered activated carbon and biomass settles to the bottom of the clarifier. A flocculating agent, preferably a cationic organic polymer may be added between the aeration tank and the final clarifier to enhance the settling characterisitics of the stream. The biomass-carbon mixture collected in the clarifier is recycled back to the aeration tank **5** via the return sludge line **7**. The addition of the powdered activated carbon to the aeration vessel enhances the bio-oxidation taking place there and substantially reduces the odor and color of the waste liquors, the color being reduced to about 20 APHA units.

During operation of the process, excess biomass builds up in the system and the carbon becomes spent after its capacity for adsorption of influent COD-BOD has been reached. A further aspect of this process resides in the concept of removing excess biomass and spent carbon from the clarifier **6** and transporting it through the excess sludge line **8** to the storage tank **1** where it is mixed with fresh sewage sludge or night soil and transported therewith to the wet air oxidation reactor. In this way the excess biomass is disposed of by oxidation and the spent carbon is regenerated. In this system where the carbon is recycled and regenerated, the liquid phase in separator **4** will contain a suspension of regenerated carbon and ash.

The ash is preferably removed before the liquor containing the regenerated carbon is returned to the aeration tank. The ash removal can be accomplished by a centrifugal device which makes use of the difference in specific gravity between the ash and the regenerated carbon. If the amount of carbon added to the system is small, the carbon in the excess sludge can be oxidized completely or removed with the ash. Virgin carbon is added to make up the original weight as needed.

Example: 78 m^3 of raw night soil having a COD of 35 g/l is collected and transferred to storage tank **1** of Figure 2.6. The night soil is then oxidized in reactor **3**, thereby reducing the COD to 15 g/l. The associated BOD_5 of the oxidized night soil is 8.5 g/l and the ammonia nitrogen content is 3.5 g/l. The reactor contents are passed to separator **4**, and the ash is selectively removed from the oxidized night soil. The liquid phase is passed to contact-aeration tank **5** containing a biomass and 630 kg of powdered activated carbon is added to the tank. The treated liquor and suspended solids are passed to clarifier **6** and the solids which settle there are recycled through sludge line **7** back to the aeration tank. Due to the biological nature of the treatment system 163 kg of excess biomass is produced.

After the carbon is spent, an amount of solid equivalent to the increased weight in the system (carbon plus excess biomass, 793 kg) is removed from the clarifier **6** and transferred to storage tank **1** (Figure 2.6). These combined solids are treated in oxidation reactor **3** with additional night soil (Figure 2.6). The regenerated carbon and the solubilized excess biomass are transferred to aeration tank **5** to begin another cycle. The system effluent has a COD of 200 mg/l, a BOD_5 of 100 mg/l, an ammonia nitrogen content of 300 mg/l, and a color value of about 20 APHA units. Without the addition of activated carbon in the flow schemes, the effluent COD, BOD and ammonia nitrogen would be about 2,000, 350, and 1,000 mg/l, respectively, and the color value about 500 APHA units.

FIGURE 2.6: WASTE OXIDATION PROCESS USING ACTIVATED CARBON

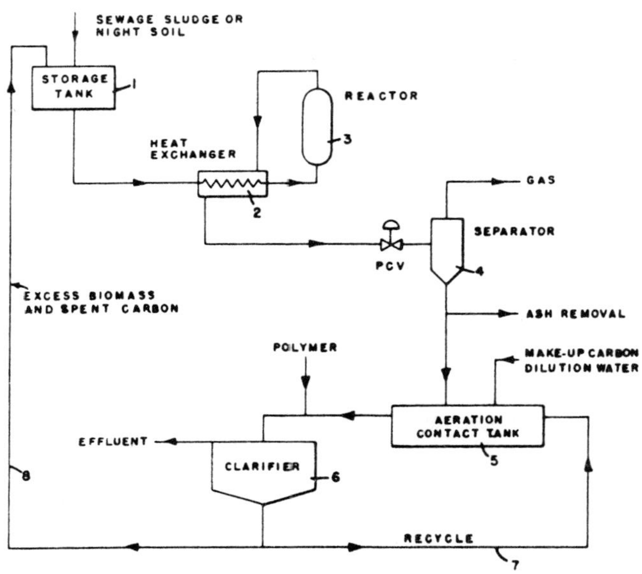

Source: U.S. Patent 3,876,536

Digestion of Fibrous Material

In a process described by *E.J.M. Berg, H.C. Deubner and L B. Nelson; U.S. Patent 3,025,151; March 13, 1962,* the solids separated in the initial primary settling stage are placed in any suitable tank or container. A biological culture is placed in the container which culture will accomplish the desired digestion or decomposition of the fibrous cellulose (or other material which generally tends to interfere with subsequent filtration treatments of the solids) while the solids are aerated.

The aeration is by any suitable means and for a suitable period of time, while maintaining the temperature of the solids within a desired range so as to encourage the growth and activity of the biological culture so as to destroy carbohydrates such as those in the form of fibrous cellulose and other constituents of the solid matter which is of a particle size that normally interferes with filtration of the solids.

The aeration of the solids accomplishes two functions; namely, it supplies the needed oxygen for the proper functioning of the microorganisms and it also intimately mixes all freshly added solids with those presently in the container. The temperature of the solids may be controlled by preheating the air used for aeration, or, if desired, the solids may be subjected to external heating while being aerated.

The sludge in the treatment tank of a sewage treatment plant, which is about 2.25% solids and consists of the solids settled out of about 60,000 gallons of primary sludge and 275,000 gallons of waste activated sludge and about 5,000 gallons of settled activated sludge from the final settling tanks of a sewage treatment plant, were placed in a container so that the depth of the sludge in the container was appproximately ten feet. In pounds of dry matter, the initial loading of the container was about 12,500 pounds of primary sludge, 9,300 pounds of activated sludge, and 400 pounds of settled activated sludge.

Air at the rate of 1,500,000 ft^3/day was blown through this mixture and the temperature of the sludge in the container was maintained by preheating the air to approximately 176°F prior to diffusing it in the container. The temperature of the sludge was maintained between 70° and 102°F. To determine readiness for processing, a sample of the sludge was brought to the pH best adapted for filtration, which in the above example was 4.5, and the sample was filtered through a Buchner funnel.

The filter cake was broken so as to expose any fibrous cellulose or other visible constituents deleterious to filtration in the filter cake and this filtration procedure and visual examination was repeated until the particle size of the constituents such as the fibrous cellulose in the sludge was reduced so as to not interfere with filtration. In the above example, the sludge was treated for a period of approximately three days prior to rendering the carbohydrates, including those present in the form of fibrous cellulose and other deleterious materials innocuous or reducing them to a particle size suitable for filtration.

Denitrification Using Raw Sewage in Oxygen-Poor Environment

P.I. Rongved; U.S. Patent 3,900,394; August 19, 1975; assigned to Activox, Inc.

describes an efficient purification and denitrification process which substantially eliminates the pollutant effect of the effluent waters resulting from the process. The denitrification step of the process comprises adding fresh raw sewage to sewage liquid in the reservoir which has undergone aerobic decomposition and is substantially devoid of oxygen.

The bacteria in the activated sludge-containing liquid thus have additional organic matter with which to react but are starved for oxygen. In order to satisfy their oxygen requirement, they act upon the nitrate compounds dissolved in the liquid and consume the oxygen resulting from the decomposition thereof. Continuous operation of the process and removal of the solid matter in the liquid provides an effluent which generally contains no more than about 10% of the nitrate compounds produced by aerobic decomposition of ammonia in the liquid.

The apparatus provided for the operation of the process comprises an aeration reservoir fitted with inlet means, at least one agitating means, and outlet means. In constructing the apparatus, it is essential that the inlet means be positioned sufficiently far from the agitating means so as to allow sufficient time for the aerobic decomposition of the sewage liquid and for the substantial depletion of the oxygen content thereof. Likewise, the inlet means should be positioned sufficiently far from the point of next aeration to allow for significant denitrification of the sewage liquid before renewed aeration of that liquid.

Matrix Filter Formed from Waste Solids

H.B. Trussell; U.S. Patent 3,655,046; April 11, 1972; assigned to Fermbionics, Inc. describes a system for disposing of municipal waste including garbage and sewage which converts the garbage portion into a matrix filter which is used to separate the liquids and solids in the sewage. The liquids are purified in a normal manner, for example by aeration, and the solids are converted into a food material suitable for livestock consumption.

In the municipal disposal system the garbage portion of the waste is pressed into a filter matrix against the surface of a screen-type conveyor belt. As the belt is moved, sewage is poured onto the matrix and solids are carried away and mixed with additional garbage. The liquid portion, which passes through the filter matrix is purified in a normal manner. The garbage and sewage solids are mixed together, pasteurized and sent to a digester, where bacteria are added to speed dissolution of solids and orient the resultant nutrients, carbohydrates, proteins, fats, vitamins and minerals. Then, the digested product may be fed directly in liquid form or the solids may be separated out and fed as meal or in pellet form.

Conditioning of Alum Sludge

J.D. Stauffer; U.S. Patent 3,720,608; March 13, 1973; assigned to Bonham, Grant & Brundage Limited describes a process and apparatus for pollution abatement by the transfer, conditioning and disposing of alum sludge. An apparatus is described which consists of (1) a holding tank which receives solids from the clarification settling tanks of a water or a wastewater purification plant; (2) a system of heat exchangers, for heat recovery and auxiliary heat application to raise the temperature of the incoming sludge; (3) a detention vessel to provide for a reaction time; (4) a decant tank; and (5) a dewatering unit for separation of the solids from the conditioned sludge.

The method comprises the subjecting of the alum sludge to a temperature of about 212°F for a period of about 30 minutes and the separation of the solids from the liquid fraction to produce a low moisture content sludge cake suitable for ultimate disposal or for use as an industrial raw material. Following are some examples of the process of conditioning alum sludge prior to dewatering.

Example 1: Heat treatment as a conditioning aid was made on an alum sludge having 5.92% total solids content. The sludge was heated at 250°F (15 psi) for 30 minutes. The sludge was allowed to cool to room temperature. The sludge was then thoroughly mixed and a test was made on a 100 ml portion to determine specific resistance. The time required to collect 70 ml of filtrate from the 100 ml was reduced to less than ⅛ of that required without treatment and the specific resistance was reduced from 13.24×10^8 sec²/g without treatment to a value of 1.52×10^6 sec²/g.

Example 2: Heat treatment as a conditioning aid was made on an alum sludge having 6.43% total solids content. The sludge was heated at 212°F at atmospheric pressure for 30 minutes. The sludge was allowed to cool to room temperature. The sludge was then thoroughly mixed and a test was made on a 100 ml portion to determine specific resistance which was 0.97×10^8 sec²/g as compared with 24.57×10^8 sec²/g without treatment. The time required to collect 30 ml of filtrate from the 100 ml was reduced from 766 seconds without treatment to 28 seconds.

pH Adjustment for Coagulant Recovery

According to a process described by *M. Ohuchi, T. Kitahori, T. Maitoko and K. Mizuguchi; U.S. Patent 3,901,804; August 26, 1975; assigned to Kanzaki Paper Manufacturing Co. and Niigata-Zimpro Ltd., Japan* the pH of the sludge resulting from coagulation-separation treatment of wastewater utilizing a coagulant consisting of a water-soluble aluminum compound is adjusted before subjecting the sludge to a wet air oxidation treatment so that the pH of the sludge obtained after the wet air oxidation treatment is either below 1.5 or above 9.0, whereby the coagulant can be recovered in the form of a soluble aluminum compound. Examples of this process follow.

Example 1: Aluminum sulfate was added by 50 ppm as aluminum oxide into the wastewater of unbleached hardwood kraft pulp (COD_{Cr}, 800 ppm). By adjusting the pH to 4.5, the resultant coagulated flocs were sedimented and separated as sludge. Polyacrylamide was added to the obtained sludge by 1 ppm as dry solid on the wastewater. Thereafter the sludge was partially concentrated by rotary screen (100 mesh wire) and the sludge concentration was adjusted to 5.0%. At this stage, pH of the sludge was 4.5, and COD_{Cr} was 52,270 ppm.

By adding H_2SO_4 in an amount of 2 weight percent on the basis of the total weight of sludge to the sludge, pH was adjusted to 1.2. The 500 ml sample of sludge adjusted to pH 1.20 was subjected to wet air oxidation treatment in a vibrating lab autoclave (capacity 800 ml) in the condition of initial pressure (air) of 30 kg/cm² and temperature of 175°C for 60 minutes. The obtained oxidized sludge was brown and the pH was 1.20. COD_{Cr} of the oxidized liquor was determined to examine the effectiveness of oxidation.

After the oxidized sludge was separated into filtrate and residue by filtration, the

		Oxidized Sludge				
		Example of this Process			Controls	
	Original Sludge	1	2	3	1	2
Conditions of pH adjustment of sludge:						
NaOH added (% on sludge)	-	-	6.0	7.5	-	2.5
H_2SO_4 added (% on sludge)	-	2.0	-	-	-	-
pH of sludge	4.5	1.2	13.8	13.9	4.5	13.4
Condition of wet air oxidation:						
Initial pressure—air (kg/cm^2)	-	30	50	50	50	50
Temperature (°C)	-	175	250	250	200	250
Retention time (min)	-	60	60	60	60	60
Properties of oxidized sludge:						
Color	Dark brown	Brown	Pale yellow	Yellow	Light brown	Pale yellow
pH	-	1.2	9.5	11.2	2.0	7.4
COD_{Cr} (mg/l)	52,270	14,685	15,991	14,008	10,630	18,991
COD removal (%)	-	71.9	69.4	73.2	79.7	67.3
Aluminum ion A (mg/l)	27	3,215	632	3,122	466	23
Soluble aluminum compounds:						
B (mg/l)	3,672	11	2,190	185	15	132
A + B (mg/l)	3,699	3,226	2,822	3,307	481	155
Recovery of aluminum compounds (%)	-	87.2	76.3	89.4	13.0	4.2

quantity of aluminum in the filtrate (A) was determined by atomic absorption photometry. After $Al(OH)_3$ in the residue was dissolved in HCl, the quantity of aluminum therein (B) was determined by atomic absorption photometry. The results are shown in the table.

Example 2: This example was similar to Example 1 except that NaOH was added in an amount of 6 weight percent on the basis of the total weight of sludge to adjust the pH of sludge to 13.8, and that the wet air oxidation treatment was carried out under the condition of initial pressure of 50 kg/cm^2 and temperature of 250°C. The results are shown in the table.

Example 3: This example was similar to Example 1 except that NaOH was added in an amount of 7.5 weight percent on the basis of the total weight of sludge to adjust the pH of sludge to 13.9, and that the wet air oxidation treatment was carried out under the condition of initial pressure of 50 kg/cm^2 and temperature of 250°C. The results are shown in the table.

Control 1: The concentrated sludge (5%) in Example 1 was subjected to wet air oxidation treatment without preliminary pH adjustment under the condition of initial pressure of 50 kg/cm^2 and temperature of 200°C. The pH of resultant oxidized sludge was 2.03. COD and aluminum were determined in the same manner as in Example 1.

Control 2: Sulfuric acid was added to the concentrated sludge (5%) to adjust the pH to 13.4. This sludge was subjected to wet air oxidation treatment under the condition of initial pressure of 50 kg/cm^2 and temperature of 250°C for 60 minutes. The pH of resultant oxidized sludge was 7.40. COD and aluminum in the sludge were determined in the same manner as in Example 1. The results are shown in the table.

Most aluminums are dissolved as aluminum sulfate in oxidized sludge with pH below 1.5. At pH 1.5 to 9.0 aluminum precipitates in the form of α-boehmite which is only slightly soluble in acid and alkali and thus impossible to recover and reuse. At pH 9.0 to 10.5 aluminum precipitates as aluminum hydroxide which can be dissolved easily and recovered. At pH above 10.5, aluminum is again dissolved as sodium aluminate. Aluminum compounds thus recovered can be reused as coagulants.

Culturing Microorganisms for Decomposition of Poisons

A. Mimura, T. Kawano and K. Yamaga; U.S. Patent 3,660,278; May 2, 1972; assigned to Asahi Kasei Kogyo KK, Japan describe the preparation of an activated sludge capable of use in biological treatment of wastewaters containing poisonous substances. The process comprises culturing in liquid medium screened microorganisms that can decompose the poisonous substances in the wastewaters and blending the cultured liquid of the microorganisms with an aqueous suspension of the usual activated sludge used in the treatment of the wastewaters.

The aqueous solution containing the poisonous substances is flowed continuously into the resulting blended liquid under aerobic conditions to force the microorganisms to combine with the usual activated sludge in commensal state. For example, among the poisons to be decomposed or assimilated are the following: hydrogen cyanide, nitrile compounds, phenols, aldehydes, petroleum oil, and other synthetic organic compounds. Microorganisms having the properties of decomposing or assimilating the poisonous substances can be obtained according to usual procedures, that is, by screening microorganisms which multiply in the culture media containing the poisonous substances as the sole carbon or nitrogen source. For instance, the microorganisms belonging to the genus Alternaria, Aspergillus, Cephalosporium, Fusarium, Trichoderma, Pseudomonas, Corynebacterium and the like are used as hydrogen-cyanide-decomposing organisms.

On the other hand, the microorganisms belonging to the genus Pseudomonas, Corynebacterium, Brevibacterium, Bacillus, Mycobacterium, Flavobacterium, and the like are used as organisms for decomposing the nitrile compounds such as acrylonitrile, acetonitrile, propionitrile, butyronitrile, succinonitrile, adiponitrile, lactonitrile and the like.

For the treatment of wastewaters containing the phenol compounds, microorganisms belonging to the genus Corynebacterium, Achromobacter, Pseudomonas, etc. are employed, and for the treatment of wastewaters containing the petroleum oils, micororganisms belonging to the genus Corynebacterium, Pseudomonas, Nocardia, Candida, etc. are employed. Furthermore microorganisms belonging to the genus Pseudomonas can be employed for the treatment of wastewater of sewage containing the synthetic cleanser ABS (sodium alkylbenzenesulfonate). The following examples illustrate the process.

Example 1: A medium consisting of 1.0% of glucose, 0.2% of peptone, 0.2% of potassium nitrate, 0.1% of monobasic potassium phosphate and 0.005% of magnesium sulfate (with its pH adjusted to 4.5) was inoculated with *Fusarium solani* and the organism was cultured at 30°C for 24 hours while aerating and stirring. 1 kg of the usual activated sludge obtained in the treatment of beer plant wastewater was suspended in water in such a way that the precipitated volume after an hour's standing of the resulting suspension liquid amounted to 50%. 1 liter of the above cultured liquid of the *Fusarium solani* was added to 10 liters of the above usual activated sludge suspension and homogeneous blending was carried out.

This blended liquid was placed in a reaction vessel having aeration equipment, and artificial wastewater containing potassium cyanide corresponding to 120 ppm of hydrogen cyanide, 0.025% of monobasic potassium phosphate and 0.0025% of magnesium sulfate (with its pH adjusted to 7.0) was continuously flowed into the blended liquid while overflowing out the liquid therefrom, under the condition of aerating, so as to acclimate and combine the microorganism cells with the usual activated sludge in commensal state, at 30°C for 8 days. Thereby, the specially activated sludge amounting to 5.6 liters was obtained, which is the volume precipitated after an hour's standing of the acclimation liquid.

1 liter of this specially activated sludge was mixed with 1 liter of artificial wastewater containing potassium cyanide corresponding to 499 ppm of hydrogen cyanide, 0.05% of monobasic potassium phosphate and 0.005% of magnesium sulfate (with its pH adjusted to 7.0), and then 500 ml/min of air were passed through the resulting liquid mixture at 30°C for 6 hours to bring about reaction, and thereby 200 ppm (initial concentration) of hydrogen cyanide in the liquid mixture were reduced to 5 ppm.

No decomposition and no removal of cyanic acid were effected in a similar treatment but using the usual activated sludge which was obtained in the treatment of beer plant wastewater.

On the other hand, placing the usual activated sludge suspension in the reaction vessel, and flowing continuously artificial wastewater which contained 0.025% of monobasic potassium phosphate, 0.0025% of magnesium sulfate and potassium cyanide corresponding to 20 to 100 ppm of hydrogen cyanide (the concentration of this hydrogen cyanide being gradually increased as time goes by) and whose pH was adjusted to 7.0, into the usual activated sludge suspension while overflowing out the liquid therefrom, under the condition of aerating at 30°C, the property of decomposing hydrogen cyanide in the resulting liquid mixture appeared gradually from the 47th day and the hydrogen-cyanide-decomposing activity corresponding to that brought about in the method of this process described as above, was achieved barely on the 63rd day after the beginning of this acclimation reaction.

This result shows that an extremely long period is needed in the conventional process to acclimate and convert the usual activated sludge into the specially activated sludge capable of being utilized to the treatment of the chemical plant wastewaters.

Example 2: Trichoderma viride and *Cephalosporium mycophilum* were separately cultured in a medium having the same composition as that of Example 1, at 30°C for 24 hours. 1 kg of the usual activated sludge obtained at a sewage disposal plant was suspended in water in such a way that the precipitated volume after an hour's standing of the resulting suspension liquid amounted to 50%.

1 liter of each cultured liquid of the *Trichoderma viride* and *Cephalosporium mycophilum* was added respectively to 10 liters of the above usual activated sludge suspension and homogeneous blending was carried out. The artificial wastewater similar to that of Example 1 was continuously flowed into the above blended liquid while overflowing out the liquid therefrom, under the condition of aerating, so as to acclimate and combine with the cells of the microorganisms with the usual activated sludge in commensal state, at 30°C for 10 days. Thereby the specially activated sludge amounting to 5.2 liters was obtained, which is the volume precipitated after an hour's standing of the acclimation liquid.

1 liter of this specially activated sludge was mixed with 1 liter of the artificial wastewater similar to that of Example 1, the resulting liquid mixture was aerated for 6 hours to bring about reaction, and thereby 200 ppm of hydrogen cyanide contained in the liquid mixture were reduced to 8 ppm. No decomposition and no removal of hydrogen cyanide were effected in a similar treatment but using the usual activated sludge which was obtained at a sewage disposal plant.

Thermophilic Microorganisms

H.K.E. Fuchs; U.S. Patent 3,864,247; February 4, 1975; assigned to The De Laval Separator Company describes a method for biological decomposition which is operable at varying outdoor temperatures.

According to the process, a treatment vessel is supplied with an aqueous liquid containing organic material and having a biologic oxygen demand during five days of at least 1,500 mg, preferably at least 5,000 mg, of oxygen gas per liter of liquid. In the treatment vessel, the liquid is brought into intimate contact with oxygen in a quantity corresponding to at least 1.1 times and at most six times, preferably at most four times, the quantity of oxygen which theoretically can be consumed by the organic material by means of the microorganisms, the liquid being kept heat-insulated and without heat addition, being kept heated by the activity of the microorganisms to a temperature of at least 42°C, preferably 50° to 58°C.

Experiments have shown that this method makes it possible to destroy, with the help of the thermophilic microorganisms, the pathogenic bacteria and various kinds of virus, worm ova and weed seeds and in addition decompose such materials as mineral oils, washing agents, hair and feathers. Furthermore, the quantity of organic material is reduced to a fraction of the starting quantity. This latter circumstance depends on the oxidation of the organic material which takes place.

Furthermore, the treated liquid is odorless and the decomposition process proper takes a time of only 6 to 10 days as compared with 30 days in prior common methods. The oxygen may be supplied by intimately contacting the liquid with air, oxygen-enriched air or pure oxygen; and the degree of utilization of the supplied oxygen is normally 50 to 70%. Thus, the oxygen percentage in the liquid during the decomposition treatment is advantageously kept at a value of only 0 to 0.5 mg O_2/liter. The specified biologic oxygen demand of the liquid is necessary to provide the heat development for maintaining the liquid at a temperature of at least 42°C.

At the start of the process, the heat development which is effected by the activity of the mesophilic microorganisms can be used for the heating of the liquid to the temperature where the thermophilic organisms are active. When the temperature rises substantially above 42°C, the activity of the mesophilic organisms decreases appreciably, and the heat development necessary in the continued process is then generated by the thermophilic organisms. If the liquid to be treated does not contain the microorganisms necessary for the process, the liquid can be inoculated with such organisms. The intimate contact between the supplied oxygen-containing gas and the liquid is preferably effected by revolving the liquid in the treatment vessel.

In those cases where a foam is developed by the liquid in the decomposition of its content of organic material, the foam can be allowed to form a heat insulation for the liquid surface. Since the foam layer has a tendency to increase in thickness, it is desirable to withdraw foam after the foam layer has reached a thickness of at least 10 cm. The foam layer thickness can be kept limited in a conventional manner, for example, mechanically by means for breaking down the foam.

The revolution of the liquid in the treatment vessel advantageously takes place in a vertical plane, in that air introduced into the liquid in the form of fine bubbles has a tendency to rise and thereby promote such revolution of the liquid. Foam formed in the biologic decomposition can be discharged from the vessel for separate use, such as the recovery of the protein content of the foam for feeding purposes.

Example 1: 10 m^3 of whey with 0.21% decomposable organic substance (= 0.12% carbon) from a dairy were supplied per 24 hours to a continuously operating plant for biologic decomposition, the plant being equipped with a well heat-insulated treatment vessel. The whey had a temperature of 42°C and a BOD_5 (= biologic or biochemical oxygen demand during five days) of 1,500 mg/liter. 1,500 m^3 of air at a temperature of +10°C (winter temperature) were blown per 24 hours into the treatment vessel. The temperature in the treatment vessel remained at 42°C. The whey content of organic material was decomposed to 98°C. The oxygen efficiency (the utilization degree of the oxygen in the supplied air) was 16.5%

Example 2: 10 m^3 of pig's manure with 5.6% of decomposable organic substance (= 3.0% carbon) at a temperature of +18°C were supplied per 24 hours directly from a pig's shed, where a room temperature of +22°C prevailed, to a plant similar to that in the preceding example. The BOD_5 of the pig's manure was 21,000 mg per liter. 31,000 m^3 of air at an average day temperature of -5°C were blown per 24 hours into the treatment vessel during a cold period. The temperature in the treatment vessel remained at 43°C. The manure content of organic material was decomposed to 95 to 98%. The oxygen efficiency was 20%.

Biotreatment of Thermally Dewatered Sludge

According to a process described by *A.H. Erickson, E.M. Pogainis and G.H. Teletzke; U.S. Patent 3,824,186; July 16, assigned to Sterling Drug Inc.* the liquors obtained by dewatering thermally conditioned organic sludge are subjected to separate biotreatment in undiluted form, thereby removing more than 90% of the biological oxygen demand from the liquors. The biotreatment is carried out in an aeration vessel containing biological sludge. The process involves carrying out the following steps:

(a) feeding the liquors to an aeration vessel containing a previously acclimated biological sludge, the loading rate of the liquors to the aeration vessel being between 1.0 and 5.0 kg of biological oxygen demand per kg of mixed liquor volatile suspended solids;

(b) agitating the mixture in the vessel for a period in excess of two hours with an oxygen-containing gas while there is maintained a dissolved oxygen content sufficient to support aerobic biological growth;

(c) passing the mixture to a settling tank to separate accumulated biological sludge;

(d) recycling the settled biological sludge to the aeration tank; and

(e) removing from the aeration tank biological sludge at a rate approximately equivalent to the rate of accumulation of the biological sludge.

Example 1: Biotreatment of Low Pressure Oxidation Liquors — A mixture of primary and biological activated sludge was subjected to wet air oxidation at 22.1 kg/cm^2 pressure and 177°C for a period of 55 minutes, such that 16.1% of the organic matter was oxidized. After allowing the mixture to cool and settle for 4 hours, decantate was drawn off having the following analysis: BOD_5, 5,580 mg/l; COD (chemical oxygen demand), 11,430 mg/l.

The decantate liquors were pumped continuously at a constant rate of 605 liters per day to a temperature-controlled 190 liter reactor held at 36.1°C. Residence time in the reactor was 7.5 hours. The reactor was equipped with a submerged agitator and a pipe which dispersed compressed air immediately below the agitator. The reactor was previously seeded with activated sludge and acclimated to sludge liquors, developing a viable biological mass. Air was continuously added at 31 liters per minute, and 26% of the oxygen in the air was absorbed by the biological solids. The resulting oxygen transfer coefficient K_La was 139 mg oxygen transferred per hour per liter per mg/l dissolved oxygen deficit.

The biological sludge and liquors were conveyed continuously from the reactor to a settling tank where the sludge solids settled out and were recycled to the aerated reactor. By removing a portion of the biological solids daily, the concentration of solids in the reactor (mixed liquor volatile suspended solids) was maintained at 4,400 mg/l. The loading rate was 4.06 kg BOD_5 per day per kg mixed liquor volatile suspended solids. The treated and settled liquors were continuously removed from the settling tank, flowing over a weir. The analysis of the liquors was: BOD_5, 276 mg/l; and COD, 3,870 mg/l. The reduction in liquor BOD_5 was 95.1%, and in COD, 66.2%. The new biological sludge which was produced amounted to 0.743 kg/kg of BOD_5 removed from the liquors.

Example 2: Biotreatment of Heat Treatment Liquors — A mixture of primary and biological activated sludge was subjected to heat treatment (no oxidation) at 22.1 kg/cm² pressure and 177°C for a period of 50 minutes. After allowing the mixture to cool and settle for 4 hours, decantate was drawn off having the following analysis: BOD_5, 3,200 mg/l; and COD, 7,070 mg/l. The decantate liquors were pumped continuously at a constant rate of 269 liters/day to the reactor of Example 1, which had been previously seeded and acclimated to sludge liquors, and held at 31.8°C. The residence time in the reactor was 16.9 hours.

Air was continuously added at 6.8 liters/minute; 31% of the oxygen in the air was absorbed by the biological solids. The resulting oxygen transfer coefficient K_La was 37.3 mg oxygen transferred per hour per liter per mg/l dissolved oxygen deficit. The biological sludge and liquors were conveyed continuously from the reactor to a settling tank where the sludge solids settled out and were recycled to the reactor. A portion of the biological solids was removed daily, thus maintaining the mixed liquor volatile suspended solids concentration in the reactor at 2,500 mg/l. The loading rate was 1.82 kg BOD_5 per day per kg mixed liquor volatile suspended solids. The treated and settled liquors which were continuously removed from the settling tank had the following analysis: BOD_5, 73 mg/l; and COD, 1,010 mg/l. The reduction in liquor BOD_5 was 97.7% and in COD, 85.7%. Biological solids production was 0.479 kg/kg of BOD_5 removed from the liquors.

Removal of Heavy Metal Ions

A process described by *E. Luck; U.S. Patent 3,915,853; October 28, 1975;* relates to a method for treating sewage so as to remove from it various organic and inorganic contaminants. The process is of a sequence of operations in which the sewage is treated to remove various impurities and to convert organic materials to gaseous or inoffensive products by biological methods.

The process comprises treating sewage before biological degradation and fermentation thereof by adding to liquid sewage an alkaline compound, such as sodium hydroxide, to precipitate out metals as insoluble carbonates, bicarbonates, hydroxides or oxides, raising the pH thereof to over 11 to destroy pathogenic bacteria and to release enzymes from bacteria in the sewage into solution, removing the insoluble metal compounds from the rest of the liquid sewage and lowering the pH of the sewage to from 3.5 to 6.5, as by addition of hydrochloric acid.

A preferred method of treating liquid sewage containing inorganic and organic components comprises adding to the liquid sewage an alkaline compound or a mixture of alkaline compounds, selected from the group consisting of sodium carbonate, sodium bicarbonate and sodium hydroxide, to precipitate out metals as insoluble carbonates, bicarbonates, hydroxides or oxides, and removing the precipitated materials from the rest of the liquid sewage.

Next is added sufficient alkaline material to the sewage to raise the pH thereof to over 11 to destroy pathogenic and other bacteria in the sewage and release enzymes from the bacteria into solution. Then the pH of the sewage is lowered to from 3.5 to 6.5 by the addition of acid, enzymatically digesting a material or materials selected from the group consisting of starches, proteins, glycerides and sugars with the enzymes released from the bacteria, adding seed bacterial, fungal or yeast organisms or a mixture thereof to the sewage to assist in decomposing organic components thereof, acidifying the sewage after substantial decomposition thereof to kill the organisms developed from those seeded, removing the dead organisms and other insoluble materials which may be present and neutralizing the liquid remaining.

The heart of the process is the concept that various sewages can be prepared for treatment by removing heavy metal ions and other precipitatable metals from them, sterilizing the sewage by treatment with alkali to raise the pH to a high level, which sterilization also results in rupturing plant and animal cells which may be present to release useful enzymes from them, and enzymatically reacting with other organic constituents of the aqueous sewage solution (or partial suspension) to further prepare such materials for bacterial, yeast or fungal decomposition reactions of known types. Additionally, after decomposition, preferably by oxidation, the remaining sewage is neutralized and ions may be removed with ion exchange resins or similarly functioning substances. The following example illustrates the process. Unless otherwise mentioned, all parts are by weight and all temperatures are in degrees centigrade.

Example: One thousand parts of municipal sewage containing about 50% normal domestic sewage, having a solids content of 0.05% of organic material after removal of large waste materials and grit, and containing about half-soluble and half-dispersed components in an aqueous medium, has a major part of the dispersed portion separated from it by settling for about a day in a quiescent lagoon or tank.

The liquid medium, containing soluble organic and inorganic materials and very finely divided organic substances, has a biological oxygen demand of about 300 ppm. It contains over 5 ppm of copper and other metal salts in similar quantities, except for greater proportions of iron and zinc salts. The organic components include the usual lipids, proteinaceous and amylaceous components, cellulosics and bacteria, many of which are pathogenic, fungi and sugars.

To the sewage is added enough sodium carbonate to increase the molarity in the sewage to 0.1, provided that none had reacted with metal ions to form insoluble salts. Instead of sodium carbonate, in some experiments equimolar parts of sodium bicarbonate and sodium carbonate are used. The pH of the sewage is raised to about 11 and heavy metal ions, including copper and alkaline earth ions, including calcium, are precipitated out, settled out and separated from the liquid medium by passing the liquid on to a subsequent treating stage. Treatment takes less than an hour with the sewage temperature at 37°C.

Settling takes about 10 hours. Subsequent to removal of the metal ions (copper concentration is decreased to less than 0.5 ppm) the liquid is treated at the same temperature with enough of an aqueous solution of sodium hydroxide to raise the pH of the sewage liquid to about 13. The caustic used is an impure material not marketable as pure sodium hydroxide. The caustic and hydrochloric acid to be used subsequently are both impure, to lower the cost of treatment. The amount of caustic used is about enough to make the sewage 0.1 molar in OH^- concentration. In about 2.5 hours substantially all the pathogenic bacteria present, including *Staphylococcus aureus* and *Escherichia coli,* are destroyed and cell structures of other organic materials present, such as cellulosics, starches and fatty materials are weakened or broken.

As a result of the caustic treatment enzymes are released from the bacteria and other microorganisms and other organic materials present in the sewage. Because of the removal of debilitating amounts of metal ions, which might inhibit enzymatic action otherwise, the enzymes are active and ready to decompose starches, cellulosics, sugars and proteins at the proper pH.

Instead of employing separate steps to treat the sewage with carbonate(s) and caustic the steps are preferably combined, with the caustic being added with the carbonate or within an hour after its addition. In some cases settling times of as little as an hour will remove almost all of the precipitated metal salts, plus a substantial proportion of the dead bacteria and other materials attacked by the hydroxyl ions present. Sometimes caustic may be used instead of carbonates.

Because the enzymes are not active at the high pH's of the sewage after caustic addition, enough hydrochloric acid is added to neutralize the caustic and carbonate alkalinity and produce a pH of 5.2. Generally about enough HCl to make the sewage 0.1 molar will be sufficient to neutralize the caustic but because of the desire to make the medium acidic and because of the presence of hydroxyl ion from the excess carbonate reagent, additional HCl is employed, enough to make the molarity of Cl^- from the HCl about 0.15. This concentration, together with HCl or Cl^- in the sewage, helps the enzymatic and microorganism actions on the organic components of the sewage liquid.

After 10 to 12 hours of enzymatic action almost all of the sewage organics are converted to a form suitable for ultimate biological conversion to compounds such as carbon dioxide, water, nitrogen, ammonia, nitrates, sulfates and, in in some cases, oxygen. In some variations of the experiment an Amberlite (Rohm and Haas) mixed anion-cation exchange agent of the 10% DVB type, containing unsatisfied SO_3H^- and quaternary ammonium groups, is used to remove harmful ions before enzymatic or microorganism action commences.

This may be effected by merely mixing the ion exchange resins of the appropriate type with the liquid sewage or by passing the sewage through a relatively loosely packed bed of the agent. Usually, however, such treatment is not required (and the treated sewage liquid remaining may be discharged to streams or lakes in neutral or substantially neutral state) because most of the potentially interfering ions are removed in the alkaline treating step, with carbonate and/or caustic.

After completion of the enzymatic treatment step, during which, if so desired, additional enzymes may be added to the liquid sewage, such as diastase, urease, papain and other amylases, proteases, lipases, oxidoreductases and hydrolases, the sewage is ready for controlled biological treatment with usefully reactive but substantially nonpathogenic organisms. The organisms added are bacteria but fungi, yeasts, viruses, algae, protozoa, etc. are also used. In this case the bacteria used include *Proteus vulgaris,* Streptomyces, *Clostridium butyricum,* Nitrobacter, *Bacillus casei* and *Streptococcus lactis* but various others may also be present.

The weight of bacteria present at the start of bacterial action may be from 0.01 to 1% and in this case is about 0.05% of the solids content of the organic component of the sewage being treated. In addition a similar proportion of fungi and yeasts is used, being mainly composed of *Aspergillus niger, Saccharomyces cerevisiae* and *Penicillium chrysogenum,* or the amounts may be halved so that the bacteria and yeasts present total 0.5% on the basis described above.

The temperature of microorganism action on the sewage is maintained at 35° to 40°C and the pH is about 6. The pH may be lower, e.g., from 3.5 to 6.5 and even lower, to 2.5 when yeasts exclusively are used as the microorganism. Instead of charging fresh bacterial cultures these may be from previous batches of treated sewage, before killing of the bacteria and other organisms used to digest the sewage. The digestion continues for 24 hours, at which time the original sewage contaminants are almost completely converted to innocuous substances, mostly in gaseous form. Then the microorganisms are killed.

Killing of the bacteria and other nonpathogenic organisms used to decompose the sewage is effected by the use of acid, again preferably HCl, and enough is employed to produce a sufficiently low pH to completely destroy the organisms. This pH will usually be lower than 4.5 and is preferably about 3 or less. Killing may be aided by the use of elevated temperatures and in this case the sewage liquid is heated to about 50°C and is held at that temperature for 4 hours. Then, the killed organisms, now less objectionable pollutants, are either settled out, as in a 48 hour settling operation used, or they may be enzymatically destroyed, in the manner previously described.. Before killing of the organisms a portion thereof, about 2%, is removed for use as the inoculum for further sewage treatments.

When settling is complete and the dead microorganisms are removed or recycled back to earlier stages of enzymatic operations in other sewage treatments, the water will usually be clear, and it is so in this operation. To make sure that there are no organisms left, the water is filtered through a sand bed and the resulting clarified liquid, containing no organic biological contaminants, is neutralized to a pH of 7 with NaOH solution and is treated with a Rohm and Haas Amberlite anion-cation exchange resin of the 10% DVB(SO_3H-quaternary) type.

The ion exchange results in a water produced which is of drinking water quality, with only a few parts per million (1 to 10) of innocuous dissolved salts, having had over 99.5% of the inorganic ions removed. The water contains no live bacteria or other harmful organisms, is sparkling clear in appearance, and can be drunk with safety. After about an hour's operation the ion exchanger is regenerated and it is found that the effluent is of increased salt concentration over that originally charged in the beginning of sewage treatment. The salt is recovered from the ion exchange effluent and is employed to generate acids and bases for use in the process. The other products of the reactions are also salvaged, with dead organisms and organic materials settled out being used for fertilizer, the metal oxides, etc. as a source of heavy metals and in some cases the gases are also recovered.

Culture Prepared from Separated Primary Sludge

A process by *N.K.G. Westberg; U.S. Patent 3,652,406; March 28, 1972; assigned to AB Vallenbyggnadsbyran, Sweden* describes biological treatment of sewage characterized by the addition of a high-activated sludge containing a culture of heterotrophic bacteria to the sewage as an inoculum, before or after the inlet of a biological oxidizing means. The high-activated sludge is produced from at least a portion of the separated primary sludge by aeration in a separate cultivating basin.

In the process, the sewage enters a pretreating part of the plant which may, for example, consist of a strainer and sand trap of conventional design. The pretreated water is then fed to a primary basin from which separated sludge passes via a homogenizing means to a separate cultivating tank where a high-activated sludge is produced. The high-activated sludge thus obtained, containing a culture of heterotrophic bacteria, i.e., bacteria adapted to high concentrations of organic matter, is then supplied to the sewage as an inoculum to one or more points before or after the inlet to the biological oxidizing means.

From the biological oxidizing means the sewage containing sludge is fed to coagulating tanks, preferably after the addition of coagulants. From the coagulating tanks the suspension is led to a sludge-separating means where the sludge is removed by sedimentation, flotation or filtration in equipment that is of conventional design, and the purified sewage is allowed to pass on. A minor amount of the suspension leaving the biological oxidizing means may be used in order to start up the cultivation of the bacteria culture in the cultivation tank.

The sludge separated in the primary sedimentation basin is used to prepare a bacteria culture which is designated as a superactive or high-activated sludge. This sludge is used according to the process instead of the return sludge used in earlier processes for inoculation of the activated sludge process. The primary sludge may suitably be homogenized before it is fed to the separate cultivating tank where the high-activated sludge is produced. The quantity of primary sludge which has not been used for this cultivation can suitably be led to a separate means for some other treatment.

Biological Treatment

H. Sontheimer and D. Heinicke; U.S. Patent 3,300,401; January 24, 1967 have found that the detrimental effects of the fined and colloidal sludge solids remaining in the sludge-water mixture after partial separation of solids from wastewater sludges can be eliminated in a simple manner by subjecting this sludge-water mixture to a biological treatment. Such a biologica treatment can be carried out in the same manner as in the activated sludge process and is applicable to sludge-water mixtures from raw and digested sludges. In all cases certain micro-organisms accumulate due to the biological treatment, which are able to act biologically on the colloidal and fine organic substances in the sludge-water mixture and to partially remove them by endogenous respiration.

The process for treating organic sludges from the wastewater purification by partial separation of the solids with centrifuges or straining apparatus followed by treatment of the resulting sludge water is characterized by subjecting the sludge water to aerobic biological digestion.

The treatment of organic raw or digested wastewater sludges by partial mechanical separation of solids followed by biological treatment of the sludge water has the advantage over the known direct aerobic treatment of the entire sludge in that with the mechanically separated organic substance those materials are separated in solid form which are difficult to attack by biological means, so that in the remaining sludge-water mixture those substances are contained which can readily be treated biologically. Consequently, the biological conditions can be controlled much better than with direct treatment of the entire sludge, particularly when the latter is carried out in combination with the biological treatment of the wastewater. Furthermore, since less organic substance is introduced into the aeration basin, the cost of operation of the biological process is lower.

The partial separation of solids from the organic sludge prior to the biological treatment can be carried out by means of any suitable apparatus, particularly decanting centrifuges of known construction or straining devices which may be followed by presses. By means of such apparatus a solids portion with very low water content is separated from the organic sludge. With decanting, a centrifuge solids contents in the centrifuged sludge of 40 to 50% can be reached without difficulty; this is entirely sufficient for subsequent preparation of compost. The sludge water originating from the centrifuge, strainer or the like containing preferably the finest and colloidal solids, is treated aerobically by aeration until it is no longer putrescible.

It has been found that the centrate of raw sludge which has been biologically treated, can be further treated by centrifuging or can be dewatered on a drying bed in a relatively short time and without any odor nuisance For this biological treatment only about half the energy is required which is used up by direct aerobic treatment and subsequent dewatering on drying beds of the entire raw sludge.

For the aerobic oxidation of such sludge waters the aeration space is proportioned so that 3 to 7 kg solids are introduced per cubic meter of aeration space per day. By introducing oxygen in an amount of about 0.6 to 1.0 kg/kg of introduced solids, about 15 to 50% of the solids are aerobically oxidized with the above load. Sometimes oxidation of 5 to 10% is sufficient. The flocculating characteristics of the sludge are thereby so much improved, that further treatment is possible without difficulties. The oxidized solids which are removed from the process by the biological treatment are particularly those which make subsequent treatment so difficult.

The efficiency of filtration, which was with an original centrate only 10 liters per square meter per hour, when adding 0.2 gram of lime per gram of solids, could be increased to 100 liters per square meter per hour by the aerobic treatment of the centrate according to the process. The settleability was also considerably increased by the aerobic treatment. With a sludge water from sludge centrifuging a substantial degree of separation could be obtained after the biological treatment. While the centrate from a decanting centrifuge has scarcely any tendency to settle even after prolonged standing, after the biological treatment about 50% of the water with a relatively low degree of turbidity can be decanted by simple sedimentation.

In a municipal clarification plant 25 m^3 of thickened raw sludge with a solids content of 6% by weight are produced per day by the mechanical and biological wastewater treatment. This sludge is treated on a conventional decanting centrifuge with an output of 5 m^3/hr, whereby 60% of the solids with a solids content of 40 to 50% are separated. There remain about 23 m^3 centrate with a solids content of 2.5% which is not suitable for further thickening by sedimentation and can be filtered only with great expense. A direct return to a point upstream of the centrifuge leads to a rapid deterioration of the efficiency of separation due to the buildup of the fine and colloidal solids in the circulation.

In accordance with the process the centrate is treated in an aeration basin having a volume of about 125 m^3 and a retention time of about 5 days. In the aeration basin aerating devices are provided which are capable of thorough mixing and of introducing about 600 kg oxygen per day under standard conditions. About 200 kg per day of the suspended substance are destroyed by aerobic decomposition.

The aerobically treated centrate which contains about 1.75% solids, can be added to the fresh sludge going to the centrifuge, after thickening it to a solids concentration of about 3 to 6%. From the thickener a portion of the thickened sludge can be returned to the aeration basin, whereby the sludge concentration is increased and the treating period shortened.

By returning the biologically treated thickened centrate to a point upstream of the centrifuge, all sludge materials can ultimately be separated on the centrifuge in solid form. This involves a circulation of solids which, in their original state, cannot be centrifuged. Due to the aerobic treatment a buildup of such solids in the circulation is prevented. With a total of 25 m^3 of raw sludge containing 1,500 kg solids, about 36 m^3 of sludge with a total of about 2,165 kg solids must be treated in the centrifuge after the circulation has become balanced. 1,300 kg thereof with a solids content of 45% are separated in solid form. 865 kg are passed as centrate to the aerobic treatment according to the process, whereby 200 kg are eliminated and only 665 kg are returned from the aeration step to the raw sludge and pass again through the centrifuge.

The centrifuge can treat the entire quantity within 7 hours per day. With an energy consumption of 13 kw for the centrifuge, and with an efficiency of the aeration of 2.5 kg O_2/kwh, an energy consumption for the entire plant of about 350 kwh per day results, which corresponds to 14 kwh/m^3 raw sludge. This low energy consumption attains a dewatering to a 40 to 50% sludge which can readily be composted.

Activated Sludge for Removal of Polyvinyl Alcohol Resin

A process described by *Y. Fujita, I. Kawasaki and H. Nishikawa; U.S. Patent 3,926,796; December 16, 1975; assigned to Nippon Gohsei Kagaku Kohyo KK, Japan* involves the preparation of an activated sludge capable of assimilating polyvinyl alcohol resins included in wastewater by subjecting an ordinary activated sludge to acclimation in an aqueous solution in the presence of a specific amount of polyvinyl alcohol resin and a specific kind of nutrient.

In acclimation, the amount of the polyvinyl alcohol resin is increased stepwise from a lesser amount to a greater amount every 85% or more drop of the chemical oxygen demand of the aqueous solution. The acclimation can be completed in a short period such as 4 to 20 days. The wastewater effluent containing polyvinyl alcohol resin can be treated by passing it through the activated sludge with high efficiency. The following examples illustrate the process.

Example 1: A 40 liter aeration tank of a treatment apparatus equipped with the aeration tank and a 5 liter settling tank was charged with 25 liters of an aqueous solution including 1,000 ppm of sodium acetate (of which BOD loading was 0.21 kg/kg MLSS/day), 69 ppm of urea, 26 ppm of potassium dihydrogen phosphate and 50 ppm of polyvinyl alcohol (degree of polymerization: 1,700; degree of hydrolysis: 87% by mol of which BOD loading was 0.01 kg/kg MLSS/day) and 5,000 ppm of activated sludge from industrial wastewater treatment.

Then the aeration was carried out at a rate of 225 liters/hour at a normal temperature, while the aqueous solution was continuously supplied to the aeration tank at a rate of 1.1 liters/hour. The residence time in the aeration tank was 8 hours. The COD of the aqueous solution after three days from the beginning of the first acclimation decreased to 2 ppm from 42 ppm of initial COD.

Then, the second acclimation was successively carried out under the same condition as the above except that the polyvinyl alcohol concentration of the aqueous solution was 100 ppm, of which COD was 84 ppm After three days, the COD of the aqueous solution in the settling tank dropped to 4 ppm. Further, the third acclimation was carried out under the polyvinyl alcohol concentration of 300 ppm. The COD of the aqueous solution supplied was 252 ppm. After four days, the COD of the aqueous solution in the settling tank dropped to 4 ppm.

Finally, the fourth acclimation was carried out under the polyvinyl alcohol concentration of 500 ppm. The COD of the aqueous solution was 420 ppm. After three days, the COD of the aqueous solution in the settling tank dropped to 5 ppm and also the polyvinyl alcohol concentration was 6 ppm. The acclimation was finished, and the activated sludge was recovered in the settling tank.

Then an aqueous solution including 500 ppm of polyvinyl alcohol (degree of polymerization: 1,700; degree of hydrolysis: 87% by mol), 1,000 ppm of sodium acetate, 70 ppm of urea and 26 ppm of potassium dihydrogen phosphate was treated in the presence of 5,000 ppm of the above activated sludge at a temperature of about 20°C for one month in the same manner as in the acclimation. The polyvinyl alcohol concentration of the treated liquid was maintained 5 to 10 ppm and the COD of the treated liquid decreased to 3 to 7% of the initial COD.

Example 2: A wastewater including polyvinyl alcohol from a polystyrene manufacturing plant (of which COD was 500 ppm and polyvinyl alcohol content was 400 ppm and in which polyvinyl alcohol was used as an emulsifier) was diluted four times, and thereto sodium acetate, calcium acetate, urea and potassium dihydrogen phosphate were added in concentrations of 250 ppm, 250 ppm, 70 ppm and 25 ppm respectively. The COD of the wastewater so prepared was 84 ppm.

The first acclimation was carried out in the same manner as in Example 1 except that the above wastewater was employed instead of the aqueous solution and the activated sludge was added to the aeration tank in a concentration of 3,000 ppm. After three days, the COD of the wastewater dropped to 10 ppm and the polyvinyl alcohol concentration was 10 ppm.

The further two stage acclimations were successively carried out with supplying into the aeration tank the wastewaters prepared by the same manner as in the above except that the wastewater was diluted to 200 ppm and 400 ppm of polyvinyl alcohol concentration, respectively. In the former stage, namely the second acclimation, the initial COD 170 ppm dropped to 5 ppm two days after and the polyvinyl alcohol concentration was 6 ppm. In the latter stage, namely the third acclimation, the initial COD 340 ppm dropped to 7 ppm three days after and the polyvinyl alcohol concentration was 8 ppm. At his stage, it was regarded that the activated sludge had been acclimated.

Successively, the wastewater was treated by supplying it to the aeration tank at a rate of 0.9 liter/hour. The residence time in the aeration tank was 9 hours. The treatment was continued for two months. During the treatment, the polyvinyl alcohol concentration in the treated wastewater was within the range of 30 to 10 ppm.

ANAEROBIC DIGESTION

High Rate Digestion

W.N. Torpey; U.S. Patent 3,259,566; July 5, 1966 describes the treatment of raw sewage sludges by anaerobic digestion to convert putrescible sewage matter into gas, while producing an inoffensive residue called digested sludge.

The process involves eliminating the digester supernatant liquor problem, while imposing extraordinarily high solids loading rates upon the digester continuously. The process revolves about the discovery that the digestion or conversion of solids to gas is operationally attainable and sustainable in only a small fraction of the time commonly found to be necessary and required for conventional digestion.

This process involves the discovery that extraordinarily high speeds of bacterial conversion of solids to gas can be effected and be maintained in continuous operation in a digester, far beyond the apparent conversion speeds that must be accepted in a conventional two-stage digestion system.

This process utilizes the discovery of high rates of solids conversion by signficantly increasing the solids loading rates on the digester far beyond the conventional limits imposed by the requirement of producing a sludge thickened to the extent that potentially releasable supernatant liquor does not enter the digester. At the same time, provision is made for continuously maintaining the gasifiable sewage solids substantially uniform throughout the contents of the digester, so that the digested sludge being continuously displaced is substantially representative of the contents of the tank.

In order to provide the required high raw sludge concentration to the digester continuously or at high frequencies there is functionally interposed between the clarification and the digestion systems, a raw sludge thickening stage, the operation of which is conducted in the manner described in U.S. Patent 2,850,449. Suffice it to say that the thickening operation is conducted so that a sludge bed is formed and maintained in a manner to provide a reservoir from which to withdraw continuously a raw sludge of the solids concentration required as well as providing an operational cushion between the preceeding clarification station and the subsequent digestion station.

Since for the purpose of such thickening clarifier sludge is drawn extraordinarily dilute, primary effluent is improved with consequent improvement of treatment plant overall efficiency. This contributes importantly to improving overall plant performance since by this thickening operation the supernatant problem is eliminated.

In the absence of the prior limitations imposed by the necessity of having to release supernatant liquor it becomes possible to increase the solids loading rates in direct relation to the rate of speed of biochemical conversion. That is to say, the loading rate can now be increased, limited only by the gas loss incurred in the displaced sludge and by that extreme loading rate of which biological failure takes place. The control of this process of sludge digestion thus is oriented towards such loading rate limits and not towards the production of supernatant liquor.

Under such conditions of sustained high efficiency digestion there is producible a concentrated digested sludge which is disposable inasmuch as a high proportion of its gasifiable matter has been converted to gas. This process also efficiently reduces the residual gasifiable matter present in the concentrated partially digested sludge being continuously displaced from the digester.

In such instance relatively greater gas losses can be accepted from the high rate digester at a further saving in digester tank volume by providing a separate chamber or residual gas extractor wherein residual gasifiable matter is converted under high efficiency. That is to say, as the sludge passes through such a supplemental chamber, the contents are maintained in a state of agitation and substantial uniform distribution, in order that such residual digestion may also proceed at the highest efficiency.

In this way, there is established a new digestion system in which the digester is operated so as to continuously sustain high rates of conversion, whereas the residual gasifiable matter is converted in the residual gas extractor under conditions of agitation and uniform distribution so that residual matter will be converted continuously in a minimum of tank volume.

In related work, *W.N. Torpey; U.S. Patent 3,220,945; November 30, 1965* describes a process where raw sewage sludge derived from a clarification operation is dewatered or thickened while having incorporated in the mass of thickened raw solids substantial quantities of digested solids derived from recirculated digested sludge. When this dewatered sludge mixture is fed to the digester, there will result a digested sludge of higher concentration than was heretofore attainable, and the sludge volume going to disposal is greatly reduced.

The process involves the following combination of operations or operating steps. First, the raw sewage is subjected to a clarification operation producing effluent and raw waste sludge containing raw putrescible solids and liquid. This raw waste sludge is subjected to further dewatering preferably in a sedimentation thickening zone to form a bed of thickened sludge in the presence of a substantial quantity of digested solids. This thickening operation is controlled to yield a sludge mixture preferably thickened to a gel state where substantially no freely releasable liquor is present in the sludge.

A body of this sludge mixture is maintained in a state of anaerobic digestion, and has the thickened sludge mixture supplied, while discharging digested sludge. Usually, this is a volumetric displacement operation in the sense that normally a volume of sludge entering the digester displaces an equal volume of digested sludge. A substantial proportion of the discharging digested sludge is diverted for recirculation, while the undiverted portion is going to disposal. The digested solids thus diverted are added to the raw sludge to form mixed sludge to be subjected to the above dewatering or thickening operation, thus effecting a marked reduction in the volume of the digested sludge going to disposal.

Methane Obtained from Coal and Sewage

G.E. Johnson; U.S. Patent 3,640,846; February 8, 1972; assigned to the U.S. Secretary of the Interior has found that a mixture of sewage and coal substantially increases the production of methane by anaerobic digestion.

In the process, a sewage sludge containing methane-producing anaerobic bacteria is mixed with particulate coal in a vessel from which air is excluded. Since sewage solids decomposition for methane production is best attained at about 95° to 100°F, the vessel is preferably maintained at this temperature, although a wider temperature range (e.g., about 50° to 150°F) can be employed.

Any methane-producing anaerobic bacteria commonly employed to produce methane from sewage can be employed in the process. A discussion of such digestion appears in *Biological Waste Treatment* by W.W. Eckenfelder, Jr. and D.J. O'Connor, MacMillan Co., N.Y. (1961) pp 248-269. Exemplary bacteria are found in the genera Methanobacterium, Methanococcus and Methanosarcina. Suitable coal-to-sludge weight ratios are about 1:100 to about 1:2. Generally, there is no limitation as to the maximum amount of coal that can be added.

The lower ranking coals such as lignite and subbituminous A, B and C are preferably employed because they contain more celluloselike material upon which the bacteria may feed. However, higher ranking coals such as low volatile bituminous (LVB) and high volatile A bituminous (HVAB) can be employed. Whatever coal is employed, at the termination of gas production the residual coal can be employed as fuel. A coal particle size of about -60 mesh is suitable although finer particulate coal (e.g., -325 mesh) is preferred since more coal surface is exposed.

The following examples illustrate the effect of coal on anaerobic digestion.

Example 1: Three 2,000 ml samples of activated sewage sludge which contained anaerobic methane-producing organisms were placed in separate glass flasks, from which air had previously been flushed from each flask system with an inert gas (helium or nitrogen). To one of the flasks was added an individual 100 gram of LVB coal (-325 mesh); to another, 100 grams of HVAB (-325 mesh) coal was added. The third flask contained no coal. The contents of each flask were slowly stirred. The product gas produced during digestion (at 95° to 100°F) in each flask was periodically quantitatively and qualitatively analyzed. The results were as follows:

	Test Results with Sludge Only			Test Results with LVB Coal			Test Results with HVAB Coal		
	Gas Analysis, vol %		Cumulative	Gas Analysis, vol %		Cumulative	Gas Analysis, vol %		Cumulative
Hours of Operation	Methane	Carbon Dioxide	Gas Volume, liters	Methane	Carbon Dioxide	Gas Volume, liters	Methane	Carbon Dioxide	Gas Volume, liters
300	–	–	–	82	18	3.9	–	–	–
400	77	23	2.4	84	16	7.3	72	28	2.5
600	–	–	–	87	13	12.8	83	17	6.5
650	82	18	4.5	83	17	13.5	78	22	7.2
800	–	–	–	86	14	14.6	80	20	11.0
1,100	79	21	5.7	–	–	–	–	–	–
1,300	82	18	7.7	87	13	15.7	83	17	12.6

Example 2: Two 2,000 ml samples of activated sewage sludge which contained anaerobic methane-producing organisms were placed in separate glass flasks from which air had previously been flushed with an inert gas (helium or nitrogen). Initially, no coal was added to either flask. Lignite was later added to both flasks at the times shown below. The contents of each flask were slowly stirred. The product gas produced during digestion (at 95° to 100°F) in each flask was periodically quantitatively and qualitatively analyzed. The results were as follows:

	Test A			Test B		
	Gas Analysis, vol %		Cumulative	Gas Analysis, vol %		Cumulative
Hours of Operation	Methane	CO_2	Gas Volume, liters	Methane	CO_2	Gas Volume, liters
100	28	72	0.6*	–	–	0.8
200	44	56	1.3	88	12	1.6
250	52	48	2.2	90	10	2.6
325	70	30	3.2	95	5	3.5
400	74	26	4.0	95	5	4.1
500	74	26	5.3	95	5	4.5
750	78	22	6.3*	–	–	4.8
900	69	31	7.8	–	–	4.9**
1,030	56	44	8.4*	9[illegible]	10	5.4
1,100	–	–	8.6	–	–	5.5

*200 g lignite (–325 mesh) added in Test A at the end of 100 hr and 1,030 hr.
**Gas production in Test B ceased after 900 hr; 100 g lignite (–325 mesh) added at the end of 900 hr which reinitiated gas production within minutes.

As can be seen from the data, the addition of coal has a pronounced effect on total gas production. The LVB coal more than doubled gas production while the HVAB coal almost achieved the same result.

At sewage facilities internal combustion engines are generally used to drive auxiliary equipment such as sewage pumps or blowers which provide air for the aeration tanks, etc. Previously, the generation of sewage gas (CO_2 + CH_4) at such facilities has generally been insufficient to completely operate those engines on sewage gas, and the fuel has to be supplemented with natural gas. The process provides a method of generating sufficient gas at such facilities to completely fuel all these auxiliary engines.

Continuous Mixing of Sludge Exterior to Digester Unit

A.L. Genter, R.R. Kennedy and R.M. Kennedy; U.S. Patent 3,105,041; September 24, 1963 describe a process and system for anaerobic sewage digestion particularly adapted to handle the waste materials of communities having a population of approximately 10,000 persons or less.

To facilitate treatment of sewage, it is common to deliver the sewage initially to a settling tank where the sewage solids are generally separated from the sewage liquids. As a result, one obtains (a) a fresh sludge, comprising the sewage solids and some liquids, which is to be processed, and (b) some comparatively clear liquid which is delivered to the settling tank with the untreated sewage, but which can be drained off to a suitable field. As shown in Figure 2.7, untreated sewage is passed through a conduit **2** into a settling tank **4**.

Preferably, the settling tank **4** is of a circular configuration, and is provided with a centrally-disposed influent well **6** provided with a series of T-shaped inlet heads **8** which cancel the velocity of streams issuing from such inlet heads as explained in U.S. Patent 3,067,878.

FIGURE 2.7: ANAEROBIC DIGESTION PROCESS

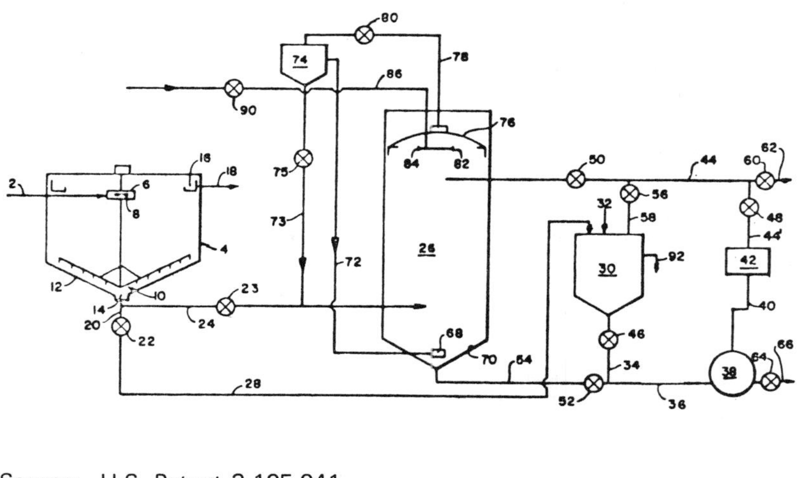

Source: U.S. Patent 3,105,041

The settling tank **4** is also provided with a sludge collector or sweeper **10** having veins extending adjacent the bottom of tapered wall **12** of the tank **4** to move settled sludge toward the tank outlet **14**. Normally the sludge settles to the bottom of the tank, and the clear sewage liquids overflow into a launder such as that designated by numeral **16** in Figure 1.3, and are then drained off for desired disposition through a conduit such as that designated by numeral **18**.

The fresh sludge which is delivered to the outlet **14** of the settling tank **4** passes into a conduit **20** having a valve **22** therein which serves to direct the fresh sludge into either conduit **24** leading directly into the processing area **26**, or to the conduit **28** which leads directly to the input of a sludge mixer **30**. Preferably, the conduit **24** includes a valve **23** which is normally maintained closed, and preferably the valve **22** is normally maintained open so that fresh sludge is passed directly to the sludge mixer **30**.

The sludge mixer **30**, or as it may be termed, the sludge receiving and distributing unit, contains partially processed sludge with anaerobic bacteria therein, and serves to mix such processed sludge with the fresh sludge entering through conduit **28** to provide a fresh sludge mixture. Preferably, two parts of fresh sludge are mixed with one part of at least partially digested sludge in the mixer **30**, and the mixture is maintained with an alkaline pH. To maintain the alkalinity of the mixture, milk of lime or any other suitable alkaline substitute may be added through the conduit **32** leading into the mixer **30**.

The fresh sludge and the partially processed sludge pass through an outlet conduit **34** and a pump inlet conduit **36** to a circulating pump **38** which pumps the mixture through a conduit **40**, heater **42**, and inlet line **44** into the enclosed processing area or digesting unit **26**. For such operation, the valve **46** in conduit **34**, the valve **48** in conduit **44'**, and the valve **50** in conduit **44** are maintained open, and the valve **52** in conduit **54** and the valve **56** in conduit **58** are maintained in closed position.

After the processing area of digesting unit **26** has been filled to a suitable level, the valve **46** in the outlet conduit of the mixer **30** is closed and the valve **52** in the outlet conduit **54** of the digesting unit **26** is opened. During all of the aforementioned operations, the valve **60** on the drain-off conduit **62** is closed and the valve **64** in the drain-off conduit **66** is closed.

After the inoculated fresh sludge mixture, or a suitable amount has been placed in the digestion unit **26**, and after valve **46** and valve **52** have been closed and opened respectively, then circulating pump **38** serves to continuously circulate the inoculated fresh sludge mixture through the process area or digesting unit **26** and also serves thereby to agitate such mixture. It will be noted that, as suggested above, a heating unit **42** is provided in the circulating path extending through conduit **54**, conduit **36**, circulating pump **38**, conduit **40**, conduit **44'**, and conduit **44**. The heating unit serves to maintain the temperature of the inoculated fresh sludge mixture at a suitable level for anaerobic bacteriological action, i.e., at a temperature from about 20° to 60°C.

Although the recirculating pump **38** provides for some severe agitation of the inoculated fresh sludge mixture, preferably agitation of such mixture is further stimulated by introducing compressed gas into the enclosed processing area or digestion unit **26** through a variable-speed gas blower **68**, disposed centrally of, and near the bottom **70** of the enclosed area of digestion tank **26**. Compressed gas is fed to the blower **68** through a conduit **72** extending from a compressed gas source **74**.

A floating gas dome **76** is provided at the upper end of the processing area of digesting unit **26** and a gas outlet conduit **78** extends to the compressed gas source **74** to feed gases exiting from the top of the unit **26** to the compression source. A valve **80** is provided in the conduit **78**, and by adjustment of the valve **80** the flow of compressed gas through the digestion unit, and thereby the degree of agitation of the sludge mixture within the unit, can be controlled.

It has been found that in most instances during violent agitation of the mixture a froth collects at the top of the sludge mixture undergoing treatment in the processing area or digesting unit **26** due to the active digestion. Such froth may be dispersed and forced to yield its floated solids by means of water sprays issuing from nozzles such as those schematically presented in Figure 2.7 and designated by numerals **82** and **84**. A suitable water line **86** leads to such nozzles, and is provided with a valve **90** so that the sprays may be cut on or off as desired.

During recirculation and processing of the sludge mixture, alkaline conditions should be maintained at all times. For this purpose, as suggested above, a suitable alkaline chemical solution may be introduced into the mixer through the conduit **32**, and the valve **46** can be opened to allow the solution so introduced to pass into the mixture circulating through conduit **54** by virtue of the pumping action of the recirculating pump **38**.

After the sludge has been at least partially processed by continuous recirculation as described above, then the same is diluted by passing water into the mixture through the conduit **86** and spray nozzles **82** and **84**. The mixture which is partially treated sludge, consists of a sludge liquor and sludge solids, and the water which is introduced is purer than the sludge liquor and serves to dilute the partially-treated sludge mixture.

After the mixture has been sufficiently diluted, the valve **90** in the water line **86** is closed, and either then, or preferably previously, the recirculation pump **38** is stopped and the partially-treated and diluted sludge mixture is allowed to remain quiescent in the enclosed processing area or digestion unit **26** whereby the sludge solids of the mixture separate from the diluted sludge liquor, and a processed sludge mixture exists in the processing area or digestion unit **26** which is suitable for delivery to sludge beds.

When the sludge mixture undergoing treatment has remained quiescent for separation as desired, then valve **48** in conduit **44'**, and valve **56** in conduit **58** are closed, and valve **60** in drain-off line **62** is opened. The valve **50** having previously been opened does not interfere with drain-off of the processed sludge liquors through conduit **62** and thus by virtue of the valve operation described immediately above, the processed sludge liquors are passed to a suitable drain field.

Either simultaneously, or after the sludge liquor has been drained off, the valve **64** in the drain-off line **66** is opened, and the valve **46** in conduit **34** and valve **48** in conduit **44'** are closed, and the sludge solids are drained out of the processing area or digestion unit **26** through conduit **66**. If necessary, the circulating pump **38** can be used to pump the solids out of the digestion unit **26**. It is important to understand that all of the processed sludge is not drained off according to the process. Instead, at least a portion of the processed sludge is stored for subsequent mixing with fresh sludge to be treated.

In the system shown in Figure 2.7, a liquid portion of the processed sludge may be stored in the mixer **30** by closing valves **60** and **48** and opening valves **50** and **56** whereby the processed sludge liquor drains into the mixer. To store a portion of the solid processed sludge, the valve **50** is closed and the valve **46** is closed, but the valves **52, 48** and **56** are opened so that the circulating pump **38** can be activated to pump the processed solid sludge into the mixer **30** through the inlet conduit **58**. Preferably the mixer **30** is provided with an overflow conduit **92** so that if too much liquid enters the mixer, the same will overflow in much the same manner as liquid overflows from the launder **16** of the settling tank **4** described initially in connection with Figure 2.7.

By again referring to Figure 2.7, it will be noted that while the processing of the sludge is carried out in the processing area or digestion unit **26**, untreated sewage can be conducted to the settling tank **4** and allowed to settle therein, so that after processing of one batch of sludge is completed in the area or unit **26**, another batch of sludge can be immediately passed thereto for processing. Of course, for this purpose, the valves **22** and **23** are maintained closed during the time that some sewage is settling in the tank **4** and other sewage is being treated in the unit **26**.

While it has been assumed that the process is initiated by conducting the fresh sewage through the conduit **28** to the inlet of the mixer **30**, it should be understood that the provision of a mixer **30**, although preferred, can be eliminated. More specifically, in instances where cost is a primary factor due to the small population, the mixer **30** can be replaced or eliminated, and fresh sewage can be fed directly from a suitable settling tank such as that designated by numeral **4**, through an inlet conduit extending directly into the digestion unit or processing area **26**.

More particularly, a conduit such as that designated by numeral **24** and provided with a valve **23** can be used for this purpose. A certain predetermined or controlled amount of processed sludge can be left in the bottom of the processing area or digestion unit **26**, and if the mixer **30** is eliminated, then mixing of the partially-processed sludge takes place during circulation of the mixture as described above and during agitation of the same.

Still another modified form of the process provides for replacing the mixer **30** by a storage tank or other storing means whereby a portion of the processed sludge can be maintained separate from the digesting unit or processing area **26**. According to this modification, fresh sludge is conducted directly into the digestion unit or processing area **26**, and as the sludge is being passed into such area or unit, the circulating pump **38** serves to gradually pump processed sludge from the storage means provided into the digestion unit or area **26**. This system is somewhat more efficient than merely leaving part of the processed sludge in the digestion unit and then passing fresh sludge therein, but, of course, is not as efficient as using a mixer **30**.

From the above discussion it should be apparent that the sludge being processed is mixed more or less on a continuous basis exterior of the digestion unit or processing area and is maintained active within the digestion unit or area during violent agitation. This process accomplishes three principal activities, all of which contribute to keeping the biological factors operating at peak capacity at all times. First, the active organisms are kept continuously in contact with the food supply; second, the food supply is uniformly distributed and made available to all of the active digestion organisms possible; and third, the concentration of the inhibitory biological intermediates and end products is maintained at a minimum level. All of these factors keep the working population of active organisms and their food supply cooperating at peak efficiency.

A process described by *R.G. Knibb; U.S. Patent 3,493,494; February 3, 1970; assigned to Simon-Hartley Ltd., England* is for a sewage sludge treatment apparatus comprising one or more units mounted externally of a sludge digester tank for heating and circulating sludge. Each unit comprises piping surrounded by heating means and having a sludge inlet at its lower end and a sludge outlet at its upper end. A conduit is provided for passing a pressurized gas into the piping at or near the lower end for escape at or near the upper end whereby the sludge is caused to circulate from the bottom to the top of the tank through the piping.

In the process, therefore, complete circulation of sludge in the tank is achieved by circulation of sludge from the bottom to the top of the tank through the external sludge carrying tubes or piping, and by circulation of sludge at the sludge level. At the same time, the external units provide for efficient operation of the system.

SLUDGE DEWATERING PROCESSES

FILTRATION

Filter Belt

A. Bähr; U.S. Patent 3,896,030; July 22, 1975 describes a filter press which is comparatively inexpensive to construct and occupies comparatively little space but which nevertheless provides a high and yet controllable filtering efficiency, the press being suitable in particular for processing sewage sludge. In order to achieve this one starts out from a filter press of the known kind with two endless filter belts which contain the sludge between them, the two belts circulating together, guided in concentric paths, around a rotary drum against which they are pressed by adjustable pressure rollers, so that the liquid phase is squeezed out of the sludge.

The process is characterized essentially in that the two filter belts travel, before reaching the rotary drum, along horizontal sections situated above the drum, the sludge being pretreated in these sections, retained between the two filter belts, before passing around the drum. Of these horizontal sections, the first can take the form of a horizontal filter belt onto which sludge is deposited by a distributor, the filter belt acting as a strainer for removing a first fraction of the liquid phase under gravity. Just downstream of this section the filter belt can, according to the process, pass over a vibrator which applies adjustable oscillations to the filter belt.

Downstream of the vibrator the two filter belts from a wedge-shaped feed inlet and just downstream of this the two filter belts, containing the sludge between them, pass along a horizontal pressure section between pressure rollers.

Downstream of the horizontal pressure section the two filter belts, containing the sludge between them, circulate together around the rotary drum, the belts preferably running onto the drum at the highest point thereof. On leaving the drum the two filter belts preferably rise over a certain distance, so as to allow the liquid which has last been squeezed out of the material to run off downwardly.

A particularly effective dewatering action is obtained by constructing the drum as a rotary roller cage. This has the advantage, in a filter press according to the process, that the sludge circulating between the filter belts around the rotary drum is worked or "kneaded" by squeezing thrusts applied sequentially in different directions between the rollers of the rotary roller cage and the external pressure rollers, producing a particularly good dewatering effect.

In the filter press the dewatering therefore takes place in a sequence of five stages, the material being worked in each stage in a manner suited to its condition on leaving the previous stage. The five stages are essentially as follows. In the first stage, the material spread out over the surface of the first filter belt is allowed to drain under gravity without the application of pressure. This is followed by a vibration stage, after which the material enters the wedge-shaped feed inlet and then passes along the horizontal pressure section.

Finally, the material is kneaded and pressed, between the filter belts, on its way around the asynchronous rotary roller cage. The raw material is assumed to be a watery sludge capable of flowing very easily. On its way through the press the material gradually loses water and becomes stiffer. Finally it is pressed to form a hard filter cake and kneaded to remove residual water.

The more intense the kneading action the less water remains in the final hard filter cake. A particularly important proposal according to the process is to support each filter belt by means of a supporting belt of greater mechanical strength, so that a higher squeezing pressure can be applied. It is assumed that the filter belts are made of a fine-mesh fabric consisting of very thin threads in order to obtain a high filtering effect.

The mesh width may be, for example, 10 to 100 microns, depending on the nature of the sludge. For example a Perlon fabric can be used, or a combination fabric made of Perlon threads and V4a wire, or it can be a purely wire fabric made of V4a wire. When a high pressure combined with a kneading action is applied to a filter belt of this kind this can involve a high rate of abrasion and a risk of rupturing the fabric.

The high applied stresses are taken by the support belts, which for example can consist of chain wires spaced 5 mm apart with woven-in steel crossbars of 1.5 to 2.5 mm diameter spaced apart so as to leave mesh openings approximately 5 x 15 mm in area.

It is preferable to drive only the two supporting belts, the filter belts themselves not being directly driven but merely entrained by the supporting belts. In regard to the amount of space occupied by the press, the five sequential process stages between them require of course a considerable length of filter belt. But the filter press is nevertheless compact in construction, because the three first sections, that is to say the dewatering by draining section, the vibrator section and the horizontal pressure section, are all situated over the rotary drum.

Furthermore, the sludge distributor which spreads the sludge rapidly right across the filter belt immediately produces a layer of sludge of even thickness extending over the full width of the filter belt. The entire available filtering surface is therefore fully utilized, giving the highest possible performance. The technical advance provided by the process is demonstrated by the fact that compared to known filter presses occupying the same amount of space the filter press according to the process provides a 50% greater effective filtering surface.

Screen and Capillary Type Belts

C.D. Beristain, B.K. Nixon and R.J. Kiefer; U.S. Patent 3,774,760; November 27, 1973; assigned to Westinghouse Electric Corporation describe a process and apparatus for sludge dewatering including gravity screening and capillary dewatering where an endless belt of screen-like material and an endless belt of a capillary material are selectively functionally utilized separately and in close proximity to effectuate the steps of the process including a repetitive dewatering of the belt as it becomes saturated.

The repetitive dewatering step proceeds with the screen-like belt supporting the sludge while the capillary belt is separately compressed, or otherwise dewatered. The process and apparatus further include a final sludge dewatering section where the screen-like belt and the capillary belt are similarly separated.

Initially the sludge which is rather dilute and contains much free water is fed to the screen-like belt alone. It has been found that with an hydraulic head of approximately 2 inches that 20 to 45% of the free water can be drained from the sludge in 20 to 30 seconds. The partially dewatered sludge is then carried by this screen-like belt to a location where the screen-like belt and the capillary belt are placed in sufficiently close proximity that the capillary belt and aqueous liquid produce a capillary force which removes further water from the sludge.

As the capillary belt becomes saturated the screen-like belt is separated therefrom and the capillary belt is treated either by compression or by the application of a vacuum to draw a substantial quantity of water therefrom. The screen-like belt and the capillary belts are then once again brought into close proximity so that the capillary force may remove still further water from the sludge.

These steps are repeated a sufficient number of times to remove a high percentage of the water from the sludge. At the end of this repetitive process the screen-like member is once again separated from the capillary belt and the capillary belt is again treated to remove water therefrom.

The screen-like belt wraps around its return rollers a foot or so along the machine length. A compression roller is associated with the return roller to further dry the sludge. The surface of the compression roller is such that the now partially dry sludge preferentially sticks to the roller surface and may be scraped therefrom into a receptacle via the usual knife-edge.

The belt and screen then proceed to cleaning locations where water under pressure is sprayed on the surface of the capillary belt and the screen-like belt, respectively, from which locations the belts are conveyed to the inlet side of the apparatus and the process is repeated.

Continuous Porous Sheet Belts

M. Adamski and D.S. Schover; U.S. Patent 3,613,564; October 19, 1971; assigned to General American Transportation Corporation describes an apparatus for removing aqueous liquid from a flowable material, containing aqueous liquid as a continuous phase and suspended solids as a discontinuous phase, to obtain a dewatered product. The apparatus has two endless, driven belts of elongated porous sheets. The belts are supported so that one belt has its path of travel inside that of the other belt and with an upper horizontal run of the outer belt overlying and abutting that of the inner belt moving in the same direction.

The sheet of the inner belt is resilient, compressible and made of cellular material capable of absorbing aqueous liquid by a wicking action, whereas the sheet of the outer belt is a fine mesh sheet with pores providing passage of the liquid through the sheet by a wicking action of the abutting cellular sheet while retaining most of the solids on the outer belt.

A feeder for delivering the flowable material onto the outer belt is mounted adjacent the beginning of the dewatering zone where there is the wicking action due to the abutment of the two belts along that zone. A device for removing dewatered product is mounted downstream of that zone. A device to compress the inner belt downstream of the upper horizontal run of that belt includes a pair of opposed rolls. Another compression device may be at the upper horizontal run but downstream of the dewatering zone at which initial wicking occurs.

L. von Reppert, W. Teske and E. Hilscher; U.S. Patent 3,527,698; September 8, 1970; assigned to Chemische Werke Albert, Germany describe an apparatus for removing water from sewage sludge including a container having a porous bottom, and either having a fabric or synthetic fiber underneath the container whereby water flowing out of the container passes through the fabric, means for tensioning the fabric and for displacing it laterally underneath the container, and a spray device for cleaning the fabric, or the container has the form of a sack or a tube or a part of it has the form of a sack and another part has that of a tube of synthetic fabric which is suspended in a tiltable support frame or in a housing open at the bottom and which may be provided with a movably mounted perforated bottom member. The apparatus also has a reservoir for collecting water flowing from the container, and means for discharging liquid at the top of the container.

Capillary Action

Y. Hanami; U.S. Patent 3,371,785; March 5, 1968; assigned to Kurita Industrial Company, Limited, Japan describes a method for rapidly dehydrating sludge by inserting into a layer of sludge containing fur or highly hygroscopic, colloidal solids formed during water-treatment procedures, a strip made of a liquid-pervious material having numerous fine passageways and spacer means with one end of the strip protruding below the bottom of the sludge layer, or with both ends protruding above and below the upper and lower surfaces of the sludge layer.

Example 1: Sludge specimens comprising aggregated sediments of mud with a moisture content of 86% collected at a water purifying plant were placed into columns having an inner diameter of 66 mm and a length of 150 mm to a depth of 140 mm. Into each of these specimens was inserted a strip of cardboard of a size of 3 mm (thickness) x 50 mm (width) internally having hollow portions each portion being of a size of 1 mm x 2 mm.

The lower end of the strip was made to protrude about 5 cm outside of the column through a slit of 3 mm x 50 mm formed at the bottom of the column.

Under these conditions, the relations between the lapse of time and the changes in the moisture content of the sludge (A) when the upper end of the strip of cardboard protruded upwardly beyond the upper surface of the layer of the sludge and (B) when the upper end of the strip was located at the surface level of the layer of sludge, are shown in the following table. For the sake of comparison, the result (C) obtained when a strip of cardboard was not inserted and the result (D) when only the upper end of a strip protruded upwardly beyond the edge of the column are also shown in the same table.

Column	Moisture Content (percent) In 2 Hours	Moisture Content (percent) In 24 Hours
A	80.5	71.6
B	80.5	75.8
C	85.9	85.0
D	85.6	83.1

Example 2: This experiment was conducted under substantially the same conditions as those for Example 1, excepting that the upper end of the strip protruded upwardly about 6 cm beyond the upper surface of the layer of sludge, that the lower end of the strip protruded downwardly 5 cm from the slit of 3 mm x 50 mm provided at the bottom of the column and also that air was passed onto the upper portion of the column at the rate of 4 meters per second.

Dehydration of the specimens was conducted under these conditions (A). For comparison, the results (B) where no strip was inserted and the result (C) where a strip of cardboard was inserted with only the upper end protruding upwardly from the upper surface of the layer of sludge and the result (D) where only the lower end of the inserted strip of cardboard protruded downwardly outside the column were also obtained. The relations between the lapse of time and the changes in the moisture content are shown in the following table.

Column	Moisture Content (percent) In 2 Hours	In 24 Hours	In 72 Hours
A	80.5	71.6	53.2
B	85.9	85.0	82.6
C	85.6	83.1	69.3
D	80.5	75.8	74.2

Eddy Flow Separator

V.R. Sparham; U.S. Patent 3,925,205; December 9, 1975 describes a method of separating colloidal or gelatinous suspended solids in a liquid phase that comprises passing the liquid through a labyrinth of eddy-forming surfaces in series with a sheet perforate member. These surfaces may be made from, for example, polyvinyl chloride, resin-bonded fibrous materials, suitably prepared metals, linoleum, timber, asbestos sheets, rubber, cork and ceramics.

The eddy-forming surfaces may in general be the surfaces of elements arranged to form the labyrinth, the arrangement being such that the liquid passes through passages formed between the elements. The passages preferably form zig-zag or nonlinear paths through the labyrinth. The direction of flow in relation to gravitational forces can be arranged to suit the type of solids in suspension to be separated from the liquid phase.

The process includes apparatus for reducing the solids content of a liquid phase containing colloidal or gelatinous suspended solids which comprises a vessel, such as a sedimentation tank, having an inlet and an outlet for the liquid, and a labyrinth having eddy-forming surfaces arranged in the vessel so that when liquid flows through the vessel from the inlet to the outlet it must pass through the passages in the labyrinth.

The labyrinth may be formed from closely spaced plate-like or tube-like members or elements which may be metallic or nonmetallic, e.g., they may be made from any of the materials mentioned above. The plate-like or tube-like members may have an undulating cross section and may also have ribbed walls. The plate-like or tube-like members of the labyrinth may be secured together to form made-up units having a plurality of passages therein for the liquid.

The passages may be inclined to one another and/or to the walls of the vessel and may be of constant or variable cross section. Preferably, the labyrinth is arranged so that when liquid flows through the vessel from the inlet to the outlet, it passes through the labyrinth in a generally upward direction.

As indicated above, a sheet perforate member is arranged in series with the labyrinth so that when liquid flows from the inlet to the outlet, it also passes through the sheet perforate member. Preferably the labyrinth is on the downstream side of the sheet perforate member, that is to say liquid flowing through the vessel from the inlet to the outlet flows first through the sheet perforate member and then through the labyrinth.

A sedimentation tank may be provided having a compartment positioned close to the top of the tank in such a way that liquid fed into the tank must pass upwardly through the compartment before it can pass out of the tank. By filling the compartment with the labyrinth, e.g., the assembly of elements, the liquid can be clarified and a large proportion of the small particles can be retained on the walls and among the cavities in the labyrinth. In addition, the labyrinth may be arranged in a tank separate from a sedimentation tank.

Filled Concrete Blocks as Filter Media

G. Spohr; U.S. Patent 3,847,808; November 12, 1974 describes a filtration system which will permit sludge removal more economically. The sludge bed is constructed such that heavy equipment can be moved thereover without disturbing the filter material regardless of the moisture content in the sludge.

In one form of the process a series of partitioned blocks are provided with cartridges filled with filter material such as sand, carbon or other appropriate chemicals. The cartridges can be removed from time to time and the filter material cleaned or replaced. Slots are provided in the edges of the peripheral blocks at various intervals so that the bound water can be drained off.

In another example tracks are provided for accomodating the wheels of a front-loader which can move along the bed to scrape off the sludge. Also, a scraper can be pulled by a winch or the like over the bed to remove the dried sludge.

Basically, the filter bed is composed of a plurality of blocks having circular openings. The blocks are composed of material such as concrete which is designed and engineered to withstand the anticipated load of motorized loading and transporting equipment having the ability to move around freely on the total area of the sludge bed.

Use of Ferrous Chloride

T.H. Oster and J.E. Cooper; U.S. Patent 3,173,863; March 16, 1965 describe a process for the preparation of sludges for dewatering by filtration to a water content sufficiently low to permit incineration without the addition of extraneous fuel. These sludges have been treated for many years by the addition of a solution of ferric chloride and lime followed by agitation. The action of this combination of chemicals on the sludge is obscure, but it does permit the mixture of sludge, lime and ferric chloride to be satisfactorily dewatered, usually upon a rotary vacuum filter.

This process is predicated upon the discovery that many sludges containing at least 3% of solids can be conditioned for filtration more satisfactorily and economically if ferrous chloride is substituted for part or all of the ferric chloride formerly used. A pressure differential of at least 3 inches of mercury should be maintained across the filtering medium. Other than the substitution of ferrous chloride for part or all of the ferric chloride the remainder of the process, that is, the addition of lime, the agitation and the filtration remains the same.

As a demonstration of this process, a sample of primary sludge containing 0.68 pound of solids per gallon was used. A solution of ferric chloride was prepared by dissolving 10 grams of $FeCl_3 \cdot 6H_2O$ in water and making the volume up to 100 cc. A solution of ferrous chloride was prepared by dissolving 17.3 grams of $FeCl_2 \cdot 4H_2O$ in water and making the volume up to 100 cc. A slurry of 10 grams of calcium hydroxide in 100 cc of water was prepared.

An arbitrary test of sludge conditioning efficiency consists of treating a 100 cc sample of sludge with ferrous chloride or ferric chloride or a mixture thereof, then adding 7.5 cc of lime slurry followed by 1 minute of gentle hand stirring. This mixture was poured quickly into an 11 cc Buechner funnel connected to a vacuum system.

The times noted extend from pouring of the sludge into the funnel until the vacuum in the system dropped from the maximum value of 28 inches Hg to an arbitary value of 15 inches Hg. Here, a satisfactorily dry cake resulted. The results obtained are summarized in the following table. In each test where both ferric and ferrous chloride were used in the same test they were added separately except in Test No. 7 where the chlorides were premixed. The 2.60 cc of $FeCl_3$ solution contains 0.156 gram of $FeCl_3$, anhydrous basis; the 1.11 cc of $FeCl_2$ solution contains 0.123 gram of $FeCl_2$, anhydrous basis.

Test	FeCl3 (cc)	Ferric Iron Fe, % Dry Solid	FeCl3 % Dry Solid	FeCl2 (cc)	Ferrous Iron Fe, % Dry Solid	FeCl2 % Dry Solid	Fe, Total % Dry Solid	Time, Seconds
1	2.60	0.66	1.92	None	None	None	0.66	41
2	None	None	None	1.11	0.66	1.49	0.66	36
3	1.30	0.33	0.96	0.55	0.33	0.75	0.66	36
4	0.90	0.23	0.66	0.50	0.30	0.66	0.56	34
5	0.60	0.15	0.44	0.50	0.30	0.66	0.59	37
6	0.30	0.08	0.22	0.50	0.30	0.66	0.30	40
7	0.90	0.23	0.66	0.50	0.30	0.66	0.53	38
8	0.90	0.23	0.66	0.50	0.30	0.66	0.53	36
9	1.30	0.33	0.96	0.30	0.18	0.40	0.51	38
10	2.60	0.66	1.92	None	None	None	0.66	38
11	None	None	None	1.10	0.66	1.49	0.66	35

Sewage disposal plants purchase ferric chloride in a 40% water solution. Operators may effect a reduction in chemical costs by contacting this solution with scrap iron or steel for a time to permit a portion or all of the ferric chloride to be reduced with concomitant dissolution of the scrap iron. This reaction proceeds more smoothly if the solution is strongly diluted with water prior to contacting with the scrap iron, cast iron, steel borings or tin cans.

Mixtures of Polyacrylamides and Ferric Salts

R.A. Lo Sasso and E.G. Rausch; U.S. Patent 3,642,619; February 15, 1972; assigned to Calgon Corporation have found that the dewatering of waste sludges is greatly enhanced if the sludge is treated prior to dewatering with a mixture of a water-soluble high-molecular weight nonionic polyacrylamide and a water-soluble salt containing ferric ions. The waste sludge treated according to this process dewaters better than sludge treated using prior art methods.

Chemical conditioning with a mixture of the polymer and ferric-containing salt yields a sludge which dewaters better than a sludge treated with the ferric-containing salt alone or the polymer alone. It is also better than a sludge treated with the polymer followed by the ferric salt or the ferric salt followed by the polymer. It has been found that the use of the mixture gives an unexpected synergistic result compared to the other methods of addition.

The method of conditioning the sludges comprises adding to the sludge prior to dewatering from 300 to 3,000 ppm of a mixture of the polymer/ferric salt. The mixture is comprised of from 0.05% by weight to about 5.0% by weight of the polymer and the remainder is the ferric-containing salt.

Addition of Ash and Lime to Sludge

G.P. Baumann and G.E.J. Thomas; U.S. Patent 3,342,731; September 19, 1967 describe a method of dewatering fresh or digested sludges or centrate resulting from centrifuging fresh or digested sludge.

Sludges may also be dewatered by filtering, centrifuging, or by a combination of the two methods. A centrifuge and filter can be arranged so that centrate resulting from centrifuging sludge is thereafter filtered. Dewatered sludge is incinerated. In this method diatomaceous earths may be added in the mechanical and biological clarification step, during sludge thickening, ahead of the sludge filtration or centrifugation, or ahead of filtration of the centrate to aid in dewatering the sludge.

Diatomaceous earths can be replaced with ash from the incineration of the centrifuge and filter cakes.

In a further development of this method it has been proposed to add ash resulting from the incineration of sludge to digested sludge or fresh sludge or to centrate resulting from centrifuging raw or digested sludge as a filter aid. In this way no additional chemicals are necessary. The filter and centrifuge cakes may be incinerated, either separately or together, depending on the particular installation, to provide the ash necessary.

Advantages of this method are that operation is inexpensive and costly supplementary chemicals are not necessary. There are certain disadvantages, however, for when it becomes necessary to add large quantities of ash (4 to 8 grams of ash per gram solids), a large quantity of heat is required to incinerate the filter and centrifuge cakes because the ash added to the sludge absorbs moisture which must be evaporated.

This method using the filter-centrifuge combination is economically feasible for sludge quantities above 50 tons per day, such as occurs from a city with 100,000 inhabitants. For smaller sludge quantities the digested sludge or the fresh sludge may be subjected directly to the ash treatment and then filtered.

It has been found in this process that the amount of ash required can be lowered substantially if lime is added, in addition to the ash, prior to vacuum or pressure filtration of the sludge or the centrate. Small line quantities can, at the same filterability, replace large quantities of ash; up to 50 to 75% of the ash requirement may be replaced by lime. These high filterabilities are not attainable through the addition of small quantities of lime alone.

For example, when filtering sludge with addition of only ash, it is necessary to add 4 to 8 grams of ash per gram of solid matter. When lime is added, less than 1 gram is required along with 1 to 4 grams of ash in order to obtain the same filterability. The ash acts as a filter aid in two ways: one, it forms a supporting porous matrix for the amorphous sludge solids, and, two, it provides reactive lime for flocculation of the sludge. Lime is generally required in amounts of 20 to 50% of the weight of the dry sludge solids for the formation of a supporting matrix.

Simultaneous use of ash and lime lessens the water quantity retained by the filter cake and lowers operational costs. With this process sludge filtration can be carried out economically even in small clarification installations. The process differs from sludge filtration using only ash in that the lime causes a flocculation reaction which is supported by finely divided ash. The ash aids in this reaction as well as aiding in development of this supporting matrix.

According to the process, employment of lime and ash for improving filterability is suitable for dewatering both fresh and digested sludges. The ash required may be provided by incinerating the solids residue from the centrifuge and the filter or can be obtained from an outside source. About 65 to 123 pounds of lime along with ash are required to treat 1,000 gallons of sludge, the sludge being approximately 10% dry solids.

During incineration of the centrifuge or filter cake after treatment with ash and lime a part of the lime in the cake remains or again becomes active; this part may be reused by mixing it with fresh lime and ash to be added to the sludge. In this way, the additional lime required in a continuous operation is less than during initial start-up of the installation.

Example 1: A digested sludge, after the addition of lime and ash obtained from incineration of digested sludge, was filtered. The quantitative ratio of lime and ash added was varied. The weight data relate to 1 g of dry solids in the digested sludge. The filter yield is stated in gallons per square foot per hour.

Addition	Filter Output
4 g. ash and 0.4 g. lime	28.8 gal./sq. ft./hr.
2 g. ash and 0.8 g. lime	28.8 gal./sq. ft./hr.

Example 2: A digested sludge was in each case filtered after the addition of ash or lime and finally after the addition of ash and lime. The weight data again relate to 1 g of dry solids in the digested sludge. The filter output is defined as in Example 1.

Addition	Filter Output
8 g. ash	16.9 gal./sq. ft./hr.
1.6 g. lime	1.18 gal./sq. ft./hr.
4 g. ash and 0.4 g. lime	19.5 gal./sq. ft./hr.

It should be observed that when the ash and lime were both added, the ash required was reduced to one-half with a corresponding change in filter output.

Amino Condensation Resins and Acrylamide Polymers

S. Mogelnicki and E.M. Gatza; U.S. Patent 3,409,546; November 5, 1968; assigned to The Dow Chemical Company have found that a mixture of a water-soluble resinous amino-condensation polymer and a small amount of a high molecular weight, cationic vinyl-addition polymer, the two polymers being present in a limited range of relative proportions, is unexpectedly efficient, especially as applied from relatively concentrated applicating solutions, for conditioning raw sewage sludges for dewatering. For best results from 0.02 to 0.3 part of the cationic vinyl-addition polymer, preferably from 0.1 to 0.2 part, are used with each part by weight of amine-condensation polymer.

Application of the polymeric blend to sludges to be dewatered is accomplished by first preparing an applicating solution of the polymeric blend which may contain from as little as 0.01% by weight up to as much as 5% by weight dissolved polymeric solids. Since large amounts of dilution water are not required, the polymer applicating solution preferably contains from 0.1 to 1.5% by weight dissolved polymer solids. The higher concentrations of polymer permitted in applicating solutions keep to a minimum the amount of dilution water required for a successful dewatering.

The aqueous suspension of organic solids, i.e., sludge, is treated with a small amount of the polymer applicating solution to provide a predetermined, desired polymer dosage, based on the weight of suspended solids. Normally, suitable conditioning of the sludges is achieved with doses of from 2 to 6 pounds of the polymer mixture per ton of sludge solids treated.

With the application of the polymer solution, the sludge is simultaneously subjected to mild agitation. Such agitation is accomplished by any convenient means such as large slowly rotating paddles, effervescing gases, cascading streams and/or turbulent flow in conduits. The agitation is designed to provide homogeneous distribution of the applicating solution throughout the aqueous sludge. After achieving suitable distribution of the solution throughout the sludge it is subjected to dewatering which may involve compaction, filtration, flotation or centrifugation. The overall process enables dewatering of the aqueous suspension with substantially reduced doses of amino-condensation polymers.

Example: A raw sewage sludge containing 7.3% by weight solids was dewatered by filtration. The raw sludge was a composite of sludge produced as a result of the primary settling of raw sewage to which was added a large proportion of aerobic trickling filter humus. The filtration was performed in a 12.5 cm Buechner funnel containing two layers of No. 1 Whatman filter paper. The funnel was seated in a vacuum flask to which was applied a standard vacuum.

Aliquots in the amount of 200 mm of the sewage sludge were treated with given amounts of a cationic polymer composition, in the form of a dilute aqueous solution, containing 0.5% by weight polymer solids. The treated sludge was then agitated by pouring the sludge from container to container 8 times. The resulting conditioned sludge was then poured into the funnel and the vacuum applied. The volume of filtrate was measured at 10, 20, 30, 60 and 120 second intervals after application of the vacuum. After 2 minutes the vacuum was discontinued and the appearance of the filter cake qualitatively determined. The results of this experiment utilizing polymer mixtures of the process as well as the polymers individually are reported on the following page.

The low molecular weight, cationic, amino-condensation resin used was a reaction product of a polyalkylene polyamine and ethylene dichloride, in the form of an aqueous solution, termed polymer A. It was characterized by a viscosity in 37% by weight aqueous solution of 2 x 10^4 cp at 25°C.

The high molecular weight vinyl-addition polymer used was a water-soluble homopolymer of acrylamide having a molecular weight of 2 million, in which 35% of the amide groups were substituted on the nitrogen atom with the dimethylaminomethyl moiety. This polymer is termed polymer B. A 0.5% by weight aqueous solution of this polymer was characterized by a viscosity of 16 cp as measured with an Ostwald viscosimeter at 25°C.

From the data on the following page, best conditioning of the sludge was achieved in Runs 8, 9 and 12. Considering these runs, it is evident that a dose of 0.75 part by weight of the high molecular weight, vinyl-addition polymer in a mixture with only 4.17 parts by weight of the condensation polymer was as effective as 9.3 parts by weight of the latter polymer used alone.

The addition polymer was ineffective as applied from a 0.5% aqueous solution. It will be noted that as more of the high molecular polymer was added beyond a certain point, the dewatering rate decreased significantly due to overdosing.

Run No.	Polymer Composition	Dose (lb/ton)*	Filtrate Volume (cc) 10 sec	20 sec	30 sec	60 sec	120 sec	Cake Quality	Residual Water (percent)
1	Polymer B	1.0	5	9	11	18	26	None	93
2	Polymer B	2.0	10	18	24	35	54	None	92
3	Polymer B	3.0	12	20	25	34	58	None	91
4	Polymer B	4.0	6	11	15	22	39	None	92
5	Polymer B	5.0	5	------	10	20	32	None	93
6	Polymer A	5.6	45	68	86	118	144	Wet	83
7	Polymer A	7.4	60	91	112	142	159	Moist	79
8	Polymer A	9.3	65	104	126	151**	162	Dry	78
9	Polymer A	11.1	88	124	142**	156	160	Dry	79
10	Polymer B	0.25	48	69	88	125	151	Moist	81
	Polymer A	4.34							
11	Polymer B	0.5	58	84	105	138	158	Moist	79
	Polymer A	4.25							
12	Polymer B	0.75	70	100	118	148	162**	Dry	78
	Polymer A	4.17							
13	Polymer B	1.0	54	74	90	120	143	Wet	83
	Polymer A	4.07							
14	Polymer B	1.5	25	40	51	75	106	Very wet	88
	Polymer A	3.88							

*The doses are the amounts of polymer per ton of sludge solids, on a dry basis.
**The vacuum broke.

Gravity Filtration Process

In a process described by *B.L. Goodman, R.M. Appleberry, J.W. Struewing and F.G. Weis; U.S. Patent 3,531,404; September 29, 1970; assigned to Union Tank Car Company* sludge is first flocculated with a coagulant and then gravity fed onto a primary, large mesh, moving screen, which is flexed intermediate its ends. Water drains from the sludges through the screen, its flow enhanced by flexing of the sludge cake which is forming.

Partially dewatered sludge is then scraped onto a secondary large mesh, moving screen, which carries the sludge between a series of pairs of compression rollers, providing successively greater compression to further dewater the sludge.

Cationic Polyalkylenepolyamines with Gravity Dewatering

A process described by *S. Kelman and C.P. Priesing; U.S. Patent 3,252,900; May 24, 1966; assigned to The Dow Chemical Company* concerns dewatering sewage sludges over filtering substrata utilizing the pressure as the result of increased sludge depth as a dewatering aid.

The process is based upon the discoveries that sewage sludges, suitably treated with a water-soluble, cationic, organic polymer containing a plurality of hydrophilic nitrogenous groups, can be rapidly dewatered with the aid of gravity over a filtering substrate and that the filter yield, while initially undergoing a decrease with increases in the depth of loading of the treated sludge, is surprisingly reversed at a critical loading which occurs at a depth of greater than two feet.

The method for dewatering a sewage sludge comprises the steps of treating the sewage sludge with a water-soluble, cationic, organic polymer containing a plurality of hydrophilic cationic groups, preferably hydrophilic nitrogenous groups, and loading the polymer-treated sludge on a filtering substrate, readily permeable to the passage of water at a depth greater than about two feet and at least the depth at which a continuous plot of the filter yield of the sludge solids on the filtering substrate against the depth to which the treated sludge is originally loaded thereon first achieves a slope of zero.

As used here, the terminology "filter yield" refers to the pounds of dried sludge solids obtained per square foot of the filtering substrate per hour. The dry solids content of a sludge is determined by procuring a weighed aliquot of a homogeneous sample of the polymer-treated sludge, drying the aliquot of the polymer-treated sludge at 110°C for four hours and thereafter weighing the residual solids. This weight that is obtained is taken to be the dry solids content of the initial aliquot.

A preferred class of cationic polymeric flocculants is represented by water-soluble polyethyleneimines of high molecular weight, N-alkyl substituted polyethyleneimines and the mineral acid and quaternary ammonium salts thereof. Some polyethyleneimines suitable for use in the process can be represented by the formula

$$\mathrm{H\overset{R}{\overset{|}{N}}CH_2CH_2\overset{R}{\overset{|}{N}}}-\left(\mathrm{CH_2CH_2\overset{R}{\overset{|}{N}}}\right)_n-\mathrm{H}$$

where R is hydrogen or methyl and n has a value of at least 400, preferably of at least 2,000. Desirable polyethyleneimines can be characterized by the viscosity of aqueous solutions. Thus, water-soluble polyethyleneimines having solution viscosities of at least 0.8 cs, and preferably at least 2 cs, for one percent by weight solutions in water, are preferred cationic polymers.

In a preferred mode of operation, which is illustrative of the process, a series of drying beds were prepared, the filtering substrate consisting of a layer of sand approximately 9 inches deep underlain with approximately 2 feet of gravel. A digested raw sewage sludge obtained from the sewage plant of a small midwestern city was treated with a flocculant grade, water-soluble, cationic polyalkylenepolyamine.

To treat the sewage, a solution containing 0.15% by weight of the active polymer solids was prepared. This polymer solution was added to a running cement mixer containing a given volume of the digested sewage sludge to provide a polymer dose based on dry sludge solids of 15 pounds per ton. Mixing was completed by operating the cement mixer for an additional 30 seconds after the addition of the polymer solution. Thereafter, the polymer-treated sewage sludge is poured into a partitioned segment of the above described drying bed in an amount sufficient to provide a layer of the treated sludge at a predetermined depth.

Subsequently, the loaded filtering beds were sampled several times during the first day of operation and thereafter once a day until effectively dewatered, i.e., having 30% by weight dry solids. The percent dry solids in the sample, which were uniform composite samples taken from a representative cross section of the sludge layer, were plotted against the depth to which the treated sludge was initially loaded on the filtering bed.

This plot is shown in Figure 3.1a. It will be observed that after a depth of 3 feet, the drying rate at the end of two days was no longer decreasing as the loading depth was increased. This was a surprising result inasmuch as one would ordinarily expect continuous decreases in the drying rate with increasing loading depths due to the likely propensity of the sludge to blind the filter in proportion to the total mass above the filtering substrate.

Upon further evaluation of the experimental data, a comparison was obtained between conventional static dewatering rates of an untreated sludge and a polymer-treated sludge in accordance with the process. The filter yield was computed in pounds of dry solids per square foot per hour.

These results are plotted in Figure 3.1b. While essentially straight lines have been used to delineate the plot, a few more points in the middle region of the plot would have defined a rounding of the curve where the downturn on filter yield was reversed toward increasing filter yields at the critical loading depth.

For the particular sludge involved in these experiments, as treated with the cationic polymer described above, the critical depth at which a reversal in filter yield occurs, i.e., the point of tangency and zero slope, will be observed to be at about 3 feet.

FIGURE 3.1: DEWATERING SEWAGE SLUDGE BY GRAVITY

a.

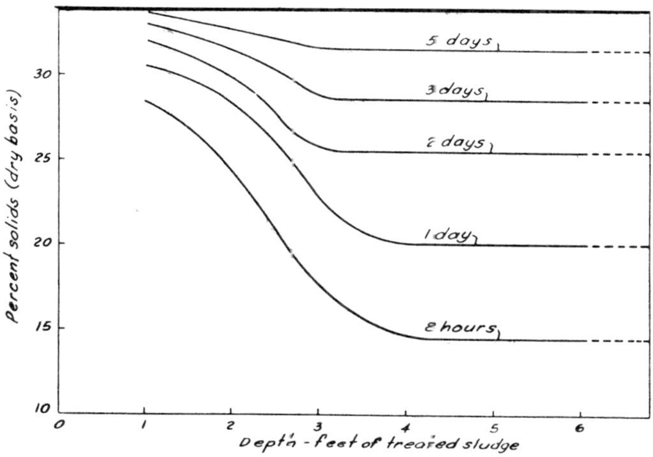

Filter Bed Performance

b.

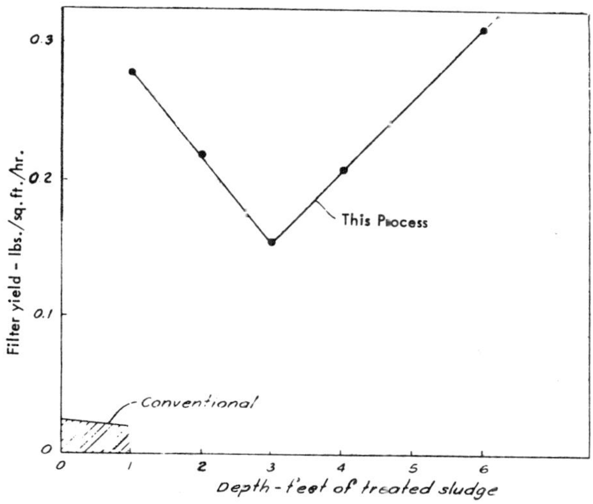

Comparison of Cationic Polymer Treated Sludge and Untreated Sludge

Source: U.S. Patent 3,252,900

Interrupted Flow

K.S. Arvanitakis; U.S. Patent 3,705,649; December 12, 1972 describes a method of and apparatus for collecting and removing contaminants from a filtration system. It involves passing unclarified contaminant bearing a liquid through a filter apparatus to collect the contaminants and discharge the clarified liquid.

At a predetermined time, the discharge from the filtration system to the filter apparatus is interrupted and the contaminants removed from the filter apparatus to the unclarified liquid where they are collected and conveyed out of the apparatus. The collected contaminants are partially solidified by exposure to heat and conveyed from the system for disposal.

CENTRIFUGAL

Solid Bowl Centrifuge

A process described by *O.E. Albertson; U.S. Patent 3,248,323; April 26, 1966; assigned to Dorr-Oliver Incorporated* provides a treatment system where dilute raw sludge, for example sludge supplied from a clarifier operation, is subjected to a continuous sedimentation thickening operation in a thickener tank, so controlled that a bed of substantial depth of thickened sludge is maintainable therein providing a reservoir of thickened sludge to be supplied to the digester contents which latter preferably is kept in a state of mechanically induced mixing and agitation.

The loading of the digester is such that incompletely digested sludge is withdrawn from the body of digesting sludge. This sludge from the digester is subjected to a treatment in a solid bowl type centrifugal machine capable of effecting the separation into a moist cake fraction and a liquid overflow fraction.

In the operation of this machine, the cake material discharging at the narrow end of the bowl contains substantially none of the undigested matter and is therefore readily disposable, while the overflow liquor containing substantially all of the digestible organic or gasifiable matter including the fats discharges from the wide end of the bowl. This latter overflow fraction from the centrifuge is then subjected to a dewatering operation, and the undigested solids along with the residual water are then returned to the body of digesting sludge for conversion into gas.

This dewatering operation is carried out preferably by way of feeding the undigested solids directly into the sludge bed in the thickening tank, causing the solids to be entrapped in the thickened sludge while separated liquor from those solids reaches the thickener overflow together with the separated liquor resulting from the thickening operation.

The undigested solids contained in the centrifuge overflow are thus recirculated into the digester through the thickener whereby the digester loading and the digestion efficiency are correspondingly increased to a point where the optimum balanced condition in the digester operation is attainable. Hence, in this digester circuit (that is the digester operating in circuit with the centrifuge and the sludge bed in a settling tank) the gas-producing efficiency of the available digester volume is thus substantially increased, even as a fully digested as well as highly concentrated sludge is delivered for disposal from the centrifuge, while a safe degree of alkalinity concentration is maintained in the digester.

In this way, higher digester efficiency and utilization of the available digester tank volume, are attainable at practically no extra cost by way of operating the digester circuit in the manner above set forth. As in Figure 3.2a the process may be used in a sewage treatment plant comprising a conventional clarification treatment system for producing raw sludge which may contain primary and secondary clarification sludge in mixture, adapted for treatment in an anaerobic digestion system.

FIGURE 3.2: TREATMENT PROCESS EMPLOYING SOLID BOWL CENTRIFUGE

a.

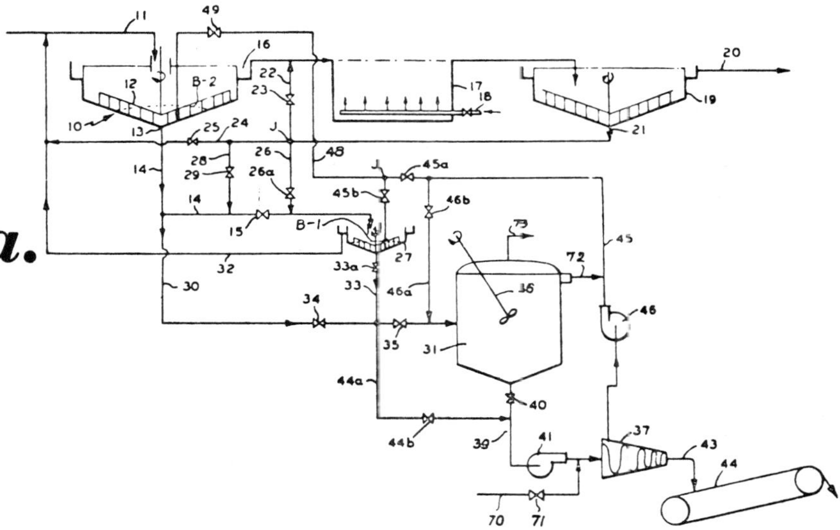

Process Flowsheet

b.

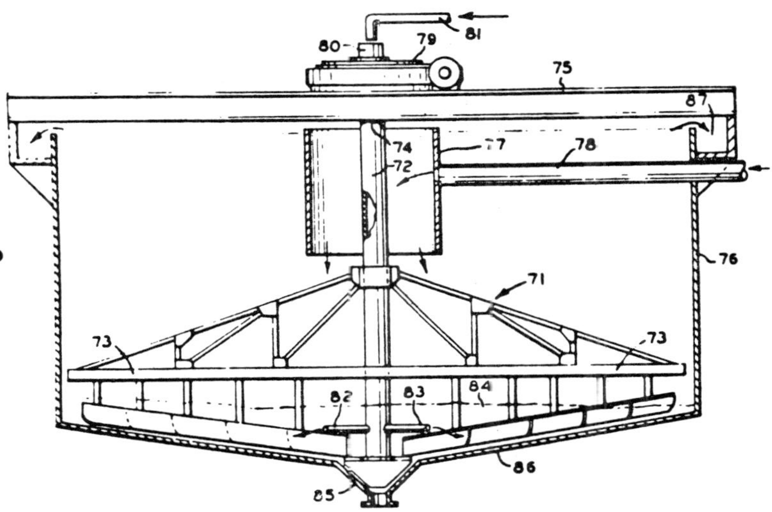

Rotary Rake Apparatus

(continued)

FIGURE 3.2: (continued)

c.

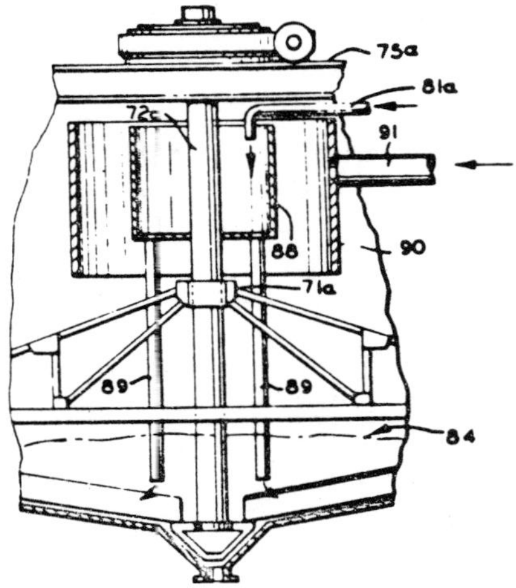

Rotary Rake Apparatus

d.

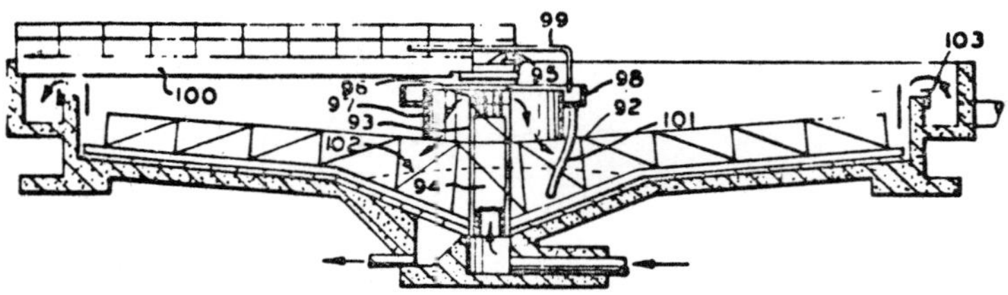

Rotary Rake Apparatus

e.

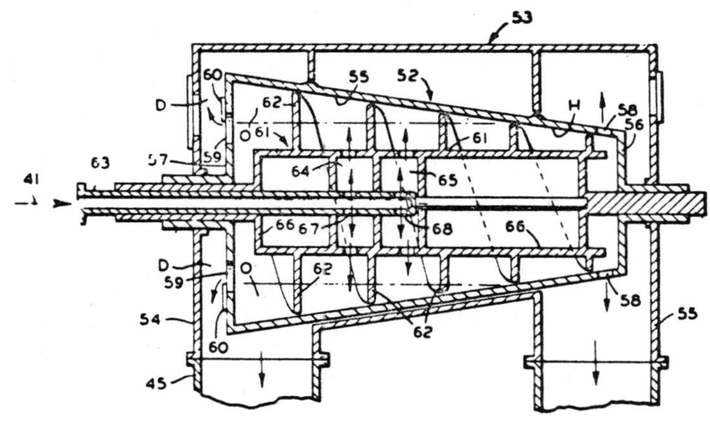

Solid Bowl Type Centrifugal Machine

Source: U.S. Patent 3,248,323

Accordingly, the clarification treatment system in this example comprises a primary clarifier tank unit **10** supplied with raw sewage as indicated by a supply conduit **11**, and equipped with the familiar rotary sediment raking structure **12** for conveying primary sludge or sediment to a bottom outlet **13** connected to a sludge discharge conduit **14** provided with control valve **15**.

The overflow launder **16** of this tank receives primary effluent containing suspended solids that will not settle in the primary tank, but may be subject to further clarification treatment. The primary effluent may be treated biologically, that is by aeration for example in an aeration tank **17** having a controlled air supply indicated at **18**.

The resulting so-called activated sludge containing the previously nonsettleable solids that are converted into settleable biologic floc structures, is transferred from the aeration tank to a secondary clarifier tank **19** which is equipped substantially the same as the primary clarifier tank, and delivering the secondary purified effluent **20** from the overflow, while secondary sludge is being withdrawn from the bottom zone as through a discharge conduit **21** leading to a juncture **J**.

From juncture **J** a branch conduit **22** provided with control valve **23** allows for recirculation of secondary sludge into the aeration tank at a controlled rate. Another branch conduit **24** provided with control valve **25** leading from juncture **J** is operable to divert a portion of the secondary sludge back into the primary clarifier tank by way of the raw sludge influent conduit **11**.

Another branch conduit **26** leading from juncture **J** provided with control valve **26a** joins the primary sludge discharge conduit **14** downstream of valve **15**, whereby mixed primary and secondary sludge may be delivered into a tank **27** which may be in the nature of a thickening tank operated in accordance with U.S. Patent 2,850,449.

The thickened sludge passes to a digestion system including a digester **31**. Another branch conduit **28** from line **24** having a control valve **29** is shown to join the primary sludge discharge conduit **14** upstream of valve **15** whereby a portion of the mixture of primary and secondary sludge may bypass the thickening tank **27** via conduit **30** for delivery to the digestion system or via line **44a** to the solid bowl machine. Effluent liquor from the thickening tank **27** may be disposed of as by conduit **32** leading back into the primary clarifier tank.

A deep sludge bed of controlled depth maintained in the thickener tank **27** and controlled in accordance with U.S. Patent 2,850,449 may serve as a reservoir for supplying the feeding sludge to the digester **31** through underflow discharge connection **33** having a control valve **33a**.

This underflow connection joins the aforementioned conduit **30** at a point between control valves **34** and **35**. The raw sludge fed to the digester **31** may be a mixture of primary and secondary clarifier sludge, which may have been subjected to a concentrating or thickening treatment such as above indicated in tank **27**.

The thickener sludge mixture is supplied preferably continuously or at short intervals to the digester **31** which in turn is operated in the special manner of this process affording significantly increased volatile solids loading rates and greatly improved digester tank utilization as will be furthermore set forth. The body of digesting sludge in the digester is kept in a state of enforced agitation as indicated by the agitator **36**. The digester **31** operates in circuit with a solid bowl type centrifugal machine **37** described below as shown in Figure 3.2e.

The digester **31** has a sludge discharge conduit **39** provided with control valve **40** and has a pump **41** operable to supply incompletely digested sludge to the solid bowl machine **37** at a rate controlled in relation to the rate of sludge feed into the digester tank. The operation of the solid bowl centrifugal machine **37** is conducted so as to selectively deliver a largely inert or undigestible solids fraction in the nature of a highly concentrated cake material having a solids concentration in the order of 40%, and a dilute fraction or suspension carrying most of the undigested gasifiable solids present in the incompletely digested sludge.

The concentrate or cake fraction representing well digested sludge discharging at **43** may be disposed of in any suitable manner for example as indicated by the conveyor **44**. A conduit **44a** having a control valve **44b** bypasses the digester enabling excess raw sludge to be supplied directly to the solid bowl machine **37**. The overflowing dilute suspension of undigested matter, however, is transferred from the solid bowl machine for further separation treatment to the thickener **27** via a transfer conduit **45** and control valves **45a** and **45b** by a pump **46**.

It will be noted that this transfer conduit delivers into the bottom zone or sludge bed **B-1** of the thickener where the solids thus become trapped, while liquor separating from these solids reaches the thickener overflow together with the overflow resulting from the thickener operation.

A bypass flow connection **46a** leads from line **45** to the digester via a control valve **46b**, whereby a portion of the centrifuge overflow may be diverted from the thickener for delivery directly into the digester for purposes to be described below. At juncture **J** the line **45** has a branch line **48** leading through a control valve **49** into a sludge bed **B-2** maintainable in the cone bottom of primary clarifier **10**. In this instance the clarifier is utilized for getting the transferred undigested solids trapped and retained in the bed of sludge while liquor separates therefrom and reaches the clarifier overflow.

A controllable supply of flocculating chemical **47** is shown for the purpose of introducing such a conditioning agent into the digester sludge entering the centrifuge. This will cause extremely fine digested matter which may be present in excess to form agglomerates or flocs large enough to be contained in the cake material leaving the centrifuge for disposal.

Escape of such fine material into the overflow of the machine, if not prevented may produce undue build-up of such fines in the digester circuit, which in turn may reduce digestion efficiency, and also by escaping with the thickener overflow reduce the effectiveness of the clarification operation.

The so-called solid bowl centrifugal machine schematically illustrated in Figure 3.2e, has a rotor structure **52** rotating in a housing **53**. The housing has a discharge neck **54** located at the wide end of the rotor structure and connected to the transfer conduit **45** for delivering the overflow fraction containing the undigested gasifiable matter to be subjected to further separation treatment. The separated solids or stabilized cake fraction **43** (see Figure 3.2a) containing mostly undigestible solids leaves the machine through a discharge neck for further disposal in any suitable manner such as indicated by the conveyor **44.**

The rotor structure **52** comprises a bowl **55** provided with an end closure plate **56** at the narrow end and an end closure plate **57** at the opposite or wide end thereof. The separated solids or cake fraction representing the digested sludge discharges centrifugally from the bowl through openings **58** provided adjacent to the narrow end of the bowl.

The opposite end closure plate **57** has overflow openings **59** associated with adjustable weir plates **60** determining the overflow level of the annular or pool body **O** of liquid centrifugal separation in the bowl. Arrows **D** at the wide end indicate delivery over these wier plates of the liquid fraction or suspension carrying the undigested matter to be further subjected to concentration in the thickener. A part or all of this liquid fraction may be returned directly to the digester through bypass line **46a** and control valve **46b.**

Within the rotor bowl of this machine there is rotatably mounted a conveyor element **61** conforming to the inner contour of the bowl and adapted to be rotated at a differential speed relative to the rotation of the bowl, so that the spirally shaped flights **62** thereof will move the centrifugally separated solids comprising most of the digested organics, out of the liquid body or pool **O** to emergence onto the narrow end portion, the so-called beach portion **H** of the bowl, finally causing the material to discharge centrifugally through the openings **58** for disposal in the form of relatively dry cake material.

The sludge from the digester enters this machine through an axially arranged stationary feed tube **63** introducing feed sludge, for example, into receiving chambers **64** and **65** formed in the hollow hub portion **66** of the flight conveyor element, and then through openings **67** and **68** into the annular liquid body **O** undergoing centrifugal separation in the bowl.

As applied in the digester operating cycle of this process, the operation of the solid bowl centrifugal machine receiving the incompletely digested sludge from the digester, may be controlled so as to remove varying amounts of the digestible solids fraction from the incompletely digested sludge. This can be done by adjusting the rate of sludge feed into the machine as by adjusting the pumping rate from the digester. As the feed rate is increased the percentage of heavy undigestible matter increases in the discharging cake, while displacing the more digestible volatile matter into the overflow for return into the digester.

Such selective displacement is effected since the undigestible volatile matter, for example, cellulose fibers, has a high specific gravity facilitating centrifugal separation, whereas the digestible fats and greases being attached to or occluded in other substances have resulting specific gravities slightly heavier than water.

Therefore, substantially all of these fats and greases may be caused to appear in the overflow of this solid bowl machine, the overflow then to be subjected to the above described dewatering operation through the sludge bed in a settling tank.

It is these greases that have the most objectionable characteristics in the digested sludge, and their potential loss accounts for a large loss of digester gas of high quality, since one pound of grease in digestion will produce 23 cubic feet of gas at 72% CH_4. Moreover, this material normally retards the digestion process as a whole because its destruction or gasification proceeds so much slower than that of the other main components, the carbohydrates and the proteins, yet will produce the bulk of the gas developing after the initial 6 to 10 days of digester detention time. But in the digester operating cycle of this process, when properly balanced, the required length of grease digestion time can be selectively increased.

The process provides for controlling the amount of suspended fines that tend to recycle with the centrifuge overflow into the thickener and into the digester. If uncontrolled the amount of fines may build up in the digester circuit so that eventually they would affect the overall treatment plant efficiency when escaping through the thickener overflow and reaching the clarifier overflow. Such buildup of fines in the circuit also would impair the digestion efficiency.

Even though the solid bowl centrifugal machine has available such control variables as regulating the pool volume in the bowl by the location of the overflow, the bowl speed, and the configuration of the bowl, there are types of feed sludges where none of the foregoing variables will provide sufficient control of the quantity of recycling fines.

In such cases, even the most significant and most effective factor available for controlling the solids fractionation between cake material and overflow, that is varying or lowering the rate of feed sludge entering the centrifuge, may be inadequate or uneconomical in view of the design capacity of the machine.

In such cases of potential fines buildup the process provides for the addition of coagulating chemicals, for example alum or organic polyelectrolytes to the feed sludge entering the centrifuge, as indicated by the supply line **70** including control means **71**. The resulting coagulation of the digested fines into larger and heavier floc-structures will cause them to be incorporated centrifugally into the cake material even with sludge feed rates corresponding to design capacity.

In this way the recovery of such solids or fines in the cake material may be increased to any extent desired. The chemical dosing may be applied intermittently or continuously, depending upon the characteristics of the treatment plant. Intermittent application of the dosing chemical means that dosing is applied only when the amount of solids in the centrifuge overflow exceeds an acceptable amount which in turn is dependent upon the design and operating characteristics of the respective treatment system as a whole. The process further provides for delivering at least a portion of the centrifuge overflow directly to the digester at a controllable rate, bypassing the thickener.

This controllable bypass arrangement makes it possible to operate the centrifuge intermittently, for example only 1, 2 or 3 days per week, even though the digester may operate on a 6 day or 7 day schedule. Hence, the volume of digester sludge to be handled by the centrifuge during its operating period may greatly exceed the volume of raw sludge normally pumped to the digester. However, that differential is compensated for through the bypass arrangement **46a** and **46b** by causing the excess of overflow volume from the centrifuge to be returned directly to the digester, thereby maintaining the sludge level in the digester at its required elevation. During the interim periods between operations of the centrifuge, overflow **72** from the digester may be returned via lines **45** and **48** to the raw sludge clarification station. The gas resulting from the operation of the digester is indicated at **73.**

Furthermore, such a bypass connection between the centrifuge and the digester, allows for a mode of operation whereby the digester contents may be purged quickly of an undesirable accumulation of relatively coarser undigested or undigestible material such as crude cellulose or the like occurring in substantial quantity in the raw sludge fed to the digester.

This purging of the digester contents is accomplished by operating the centrifuge alternatively for long periods in circuit with the thickener and with the digester, and for short periods in circuit only with the digester. During the long operating periods when nondigestible cellulosic matter or the like may accumulate in the digester contents, the centrifuge is operated in the normal manner, that is with a feed rate as close as practical to its design capacity, whereby a major portion of the feed solids substantially free of digestible matter is recovered in the discharging cake material, while the remainder comprising largely the undigested matter in the overflow is recycled into the sludge bed of the thickener.

The aforementioned coagulating treatment with a chemical may or may not be required during these periods of normal operation. However, during the intervening short periods the centrifuge is operated at a feed rate which greatly exceeds the rate necessary for removing the organic solids. Accordingly, with quantities of the fines digested or inert matter thus being displaced into the overflow of the centrifuge, the solids recovery in the cake material will be only about 20 to 50% of normal, with the cake material containing mainly the solids of relatively high specific gravity, namely those of high ash content and of cellulosic character.

Such periodical removal or purging in the digestion process, is effected without impairing the required degree of alkalinity in the digester contents. During the relatively short purging periods, the supply of raw sludge from the thickener to the digester may be continued, or it may be discontinued while allowing thickened sludge to be stored in the thickener.

Furthermore, by thus temporarily recycling the additional fines from the centrifuge directly into the digester, the thickener and/or the clarification station are protected against loads that might be detrimental to their efficiency. Figures 3.2b to 3.2d illustrate a practical manner of introducing the centrifuge overflow suspension into the sludge bed of a sedimentation tank or thickener for effecting the entrapment and retention of the solids, by way of induction conduit means incorporated in the rotary rake structure.

In Figure 3.2b, such induction means are embodied in a type of settling tank which may be the thickener **27** (see Figure 3.2a) where a rotary rake structure **71** has a vertical hollow shaft **72** and rake arms **73** connected therewith. The top end portion **74** of the shaft is mounted for rotation upon a horizontal beam construction **75** extending transversely of a tank **76** which has a feedwell **77** supplied by a feed conduit **78** (corresponding to conduit **14** in Figure 3.2a). The vertical shaft is connected to a conventional drive head **79** whereby it is supported from the beam construction. The hollow shaft has an upper exposed end **80** into which a supply or transfer pipe **81** delivers the overflow suspension from the centrifuge for discharge at points **82** and **83**, for example, sufficiently deep in the sludge bed **84** for which there is a discharge **85** provided in the low point of the conical tank bottom **86**. In this way the solids of the suspension get entrapped in the sludge while liquid therefrom finds its way upwardly to the overflow **87**.

According to Figure 3.2c, a rotary rake structure **71a** which may be similar to the one in Figure 3.2a, is provided with induction means for the centrifuge overflow suspension, comprising a special feedbox **88** surrounding the shaft **72a** and located below the transverse beam construction **75a**. A transfer pipe **81a** delivers the suspension from the centrifuge into this feedbox for discharge through down pipes **89** terminating at a suitable depth within the sludge bed **84a**. A stationary feedwell cylinder **90** or the like surrounds the rotating feedwell **88**, and is supplied by a feed conduit **91** (corresponding to feed conduit **14** in Figure 3.2a).

According to Figure 3.2d, the induction means are embodied in a type of settling tank which may be the clarifier **10** (see Figure 3.2a), where a rotary rake structure **92** has a central cage construction **93** surrounding a pier **94** and mounted for rotation of the rake structure on the pier. A conventional drive mechanism is also mounted on the pier, indicated at **95** for rotating the rake structure on the pier.

The regular feed supply to this settling tank is shown to be by way of a supply duct underneath the tank bottom and up through the hollow pier for discharge through passages **96** for delivery into a feed well **97** surrounding the cage construction and unitary with the rake structure to rotate.

Surrounding this feedwell **97** and in turn unitary therewith is an annular feed trough **98** to which a transfer pipe **99** (corresponding to line **48** in Figure 3.2a) supported upon a stationary radial bridge construction **100**, delivers the centrifuge overflow suspension for discharge through a down pipe or pipes **101** terminating sufficiently deep within the sludge bed **102**. In this way, the solids of the suspension become entrapped in the sludge while liquid separates finding its way upwardly to the overflow **103**.

In summary, it will be seen that the digester operating circuit as a net result delivers a well digested sludge, even though only incompletely digested sludge is discharged from the highly loaded digester itself, and that these results are attainable by selectively increasing the solids detention time relative to the liquor detention time in the digester.

More in particular, this is accomplished by having the digester operating in circuit with the solid bowl centrifuge and with the thickener or settling tank, whereby the centrifuge discriminates between the digested sludge solids contained in a disposable cake material and the undigested matter contained in the overflow suspension returning to the thickener, and whereby the sludge bed in the thickener traps and retains the solids from the centrifuge overflow suspension for return into the digester together with the thickened sludge for conversion into gas.

Among the advantages resulting from the foregoing improvements are increased utilization of the available digester tank volume relative to the conventional, lower heat requirements for the digester contents, lower construction costs, reduction of labor and space requirements along with the elimination of sludge filters and/or sludge drying beds for digester sludge.

In related work *O.E. Alberston; U.S. Patent 3,226,317; December 28, 1965; assigned to Dorr-Oliver Incorporated* describes a digestion system, the operation of which is so controlled that the effective detention time in the digester of the gasifiable organic substance is greatly increased relative to the detention of the liquor, that is, the solids detention time is rendered independent of the liquid detention time.

More particularly, the slow digesting raw sludge components, such as the grease and fats, are thus selectively given a longer detention time needed for their eventual digestion and conversion into gas. The net result is that a significantly higher organics loading rate with greater bacterial feed concentration and greater bacterial density is maintainable in the digesting mass along with an adequate alkalinity concentration in the liquor necessary for digestion. At the same time there is a maximum recovery of gas in a minimum of digestion space.

C.L. Amero; U.S. Patent 3,468,794; September 23, 1969; assigned to Bird Machine Company describes a process for improving sludge digestion in anaerobic digesters. Sludge is withdrawn from a digester, centrifuged to remove a fraction of the mother liquor low in solids and acidic, the remainder being returned for further digestion usually to the same digester. By replacing the removed liquor with less acid liquor and by suitably regulating the process undesired acid build-up in the digester is minimized.

By selectively utilizing as the source of replacement water raw sludge or water essentially free of solids, the solids consistency in the digester is increased or maintained as desired. By a like centrifuging operation the scum layer which forms at the top of the digester contents is broken down and returned as a manageable slurry.

Centrifugal Separation Combined with Lagooning

J.M. Valdespino; U.S. Patent 3,625,883; December 7, 1971 describes a method for the advanced treatment of liquid waste. The liquid waste is first fed to a comminutor for chopping and mixing the solid waste with the liquid to give a more homogeneous liquid waste which is then fed to a surging tank for providing an intermittent flow into a centrifuging step for centrifuging the waste through a filter medium such as sand, or diatomaceous earth for separating the solids from the liquid effluent.

Burning of the separated solids may take place within or without the centrifuging filter and the separated effluent fed to an oxidation lagoon for aerobic biological degradation of solids in the waste which reduces the biological oxygen demand of the effluent. The generation of algae through photosynthesis or plant growth may also be provided in the lagoon to remove phosphates, nitrates and other nutrients. Effluent containing algae, microbiological agents, and the like, from the lagoon is fed to a second centrifuge filter for filtering the effluent through a filter medium to separate the remaining solids from the effluent, and the remaining solids are burned and the separated effluent discharged. The effluent may then be sterilized, such as by chlorination, and the ash in each centrifuge removed.

Grease Handling and Disposal

In a process described by *O.E. Albertson; U.S. Patent 3,252,896; May 24, 1966; assigned to Dorr-Oliver Incorporated* the problem of grease handling and disposal is met in an extremely simple and effective manner by introducing the previously separated and collected scum or the grease fraction directly into the sludge strata in a respective settling tank.

This operation is conducted in such a manner that the scum or grease fraction becomes dispersingly incorporated or entrapped in a relatively large volume of the sludge containing the settled combustible organic matter. Thereby, the slugging effects and upsets are dissipated and eliminated, allowing the actual destruction of the organic sewage matter including the grease and fats to proceed under desirable and uniform conditions, with the sludge volume acting as a buffer or an equalizer or a cushion.

Taking sewage treatment, for example, this process may be practiced by the operation of a single sedimentation unit, as exemplified in Figure 3.3, or by the operation of one or more clarifier tanks delivering their underflow sludge to a sedimenting thickener.

FIGURE 3.3: SOLID BOWL CENTRIFUGAL PROCESS—GREASE HANDLING AND DISPOSAL

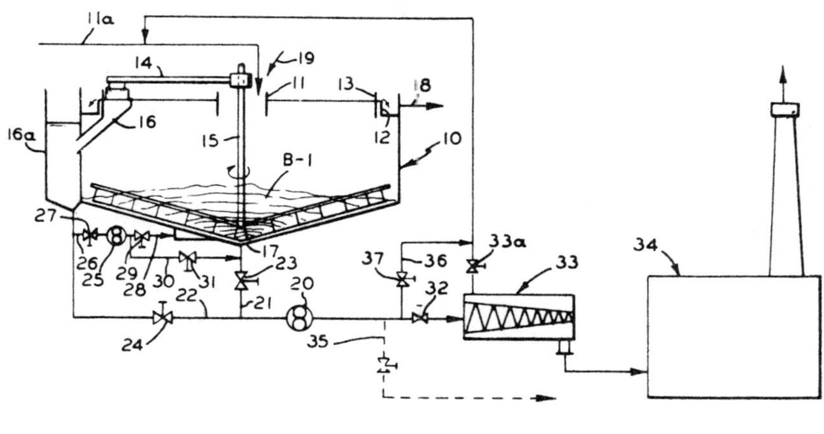

Source: U.S. Patent 3,252,896

The process is concerned with special arrangements whereby previously separated floating scum material or the grease fraction is introduced into the sludge bed in a respective sedimentation tank in such a manner that the grease becomes dissipated and entrapped in a relatively large volume of sewage sludge. Such dissemination of the high caloric grease into a large volume of sludge of relatively low caloric value, provides a cushion or buffer against the load shocks or the like in the subsequent disposal operation whereby both the organic matter and the grease are subjected jointly to destruction.

In the single tank example of Figure 3.3, there is shown a continuous sedimentation tank **10** having a feedwell **11** supplied with waste liquid or sewage **11a**, an overflow launder **12**, a scum baffle **13** concentric with and spaced inwardly from the launder, and a scum skimming device **14** rotating together with a rotary sludge raking structure **15**. A scum trap **16** located adjacent to the scum baffle receives the floating scum moved by the skimmer, for delivery into a scum pit **16a** provided marginally of the tank. An example of a scum skimming device and trap is found in U.S. Patent 2,611,489.

The settling tank has a conically shaped bottom where the sewage sludge may undergo a degree of thickening as indicated by sludge bed **B-1**, while being kneaded by the blades of the raking structure, and there is a centrally disposed bottom outlet means **17** for sludge discharge. The separated overflowing sewage liquor discharging at **18** may be sent to waste or it may be subjected to any of the forms of conventional secondary aerobic biologic treatment and part or all of the resulting secondary sludge may be supplied to the tank **10** as indicated at **19**.

A main pump **20** has a first intake connection **21** with the sludge outlet means **17**, and a second intake connection **22** with the bottom of scum pit **16a**. Control valve means are provided at the intake side of the pump as exemplified by control valves **23** and **24** in the respective intake connections of the pump. These valves may be manipulated in such a manner that the pump will normally draw underflow sludge along from the tank until such time that the scum pit will be full or will have accumulated a substantial inventory stratified into a volume of water and sludge subjacent to a top layer of zone of grease.

At that time the valves may be set so that the pump will temporarily draw water alone from the pit, or so that underflow sludge and the subjacent volume of scum material are drawn simultaneously. After the nongrease fraction has thus been drawn down in the pit, the initial valve positions may be restored so the pump will again draw underflow sludge alone.

But then, the remaining grease fraction is pumped separately from the scum collecting pit into the body of sludge bed in the tank. For that purpose there is provided an auxiliary pump **25** of relatively small capacity as compared with pump **20**. This auxiliary pump has an intake connection **26** with the bottom of the pit, provided with control valve **27**.

A discharge connection **28** provided with control valve **29** leads from the auxiliary pump to the tank bottom for delivery of the grease fraction in a region spaced from the sludge outlet means of the tank. With valve **24** closed, and valves **27** and **29** open, the pump **25** will introduce the grease fraction from the pit relatively slowly into the body of sludge, causing the grease to be dispersed, even as main pump **20** may continue drawing sludge through valve **23**. Then, after the auxiliary pump is stopped and valves **27** and **29** are closed, the pit will again start refilling with scum supplied thereto by the skimming device.

The auxiliary pump **25** is shown to have an alternative discharge connection **30** provided with a control valve **31**, leading directly to the sludge outlet means of the tank. Such a connection may be operated by the manipulation of the respective valves. The main pump **20** in this example delivers the underflow sludge interspersed with the grease, through a control valve **32** to a dewatering station **33** schematically indicated by a centrifugal machine of the solid bowl type, although for instance a continuous vacuum filter unit or rotary drum filter may also be employed in lieu of the centrifuge.

These machines are capable of dewatering the underflow sludge to a solids concentration of from about 25 to 35% which is the consistency of a moist cake material. In the instance of the centrifuge, the conically shaped centrifuge rotor bowl delivers from its wide end the separated sewage liquor here shown to return through a control valve **33a** to the feed inlet of the tank.

The moist cake material is delivered from the narrow end of the rotor impelled by a conically shaped conveyor screw rotating at a differential speed within and relative to the solid rotor bowl, which cake material is then sent to a combustion station **34**. A combustion system for moist cake material is exemplified in U.S. Patent 2,171,335 where the cake material is subjected to a drying operation before being subjected to incineration proper.

An alternative connection **35** shown in dotted line indicates that the destructive disposal of the sludge with the grease dispersed, may also be effected by way of anaerobic digestion in a digester tank A further discharge connection **36** with control valve **37** from pump **20** bypassing the centrifuge, allows the pump delivery to be shunted directly into the settling tank while pump **20** is drawing down the nongrease or water fraction from the pit.

This can be arranged by closing valves **32** and **33a**, and opening valve **37**. The process may also be advantageously employed in a sewage treatment operation where dilute underflow sludge from a primary clarifier tank is pumped to a much smaller but highly loaded sedimentation thickener to effect further substantial solids concentration by applying thickening controls such as described in U.S. Patent 2,850,449.

Batch Dewatering with Continuous Feed to Incinerator

P.J. Cardinal, Jr.; U.S. Patent 3,623,438; November 30, 1971; assigned to BSP Corporation describes a process for dewatering sewage sludge obtained from a raw sewage tertiary treating procedure. The apparatus includes a batch type dewatering device for sewage sludge, a collector for accumulating successive batches of sludge discharged from the dewatering device, and means for feeding a continuous stream of dewatered sludge from the collector into an incinerator, such as a multiple hearth furnace.

Figure 3.4 is a schematic view of the apparatus combination utilized in effecting dewatering and incinerating of sewage sludge by the process. As shown, a quantity of sewage sludge, obtained from a tertiary or like raw sewage treating procedure, is introduced through conduit **1** into a dewatering device **2**. Preferably, the dewatering device is a centrifuge unit designed to operate upon successive batches of sewage sludge, each of predetermined quantity, introduced thereinto in sequence from the sludge supply. The successive batches of sludge are fed into, processed in, and discharged from the dewaterer **2** in a series of successive steps dependent upon the capabilities and incinerating requirements of the sludge dewatering and incinerating system.

FIGURE 3.4: DEWATERING OF SEWAGE SLUDGE

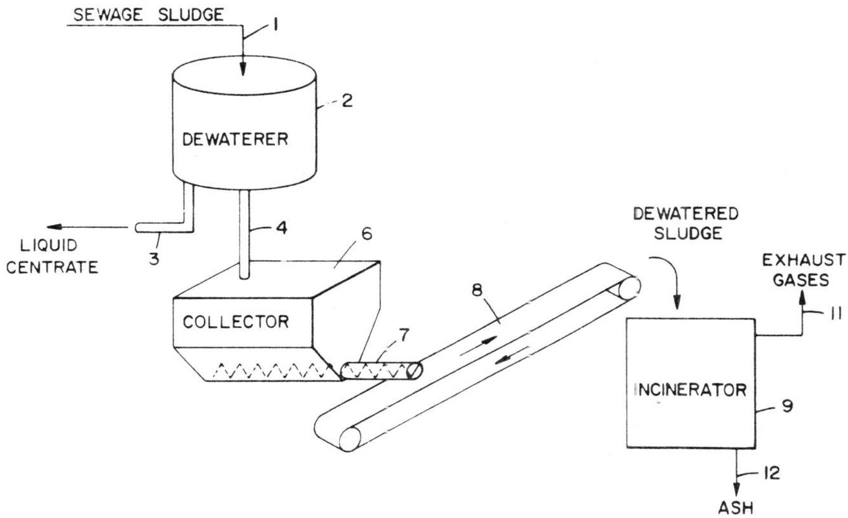

Source: U.S. Patent 3,623,438

The liquid centrate separated from the sludge solids in the dewaterer is discharged through conduit **3**, for reintroduction into the tertiary treating system or for disposal as may be desired. The dewatered sludge solids are discharged as a cake through conduit **4** for subsequent handling. Upon discharge of each batch of dewatered sludge solids from the dewaterer, a subsequent batch of sludge is introduced into the dewaterer for treatment.

Each batch of sludge dewatered in centrifuge **2** is introduced into a bin or like collector **6** from which the dewatered sludge may be selectively withdrawn and transferred for subsequent incineration. In that connection, the collector is utilized in conjunction with means for continuously withdrawing sludge cake from the collector and introducing the same as a continuous stream into the incinerator.

The continuous feed means utilized includes a mechanism for withdrawing dewatered sludge from the collector **6** and for thereafter transferring the withdrawn sludge into the incinerator. To that end, a screw conveyor **7** is positioned to extend along the bottom of the collector for withdrawing a continuous stream of sludge therefrom in selective quantities as desired. Alternatively, other mechanisms for withdrawing sludge from the collector may be utilized, such as a reciprocating piston feeding device.

Sludge withdrawn from the collector by the illustrated screw conveyor is transferred by a continuous belt conveyor **8** and is discharged from the belt conveyor into the incinerator **9** in which the sludge is burned. After sludge is burned in the furnace, the exhaust gases produced during such burning are carried from the incinerator through conduit **11** for subsequent treatment, disposal or selective reuse, in known fashion. The incinerated ash produced in the furnace is discharged through conduit **12** for disposal or selective treatment and reuse.

The simplified dewatering and incinerating procedure described, depending upon the capacity of the furnace **9** utilized, may be employed in conjunction with second and third or more dewatering devices of the batch type described. Each such dewaterer may be positioned to discharge its dewatered product directly into a single collector **6**, or additional collectors may be utilized in conjunction with the additional dewatering devices employed.

If the incinerating system utilizes a furnace of large size, it may be desirable to utilize two or more dewatering devices in sequence as noted to insure that a continuous stream of dewatered sludge is available for feeding into the incineration furnace in a continuous burning operation.

The batch type centrifuge utilized has been found to produce a drier sludge cake than a continuous centrifuge, thus permitting a smaller and less expensive incinerator furnace **9** to be utilized which operates with less fuel consumption or no fuel consumption at all depending upon the nature and moisture content of the dewatered sludge.

Previously, continuous centrifuge and continuous vacuum filter operations carried out on sewage sludge resulted in a sludge cake containing about 16 to 22% solids. With the improved apparatus and process, utilizing a batch type dewatering procedure, a sludge cake containing about 22 to 35% solids is producible, with the exact solid percentage obtained depending upon the nature and origin of the sewage sludge introduced into the dewatering device.

It should be noted that sludges from the primary treating step and the secondary treating step of a tertiary raw sewage treating procedure will vary substantially in their moisture content, thereby resulting in dewatered sludge cakes of varying moisture contents.

By way of specific comparative example, previously sludge dewatered in a continuous operation contained typically 17% solids. At a continuous centrifuge rate producing 1,000 pounds of dewatered sludge per hour, the furnace infeed rate required is 5,800 pounds per hour. However, using a batch type centrifuge, a drier dewatered product containing 25% solids may be readily produced, thereby requiring only 4,000 pounds per hour furnace infeed rate from the same dewatering rate of 1,000 pounds per hour.

As a result, for the same dewatered sludge centrifuge outflow rate, an incinerator 30% smaller than that required for treating sludge from continuous vacuum and/or centrifuge filtration may be utilized. The drier dewatered product produced results in a substantial savings in fuel costs in the furnace due to the lower moisture content in the sludge. Such fuel costs may even be eliminated entirely in certain cases due to the self-burning characteristics of the drier sludge produced in the batch process described.

Lime and Carbon Dioxide Treatment of Centrifuged Sludge

H. Sontheimer; U.S. Patent 3,345,288; October 3, 1967 describes a process for improving the dewaterability of organic sludges from wastewater treatment. In the process raw sludge to be dewatered is first thickened by simple sedimentation or flotation. The water removed is returned to a clarification basin. Thickened raw sludge is fed to a decanting centrifuge. The centrifuge usually removes 50 to 80% of the solids content of the raw sludge with a cake moisture of 40 to 60%. The centrate contains 20 to 50% of the solids content of the raw sludge and has a solids content of about 15 to 45 grams per liter.

The centrate containing 20 to 50% of the original solids contained in the raw sludge is mixed with a quantity of either dry or aqueous $Ca(OH)_2$ slurry of at least 20% concentration. After good mixing the lime-containing centrate is treated in a reactor with carbon dioxide-containing gas and mechanically agitated until the pH value is below 10, in the range of 9.0 to 8.0.

The mixture is then thickened by sedimentation, flotation or by other known methods. The resulting sludge has a solids content of about twice that of the centrate. This sludge is then mixed with incoming raw sludge to be dewatered and centrifuged to remove the suspended solids. Thickening of centrate treated with lime and carbon dioxide and thickening of the raw sludge to be dewatered may be carried out together.

The dewaterability of industrial sludge is particularly improved when both the raw sludge and centrate are treated with lime and CO_2. The thickener in such an arrangement must be sized correspondingly. Raw sludges consisting of primary sludge or mixed sludges containing a relatively high proportion of primary sludges are difficult to centrifuge with efficiency; therefore, a large recycle or a relatively great lime addition is necessary in order to obtain good solids recovery in the centrifuge. Raw sludges when treated in this manner with hot off-gases for neutralization are heated from 60° to 80°C, whereby some sterilization of the raw sludge is achieved. The dewatered sludge may be composted with no health objections and digestion in the thickening process will be prevented.

It was found that the gas necessary to neutralize the lime treated centrate could be any gas containing carbon dioxide, for instance, combustion off-gases. It was also found that the carbon dioxide necessary to neutralize the lime treated centrate could in some systems be produced within the system itself. It is known that carbon dioxide is produced in the aerobic or anaerobic digestion of sludge.

This carbon dioxide, present in statu nascendi, can be used as a neutralizer. If the process is carried out in conjunction with an aerobic sludge treatment, lime can be added to the centrate and after a sufficient holding time, which depends on the characteristics of the sludge, introduced into the digester. The carbon dioxide-containing digester gas not yet combined in statu nascendi with the lime-containing centrate can be further utilized by recirculating the digester gas to the digester so that the contact time is increased. Most standard sludge digesters are equipped for such an operation.

EVAPORATIVE

Rotating Contactor Drum Containing Preheated Aggregate

According to a process described by *H.H. Neukamm; U.S. Patent 3,909,410; September 30, 1975* sludge in the form of a slurry is continuously distributed onto preheated aggregate pieces which are continuously fed into a lower region of a rotating contactor drum. A portion of the slurry substantially instantaneously adheres to the heated aggregate and the nonadhering slurry passes through perforations in the drum for recycle.

The aggregate with adhering sludge is carried upwardly by means of baffles in the drum and is dumped into a collector trough in an upper region of the drum. This coated aggregate is subjected to further heating as it travels down an inclined path to dry the coating into a hard encrustation and enters a tumbler drum which breaks the sludge coating away from the aggregate and pulverizes the sludge into a dust or powder. This dust is collected as product and the decoated aggregate is recycled.

The general layout of the system can best be seen from Figure 3.5 with sludge slurry **S** being received from a sewage treatment plant through pipeline **11** for collection in a holding tank **10**. In the bottom of tank **10** is an agitator **12** and the slurry is pumped from holding tank **10** via pump **13** through line **14** into a rotating tumbler drum **15**.

This tumbler drum is the main feature of the process and has a cylindrical portion **16** with the ends being partially closed by means of the truncated conical end walls **18** and **19** leaving axial end openings **20** and **21** respectively. A portion of the cylindrical wall **16** contains perforations **17** and this perforated region is surrounded by a truncated conical skirt **22** leaving an annular space between the skirt and the cylindrical wall **16**.

A pair of steel rings **23** and **24** are mounted on the drum and these rest upon support wheels **25** and **26** respectively to support the tumbler drum for free rotation. An electric motor is utilized for rotating the drum at a slow speed.

The sludge slurry arriving in line **14** is distributed downwardly in the drum through a distributing head **27**. This slurry in a lower region of the drum comes into contact with previously heated aggregate pieces **A** which enter the drum through chute **28**. These aggregate pieces are at a temperature of about 212° to 325°F. The slurry in coming into contact with this hot aggregate immediately adheres to the aggregate forming a coating **C** thereon.

The drum contains a series of longitudinal baffles **30**, which may be angularly mounted with respect to the longitudinal axis of the drum, and these baffles divide the periphery of the interior of the drum into a series of compartments, which carry the aggregate upwardly as the drum rotates. This means that immediately after being contacted by the downwardly falling sludge slurry, the hot aggregate is carried away from contact with the slurry and up into an upward region. The slurry which does not adhere to the aggregate during its contact time passes through the perforations **17** and into the surrounding skirt **22** for collection.

FIGURE 3.5: PROCESS EMPLOYING ROTATING CONTACTOR DRUM

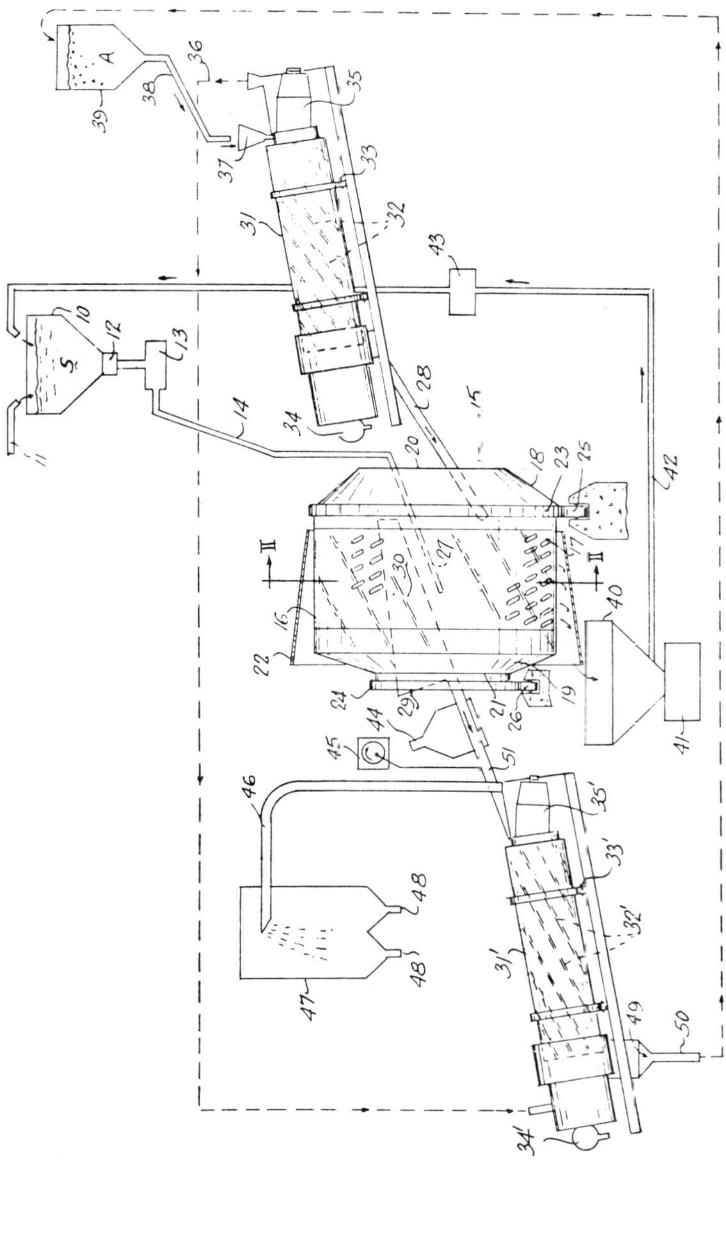

(a) View illustrating the process and apparatus.

(continued)

FIGURE 3.5: (continued)

b.

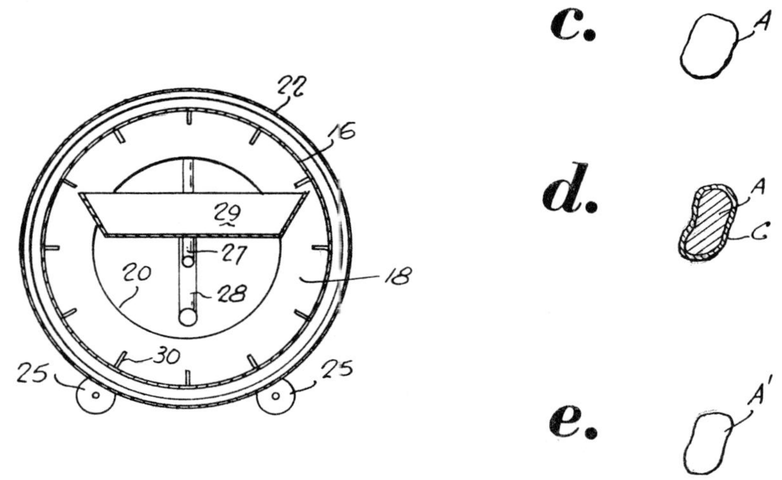

(b) Perspective view showing details of the contactor drum.
(c) View of an aggregate piece before being subjected to the slurry.
(d) Cross-sectional view of the aggregate piece encrusted with slurry.
(e) View showing the aggregate pieces after the encrustation has been removed.

Source: U.S. Patent 3,909,410

In the upper region within the drum cavity is mounted a collecting trough **29** so that the coated aggregate pieces in each peripheral compartment as they reach a certain point during the rotation of the drum are dumped from the compartment into the trough **29**. From there they travel down an inclined chute **51**.

The preheated aggregate particles which enter the drum **16** through chute **28** are heated in an inclined rotating drum **31**. Angled ribs **32** are mounted within the drum to cause a tumbling action of the contents as the drum rotates. The drum is mounted on rollers **33** for free rotation and is rotated by means of an electric motor.

The drum is heated by means of a burner **34** mounted at the lower end thereof and a fan **35** is provided at the upper end and for drawing the air heated by burner **34** through the drum in counter-current flow to aggregate particles entering the drum through inlet hopper **37**. The aggregate particles arrive in hopper **37** from a large hopper **39** through chute **38**.

The excess slurry which is collected at the bottom of drum **15** by means of skirt **22** flows into a collecting hopper **40** having a vibrator **41** at the bottom thereof and this slurry is recycled through line **42** and pump **43** back into collection tank **10** for mixing with fresh sludge from the treatment plant. Some small pieces of broken aggregate pass through the perforations **17** and can be screened out of the sludge being recycled. The coated aggregate collected in trough **29** as it moves down chute **51** passes through a heater **44** which dries the coating on the pieces. The coating is simultaneously dried from the interior by the heat previously applied to the pieces in heating drum **31**. To aid movement of the coated pieces down the chute a vibrator **45** is connected to the chute.

The dry coated particles from chute **51** enter a second inclined heating drum **31'** similar in design to heating drum **31**. It also has a burner **34'** in the lower end and a fan **35'** in the upper end for drawing hot air through the drum in countercurrent flow to the coated pieces. This hot air is passed through conduit **36** into the lower end of drum **31'** to assist in heating drum **31'**. It completely dries the coating on the pieces. Rotation of the drum with the continual tumbling action caused by the ribs **32'** breaks the coating (a hard encrustation) away from the particles in the form of a dust or powder.

This dust is entrained in the emerging hot air stream whereby it is separated from the aggregate and carried from the upper end of drum **31'** through conduit **46** into collector tank **47**. The dust or powder can be withdrawn from the hopper **47** through outlets **48** for further use. The complete drying and pulverizing in drum **31'** also aids in disintegrating any particulate material in the sludge, e.g., seeds, etc.

The decoated aggregate pieces A' collected in a hopper **49** in the lower end of drum **31'** are transferred by means of a conveyor **50** back into hopper **39** for collecting aggregate pieces to start the cycle once again. The entire system is operated with all of the chutes, etc., being closed from the atmosphere to conserve heat. Thus, even the decoated aggregate pieces being collected in hopper **49** still retain a considerable amount of heat and all of this partially heated decoated aggregate is reused in the system.

Only sufficient fresh aggregate is added into hopper **39** to make up any losses from pulverization. Also, no unnecessary heat losses take place in drum **16** since the only heat applied to that drum is within the aggregate pieces themselves and entirely transferred to the sludge forming the coating. The sludge which does not form any coating is quickly out of the influence of these heated pieces so that the recycled sludge picks up very little heat from the heated pieces. In other words, all of the stored heat in the aggregate pieces is used to form the sludge coating thereon without any wasted in needless heating of large amounts of sludge.

A material must be selected as aggregate which will withstand the tumbling action without disintegration. The aggregate pieces typically have a major dimension of ¾ to 1 inch and may be granite gravel, raw agricultural limestone, steel balls or fragments, open-hearth furnace slag or combinations thereof. Since the sludge powder is for use as an agricultural fertilizer, the aggregate material can advantageously be selected to compensate for soil deficiencies. For instance, if the particular soil being treated has a lime deficiency, limestone aggregate is advantageously used; if phosphates are deficient, then certain forms of slug aggregates are preferred.

Fluid Bed Heating and Off-Gas Incineration

According to a process described by *F.R. Keller; U.S. Patent 3,805,715; April 23, 1974; assigned to the U.S. Atomic Energy Commission* sludge is dried at extremely high thermal efficiencies and the odor bodies emanating from the sludge during drying are eliminated by incineration. Both the drying of the sludge and the incineration of the odor bodies are carried out in fluidized beds.

Referring to Figure 3.6, there is shown a fluidized bed dryer indicated generally at **10** and a fluidized bed incinerator indicated generally at **11**. A blower **12** takes in air and forces an air stream through line **13** to the fluidized bed dryer **10** where the air stream from line **13** serves as the fluidizing air for the fluidized bed dryer **10**. A section of line **13** lies in the upper portion of the fluidized bed incinerator **11** above the bed media level so that this section of line **13** lies in the stream of hot exit gases **16** from the incinerator **11**.

The fluidizing air stream for the dryer **10** passing from the blower **12** to the dryer **10** through line **13** is preheated in this section of line **13** by the stream of hot exit gases **16** from the incinerator **11**. Sewage sludge to be concentrated and dried in the dryer **10** is fed by pump **14** through line **15** to the fluidized bed dryer **10**. Similar to line **13**, a section of line **15** lies in the upper portion of the fluidized bed incinerator **11** above the bed media level of the incinerator. The sewage sludge feed being fed to the dryer **10** by the pump **14** is preheated by the hot exit gases **16** of the incinerator **11** as the sludge passes through this section of line **15**.

FIGURE 3.6: FLUID BED/INCINERATION PROCESS FOR REMOVAL OF ODOR BODIES

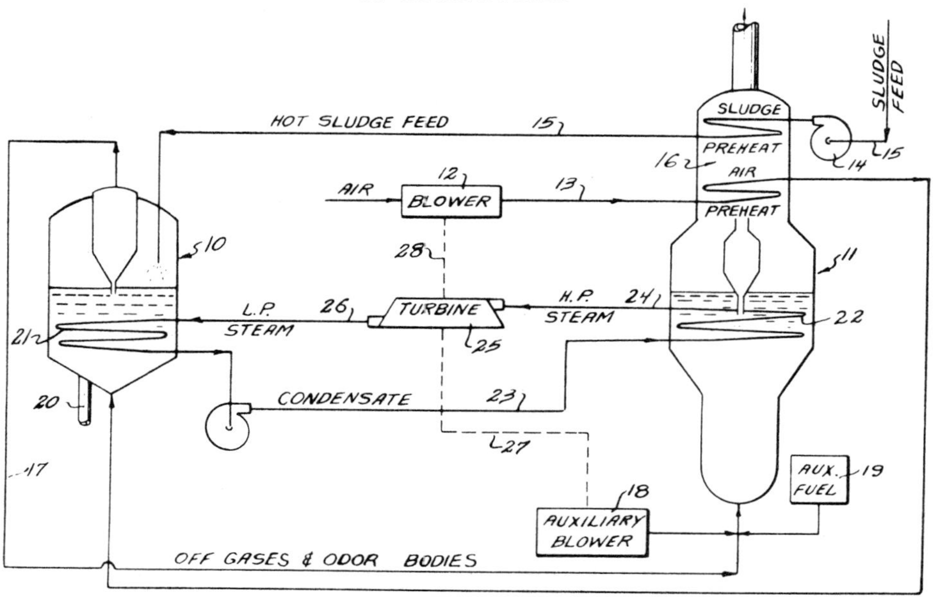

Source: U.S. Patent 3,805,715

As the sludge is dried in the fluidized bed dryer, various volatiles and gases including noxious and obnoxious odor bodies emanate from the sludge. The stream of off-gases from the dryer including these odor bodies passes through line **17** to the incinerator **11**, this stream of air and gases acting as the fluidizing air stream for the fluidized bed of the incinerator. The fluidizing stream of off-gases from the dryer can be supplemented as is necessary with an air stream from an auxiliary blower **18** to insure proper fluidization of the bed media of the fluidized bed incinerator **11**.

The odor bodies are eliminated by incineration in the fluidized bed prior to venting the exit gases **16** from the incinerator **11** to the atmosphere. While the fuel values of the off-gases and odor bodies are effectively used for producing heat in the incinerator **11**, an auxiliary fuel source **19** provides such auxiliary conventional fuel as is necessary to produce high temperatures and insure complete combustion of the odor bodies in the incinerator **11**. After the odor bodies have been eliminated by incineration, the hot exit gases **16** from the incinerator pass over sections of line **13** and line **15** before they are vented to the atmosphere, thereby preheating the sewage sludge feed and fluidizing air for the dryer as described previously.

Concentrated and dried sludge can be removed from the fluidized bed dryer **10** by conventional means, means for removing the dried sludge being represented in the drawing by a tap **20**. The concentrated and dried sludge can then be disposed of in a more acceptable manner and with less harm to the environment or put to some beneficial use.

Heat to promote the drying of the sludge is derived from steam passing through a heat exchanger **21** submerged in the bed of the fluidized bed dryer. The steam is generated in a heat exchanger **22** submerged in the bed of the fluidized bed incinerator, the condensate and exhaust from the heat exchanger in the dryer being recycled through line **23** to the second heat exchanger **22** in the incinerator for regeneration of steam. Both generation of steam in a heat exchanger submerged in a fluidized bed and the liberation of heat from a heat exchanger submerged in a fluidized bed are known and practiced in the art.

In the preferred case, high-pressure steam is generated in the incinerator heat exchanger **22**, and after passing through line **24** to an extraction turbine **25**, the high-pressure steam is employed to drive the turbine. The low-pressure exhaust steam from the turbine **25** passes through line **26** to the heat exchanger **21** in the dryer where the low-pressure exhaust steam provides the heat for drying the sludge. The output of the turbine **25** can be used to drive the blower **12** and the auxiliary blower **18** for the fluidizing air for the two fluidized beds, turbine **25** being appropriately connected to the blowers **12** and **18** as represented in the drawing by dotted lines **28** and **27** respectively.

Heat Treatment

According to a process described by *G.H. Teletzke, P.V. Knopp and A.H. Erickson; U.S. Patent 3,697,417; October 10, 1972; assigned to Sterling Drug Inc.* sewage sludge is briefly heated in a continuous process to a temperature between 190° and about 230°C for a period of time, up to about 240 seconds, which is inversely proportional to the temperature to which the sludge is heated and which is just sufficient to substantially improve the dewatering characteristics of the sludge. The sludge is immediately cooled thereby reducing color and BOD (biological oxygen demand) formation in the supernatant liquid.

Heat Transfer Process Utilizing a Carrier Substance

R. Rummel; U.S. Patent 3,619,422; November 9, 1971; assigned to Von Roll AG, Switzerland describes a process for the transfer of heat and/or matter between two or more gaseous or liquid agents, the process utilizing a carrier substance continuously circulated in closed cycle successively through at least two working spaces where heat and/or matter is respectively taken up and yielded. The carrier substance is present in a thin layer and the agents are caused to flow across the layer of carrier substance at both working spaces.

Referring to Figure 3.7, the simplest form of the process is shown, with an upper working space **1** and a lower working space **2**, both shaped as narrow pits in vertical arrangement. The sidewalls **3** and **4** of both working spaces are permeable to the streams of receiving and yielding agents, which flow through the working spaces **1** and **2**, entering and leaving in the direction of the arrows, through the ducts **5** and **6** and, respectively, **7** and **8**. In both working spaces **1** and **2**, the carrier substance is present as a thin layer from a hopper **9** and through a channel **10**, and moves downwardly through the two working spaces in succession, the channel **11** serving as connection.

The channel **12** leads into a cellular rotating block **13**, serving for the control of the rate of recirculation, while a conveyor **14** ensures the recirculation as indicated by the arrow, the carrier substance being brought back into the hopper **9**, thereby completing its closed circuit.

From the arrows in Figure 3.7a, it is apparent that both agents are flowing through the thin layer of carrier substance, crosswise to the vertical moving direction of the latter. When using the assembly described above for preheating air, hot combustion gases are led through the duct **5** into the working space **1**, to heat up the carrier substance by yielding at least part of their sensible heat, and then to leave through duct **7**.

The hot carrier substance then moves through the channel **11** into the lower working space **2**, wherein fresh air is fed through duct **6**. By flowing through the hot carrier substance, this air is preheated and leaves as such through duct **8**. The rate of movement of the carrier substance through the working spaces **1** and **2**, which is that of its recirculation, is governed by the speed of rotation of the cellular lock **13**. A possible passage of air from the working space **2** into the working space **1** is prevented by the length of the connecting channel **11**, which is filled with the carrier substance. Even at high air pressures in the working space **2**, only very small amounts of air may surmount the resistance of the column of carrier substance in the channel **11**. The same is also true for the channels **10** and **12**.

In Figure 3.7a, the channels **10, 11** and **12** are shaped as vertical pits, coaxial with the working spaces **1** and **2**. At higher working pressures, for instance in the working space **2**, the cross section of the channels **10, 11** and **12** may be reduced by means of internal partitions, equally spaced, thereby opposing a higher resistance to the possible overflow of an agent from one working space into another.

FIGURE 3.7: HEAT TRANSFER PROCESS UTILIZING A CARRIER SUBSTANCE

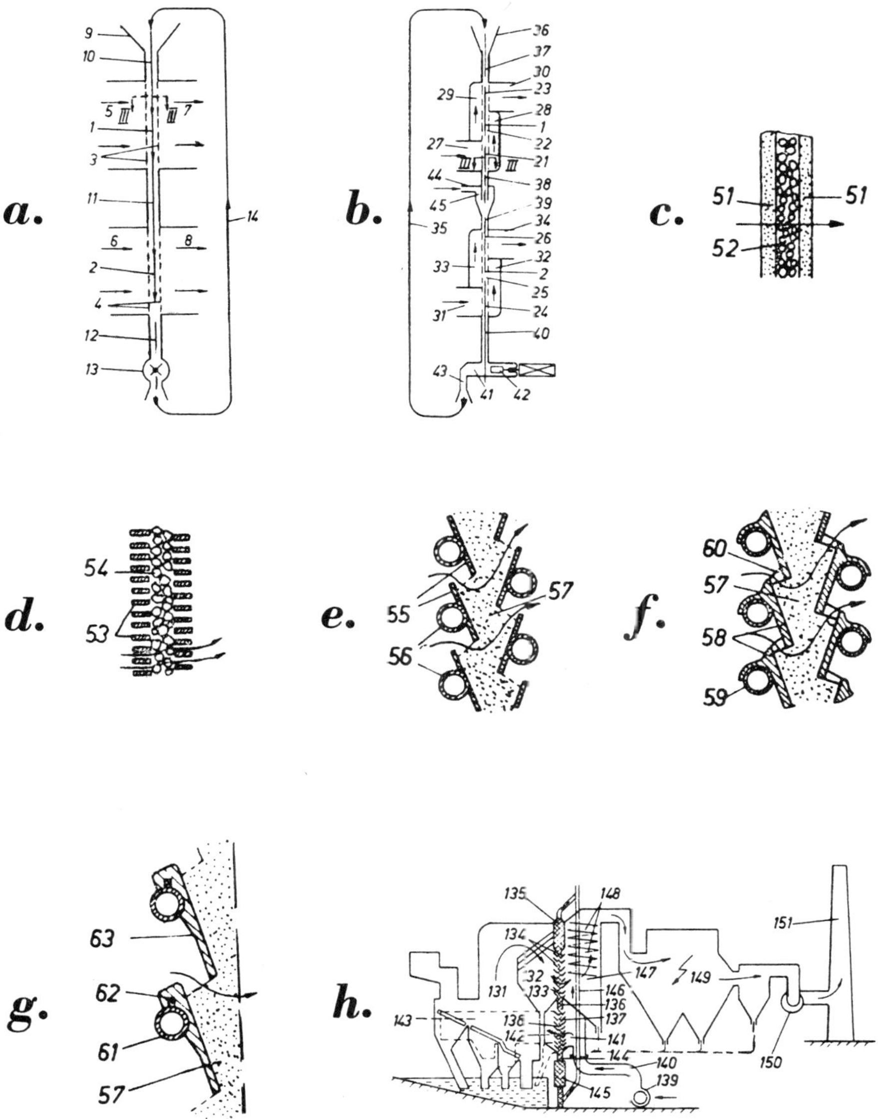

(a)-(g) Heat Transfer Surfaces and Assembly
(h) Refuse Incinerator Plant

(continued)

FIGURE 3.7: (continued)

i.

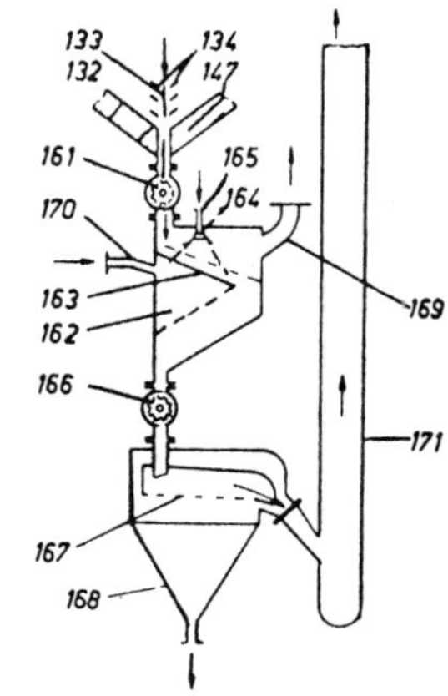

j.

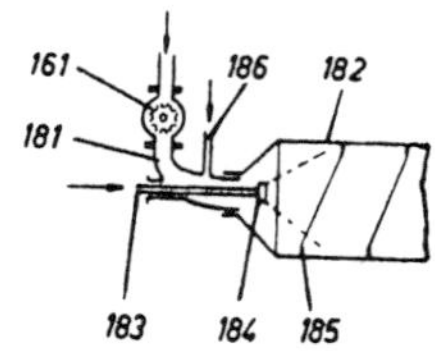

(i) Sludge Drying Unit with Stationary Dryer Space
(j) Sludge Drying Unit with Rotating Dryer Space

Source: U.S. Patent 3,619,422

In Figure 3.7b, another form is shown, also having two working spaces **1** and **2** in vertical arrangement, the working spaces, however, being divided each into three sections **21, 22, 23** and respectively, **24, 25, 26.** The agent entering through the duct **27** into the working space **1**, as indicated by the arrow, flows across the section **21** and is then led through the duct **28** across the section **22** and through the duct **29** across the section **23**, to leave eventually through the duct **30**. The other agent follows an analogous path through the three ducts **31, 32** and **33,** to cross in three steps the lower working space **2** and leave eventually through the duct **34**.

The carrier substance is recirculated by means of a conveyor **35**, into the working space **1** through a hopper **36** and a channel **37** and moves successively through the upper working space **1**, then channels **38** and **39**, the lower working space **2** and the channel **40** to a discharging device. This device essentially comprises a horizontal channel **41**, wherein a plunger **42** is moved to-and-fro by a hydraulic drive, thereby propulsing the carrier substance through the channel **43** into the recirculating conveyor **35**. To prevent the overflow of any agent from one working space into the other, an occluding agent is led through the duct **44** into a chamber **45**, between the channels **38** and **39**.

The permeable sidewalls of the working spaces in Figures 3.7a and 3.7b, symbolized by dotted lines, may be of various structures and configurations to best ensure the free passage of the agents and the free movement of the carrier substance. Figure 3.7c depicts an enlarged section of a pitlike working space, with two sidewalls **51** of porous structure, permeable to clean gases, and between the sidewalls a thin layer **52** of a coarse-grained carrier substance, for instance ceramic pebbles or metallic balls, easily sliding downwards and offering little resistance to the agent's through-flow.

Figure 3.7d depicts a vertical section of a pitlike working space with sidewalls consisting of metallic trellis, convenient for a coarse carrier substance **54**. According to the operating temperatures and to possible chemical influences, suitable materials are used, for both the trellis **53** and the carrier substance **54**. In Figure 3.7e, a vertical section through a pitlike working space is partially represented.

Here, the working space is limited by sidewalls consisting of oblique steel plates **55**, arranged as in Venetian blinds. For greater rigidity and for cooling, tubes **56** are welded to the plates **55**. In the two opposed sidewalls the plates **55** are displaced with regard to each other, thereby ensuring a homogeneous layer **57** of carrier substance between them. The tubes **56** and the plates **55** may be incorporated in the heating surface of a steam boiler, for instance, as a vaporizer or as a superheater. In this form with the sidewalls as described, both fine and coarse carrier substances may be used.

Similarly to Figure 3.7e, Figure 3.7f shows an enlarged vertical section of a working space with sidewalls consisting of ceramic plates **58**, supported by tubes **59**, as is the case with roof tiles. Lateral noses **60** protruding at the lower ends of the plates **58** ensure equal spacing between consecutive plates. Here again, the tubes **59** may be used as a heating surface in a steam boiler. The form with ceramic plates **58** has been found particularly suitable and is recommended for operation at higher temperatures.

In Figure 3.7g, a form similar to that of Figure 3.7f is described, with the tubes **61** having longitudinal fins **62**, fitting into corresponding grooves of the plates **63**, for higher rigidity of the assembly. At very high operating temperatures, the sidewalls may consist entirely of ceramic material, especially when heat transmission through the supporting tubes is not possible or not desirable. Figure 3.7h is a schematic representation of a refuse incinerating plant with a steam boiler for the recovery of heat, where the combustion air is preheated by means of an assembly according to the process, serving as a heat exchanger, the assembly being incorporated in the steam boiler.

Combustion gases effluent from the combustion chamber **131** pass into a first flue duct **132**, the right-hand sidewall of which constitutes the working space **133** of the heat exchanger. The two permeable sidewalls **134**, defining the working space **133**, are of the kind represented in Figures 3.7c, 3.7d and 3.7e and described above. Through the tubes carrying the plates steam would be flowing for superheating. Above the working space **133**, a bunker **135** is provided for the carrier substance.

A narrow channel connects the first working space **133** to a second one **137**, placed coaxially under the first, with sidewalls **138** similar to **134**, permeable to the stream of air to be preheated. A fan **139** drives the air through a duct **140** into a chamber **141**, from where it flows across the carrier substance in the working space **137** and then into a chamber **142**, to be ducted through **143** below the grates, as underblast for the burning of refuse. The carrier substance passes from the lower working space through a channel **144** into an intermediate bunker **145**, from where it is recirculated by an elevator **146** into the bunker **135**.

When flowing across the carrier substance in the upper working space **133**, the combustion gases yield a part of their sensible heat to the carrier substance, and pass into a second flue duct **147**, where their residual sensible heat is transmitted to the heating surfaces **148** of the steam boiler. Eventually, the combustion gases pass through the dedusting equipment **149** and the fan **150** to be expelled through the stack **151**.

In this example, molten and granulated slag resulting from the incinerating process, in sizes of 3 to 6 mm may be used as the carrier substance. At practically no cost, this material is resistant to high temperatures and to chemical action by the combustion gases as well as to superheating which may occur during a standstill of the circulation, due to some disturbance. A further important advantage lies in the fact that, in the case of dust-loaded combustion gases, the performance of the carrier substance is not hindered, and the dust retained in the carrier substance may easily be separated by sieving.

A further important advantage is that, because of the large cross section for the flow of the combustion gases in the upper working space, the thickness of the layer therein may be kept at some 30 to 40 mm, with a drop of pressure head of only a few millimeters of water column for the stream of combustion gases. The arrangement described above may also be used when aggressive or poisonous substances have to be removed from combustion or other gases. In such cases, carrier substances are used which are able to bind chemically such noxious substances as nitrous gas or combinations containing chlorine, fluorine or phosphorus.

For the drying of sewage sludge, metallic balls such as steel balls of about 25 mm in diameter are used as the carrier substance. In this case also, the working space for the heating up of the steel balls may be incorporated in a steam boiler as already described with respect to Figure 3.7h.

Figure 3.7i depicts a schematic representation of a drying plant for sewage sludge. The first working space **133** for the heating up of the steel balls used as the carrier substance is located between the flue ducts **132** and **147** of the boiler, and are defined by the permeable sidewalls **134**. The hot carrier substance moves from the working space **133** and through the cellular lock **161** into the second working space **162**. Here, the carrier substance slowly slides over an inclined chute **163**, while the liquid sludge is being sprayed over it by means of the nozzles **164** fed through the pipes **165**.

In contact with the hot carrier substance, the sewage sludge is dried, leaving the working space **162** through the cellular lock **166** together with the carrier substance from which it is separated by means of a vibrating sieve **167**. The dry sludge is collected in a receptacle **168** from which it is pneumatically conveyed into the furnace. The water vapor resulting from the drying process is drawn from the working space **162** through a duct **169** and caused to commingle with hot combustion gases for deodorization. A stream of exhaust gases from the boiler is led into the working space **162** through duct **170** for continuous scavenging of the working space.

From the vibrating sieve **167**, the carrier substance is directed into an elevator **171**, to be brought back into the working space **133** and thus complete the cycle. With some sewage sludges, solid crusts may adhere to the steel balls during drying which, however, peel off as small shells dispersed in the carrier substance and are gradually crumbled and ground into fine dust and eventually separated in the vibrating sieve **167** during the following cycle, all without hindering the normal operation. When such shells are very hard and subsist as such in normal operation, it is recommended that another form of the process be used as will be described below with reference to Figure 3.7j.

In Figure 3.7j, the hot carrier substance issuing from the cellular lock **161** is led through the channel **181** into a revolving cylinder **182**, where the liquid sludge is sprayed by a nozzle **184**, fed through the pipe **183**. By means of helicoidal ledges **185** internally welded in the cylinder, the carrier substance is caused to travel along its axis. During this travel and because of the tumbling and reciprocal friction of the steel balls, the adhering crusts are detached and ground into fine dust, which is thereupon separated in the usual way. Scavenging gas may be blown into the cylinder through duct **186**, to accelerate the effluence of the water vapor resulting from drying.

A refuse incinerating plant as represented in Figure 3.7h and equipped with an assembly according to Figure 3.7i for the drying of sewage sludge, is presumably built for a town of, for instance, 140,000 inhabitants. The assumable daily amount of sludge for this town is about 19 tons of dry substance, corresponding to some 95 m^3 of liquid sludge with 80% water contents. For drying this sludge down to water contents of only a few percent, some 2.2×10^6 kilocalories of heat are necessary, which according to the process, have to be supplied by the hot carrier substance. The carrier substance consisting of steel balls of 25 mm diameter is heated in a first working space, incorporated in the refuse incinerator, up to some 450°C.

During drying, the carrier substance yields part of its heat contents, leaving the drying space at a temperature of about 150°C, together with some 800 kg/hr of solid, dry sludge. For this some 5.3 tons of carrier substance must be recirculated 10 times per hour. The dried sludge is separated from the carrier substance by sieving and then conveyed into the furnace, there to be incinerated.

The water vapor resulting during drying is led into the stream of hot gases issuing from the combustion chamber. A volume of effluent gas of some 23,000 m^3/hr consisting of combustion gases from the burning of refuse and of dry sludge commingled with the water vapor from the drying process is then led through the upper working space, to flow across the layer of carrier substance therein at a velocity of about 1 m/sec. Through this, the temperature of the gaseous mixture is reduced from 750° down to about 500°C. With a thickness of 50 mm for the layer of carrier substance, the drop of pressure head for the gas stream is only 2 mm of water column. The residual heat contents of the effluent gases is recovered by means of the heating surface of the boiler.

Addition of Nonvolatile Oil in Multieffect Evaporation Process

C. Greenfield; U.S. Patent 3,323,575; June 6, 1967 describes a waste solids concentrate dehydration process which comprises the steps of mixing sewage sludge with an oil to obtain a mixture which will remain fluid and pumpable even after the removal of essentially its entire water content; subjecting the resulting mixture of sewage solids, water, and oil to a plurality of sequential dehydration steps by heat evaporation where each of the successive dehydration steps is at a successively lower temperature and the resulting oil-containing mixtures are of successively higher concentration because of increasing dehydration, the evolved vapors from each heat evaporation step except the last one supplying a substantial portion of the heat requirements of the following heat evaporation step; and withdrawing a substantially anhydrous slurry of sewage sludge, or sewage solids, and oil.

Thereafter, mechanical pressure of either a static or dynamic variety or both is exerted upon this anhydrous slurry to express the substantially greater part of its oil content to leave the sewage solids in a largely dry and oil-free or fat-free condition. Initially the solids particle size should not be in excess of ¼ inch, and a grinding step may be necessary for particle reduction to make sure that the later formed mixture or slurry of sewage sludge and oil will flow through pipelines and pipe fittings without impediment.

Sewage sludge itself usually contains an appreciable amount of oil independent of oil which may be added to it in the above mixing step. This oil will be carried through the dehydration steps along with the sewage solids and the added oil, and be subjected to being pressed out of the dehydrated slurry along with the added oil. If the dry or substantially water-free slurry is pressed sufficiently vigorously it can thus be made to yield oil in a quantity or at a rate equal to or in excess of that in or at which oil was previously added to the sewage sludge in the mixing step.

Generally it is desirable that the pressing step yield enough oil for the mixing step so that the process will be self-sufficient in respect to oil requirements. More desirably in most cases the pressing step will provide somewhat more oil than is needed for the mixing or pumpability maintenance step so that the process will provide a net oil yield or product.

The dry sewage solids left after the pressing operation can be utilized as fertilizer or possibly as animal feed or for other purposes outside the process itself, and thus constitute a process product. Further, being burnable, they can be used as fuel for the generation of steam needed to run the evaporator component of the apparatus for the dehydration steps, and also the steam needed to run auxiliary equipment such as pumps, either directly if they be steam driven pumps or indirectly if they be motor driven pumps and the steam is used to run a turbogenerator directly. The process can thus be self-sufficient in respect to fuel requirements.

With sewage oil being cycled back for the mixing step, sewage solids being burned for process fuel, and water evaporated from the sewage sludge being discharged after condensation at sufficient purity to enter a natural stream or river, the process thus provides useful, compact and complete disposal of all of the components of sewage sludge except to the extent that some net quantities of commercially useful products such as dry sewage solids and sewage or sewage-like oil may be generated, and further that there may be some disposable ash left after the burning of pressed sewage solids for steam generation purposes.

The oils which are utilized with the sewage sludge are inert, relatively nonvolatile oils or fats, or other oil-like materials. Typical of these are tallow, other animal fats, and vegetable oils all of which often can be derived directly from the process operation; fatty acids; petroleum oils and their fractions and derivatives including fuel oils, glycerins, glycols, and mixtures thereof, and miscellaneous liquid wastes from industrial plants, generally wastes of an organic nature. It is desirable to employ an oil that imparts process credits, i.e., one that can add value to the sewage product, such as waste oils normally found in sewage or industrial waste, or fuel oils, or, as suggested above, utilized oils derived in the practice of the process itself so as to minimize cost factors. The quantity of oil is such that its ratio in the system is in the range of 2 to 20 parts by weight, based on the nonfat solids. This refers to total oil, i.e., that added plus that derived from the process for reuse. This amount of oil gives a fluid pumpable mixture even in the absence of sewage water, indeed a mixture of even improved fluidity and pumpability.

In related work *C. Greenfield; U.S. Patent 3,251,398; May 17, 1966* also describes a multistage, multieffect process which comprises mixing the sewage sludge with an oil to obtain a mixture which will remain fluid and pumpable after the removal of the water content therefrom; subjecting the resultant oil-containing mixture to a plurality of sequential dehydration steps by heat evaporation wherein each of the successive dehydration steps is at a successively higher temperature and the resulting oil-containing mixtures are of successively higher concentration because of increasing dehydration, the evolved vapors from each heat evaporation step supplying a substantial portion of the heat requirements of the preceding heat evaporation step; and withdrawing a substantially anhydrous sewage sludge oil slurry.

Additionally, the oil can be separated from the sewage slurry to provide a substantially oil-free sewage solids product. The latter can be utilized for fertilizer, possibly animal feed products, or as fuel for the process.

C. Greenfield; U.S. Patent 3,304,991; February 21, 1967 describes a waste solids concentrate dehydration process which comprises the steps of mixing sewage sludge with an oil to obtain a mixture which will remain fluid and pumpable even after the removal of essentially its entire water content; subjecting the resulting mixture of sewage solids, water, and oil to one dehydration step by heat evaporation, and withdrawing a substantially anhydrous slurry of sewage sludge, or sewage solids, and oil.

As described in U.S. Patent 3,323,575 above, mechanical pressure of either a static or dynamic variety or both is exerted upon this anhydrous slurry to express the substantially greater part of its oil content to leave the sewage solids in a largely dry and oil-free or fat-free condition. Initially the solids particle size should not be greater than ¼ inch, and a grinding step may be necessary for particle size reduction to make sure that the later formed mixture or slurry of sewage sludge and oil will flow freely through pipelines and pipe fittings.

In comparison with the multieffect evaporator process described in U.S. Patent 3,251,398 as described above, this process is generally at a disadvantage with respect to fuel consumption whether or not the process is capable of generating some or even all of its own fuel. The outstanding advantages of this process and apparatus are their comparative simplicity which makes for low first cost of installation and their amenability to reduction to small scale or package installation and operation in cases where lack of complication and small space requirements for equipment may be more important for the sake of convenience and indeed of more economic benefit than the lower fuel requirement of a multieffect evaporator installation.

Multieffect Evaporator Installation

A.P. Aldridge; U.S. Patent 3,650,403; March 21, 1972; assigned to Simon-Carves, Limited, England describes a method of treating an aqueous biological slurry comprising the steps of feeding the slurry through an evaporator plant where it is subjected to heat treatment and where a part of its water content is removed by evaporation, filtering the slurry which has passed through the evaporation plant and recycling the aqueous filtrate for further treatment with the slurry in the evaporator plant.

Referring to Figure 3.8, it will be seen that the installation includes a processing retort or evaporator generally indicated at **10**. The sludge to be treated is passed into the evaporator by a line **11** and emerges from the evaporator through a line **12**. The temperature in the evaporator should be not less than 350°F in order to ensure that the biological content of the sludge is suitably conditioned to permit ready filtration. The sludge emerging from the evaporator **10** is passed by the line **12** to a filter plant generally indicated at **13**, which may conveniently be comprised by a rotary vacuum filter, or filter press, for example.

The aqueous filtrate from the filter plant is recycled by a line **14** and mixed with the feed of sludge in line **11** passing into the evaporator. The filtered solids from the filter plant in the form of filter cake are passed to a furnace **15** which fires a waste heat boiler **16** which provides at least a portion of the heat requirement for the evaporator. Steam is drawn off from the evaporator through a line **17** and may be condensed by heat exchange with the ingoing sludge in line **11**. The condensate may be returned to the inlet of the sewage plant and will not cause deterioration of the quality of the final effluent because it is readily biodegradable.

Typical operating conditions for an installation such as described above and fed with sludge at 5% suspended solids may be as follows. One pound of sludge at 5% suspended solids entering the evaporator through the line **11** will be combined with 0.4 pound of filtrate from the filter making a total feed to the evaporator of 1.4 pounds. In the evaporator 0.9 pound of water is evaporated and discharged through line **17** and 0.5 pound of blowdown at 10% suspended solids is discharged to the filter. Part or all of the evaporator operates at 350° to 400°F, thus heat conditioning the sludge and making it pumpable and filterable at 10% suspended solids.

It should be noted, however, that heat treated sludges are pumpable with a solids content of up to as much as 18%. The blowdown passing through the filter produces 0.4 pound of filtrate which is recycled as has been described, and 0.1 pounds of filter cake at 50% solids. The step of recycling the filtrate ensures that those compounds in solution and colloidal suspension are retained in the system and are eventually rejected with the filter cake. Such compounds are not readily biodegradable and the filtrate should not therefore be discharged to the inlet of the sewage works.

Characteristically the 0.1 pound of filter cake may be incinerated with 100% excess air to produce approximately 0.75 pound of waste gas at 1500°F. Thus the waste heat recovery is of the order of 200 to 224 Btu per pound of original sludge intake and thus comprises an important heat source for the evaporator. Obviously additional fuel is required to fire the boiler to provide sufficient heat for the evaporator to operate as here described.

FIGURE 3.8: EVAPORATION TREATMENT OF BIOLOGICAL SLUDGE

a.

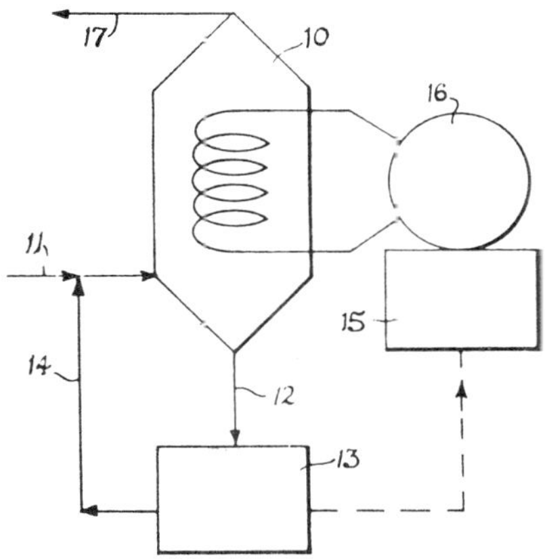

Single Evaporative Unit

b.

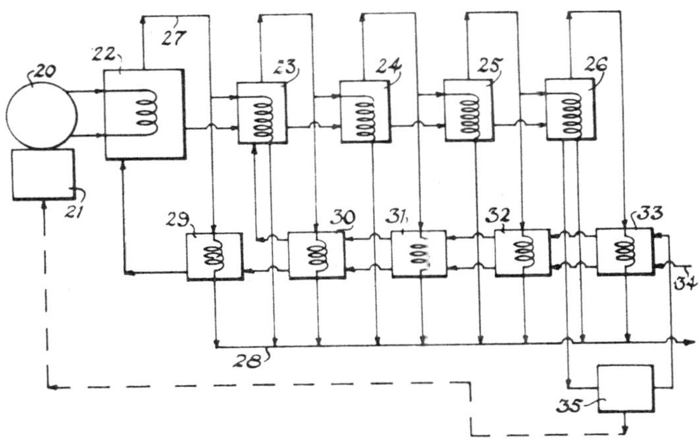

Multieffect Evaporator Plant

c.

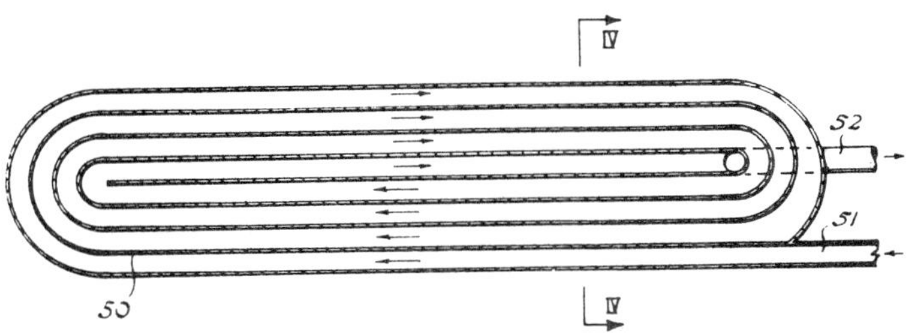

Horizontal Cross Section of Retort

(continued)

FIGURE 3.8: (continued)

d.

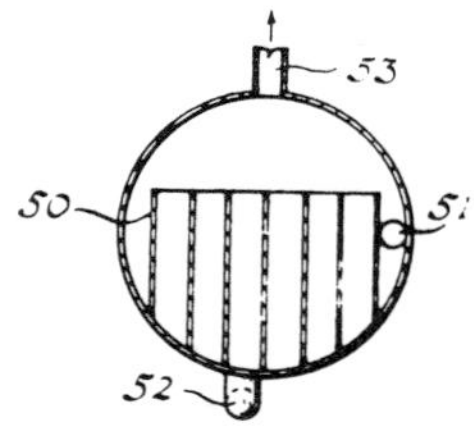

Cross Section of Retort Along Line IV—IV of Figure 3.8c

Source: U.S. Patent 3,650,403

Referring to Figure 3.8b it will be seen that a typical installation having a multi-effect evaporator plant comprises as before, a waste heat boiler **20** fired by a furnace **21**. The evaporator plant comprises a processing retort **22** which also forms the first effect of the evaporating plant, and four further evaporator vessels **23, 24, 25** and **26** forming the second, third, fourth, and fifth effects of the multieffect evaporator plant, respectively. The processing retort or first effect vessel is heated by steam from the waste heat boiler. Steam produced in the processing retort is discharged through a line **27** which divides to supply the steam after it has been condensed to a line **28** via a vapor heat exchanger **29** and the second effect vessel **23**, respectively.

Likewise, the steam produced in each of the vessels **23** to **25** inclusive is supplied to the line **28** via heat exchangers **30, 31** and **32** and the vessels **24, 25** and **26,** respectively.

Steam produced in the vessel **26** is supplied to the line **28** via a final vapor heat exchange unit **33**. It will be noted that each of the vessels **23** to **26** inclusive is heated by a portion of the steam produced by the preceding vessel in the multi-effect evaporator plant, and that each of the evaporator vessels operates at successively lower temperatures and pressures. The incoming sludge enters the apparatus through a line **34** and passes to the processing retort via all of the vapor heat exchange units **29** to **33** inclusive. The sludge is passed successively through the vessels **22** to **26** inclusive and the blowdown from the vessel **26** is passed through a filter **35**, the aqueous filtrate being passed through the vapor heat exchangers **30** to **33** inclusive before entering the second effect vessel **23** there to mix with the sludge which has been pretreated in the processing retort.

The filter cake from the filter **35** is incinerated in the furnace **21** to fire the waste heat boiler **20**. The condensate from the line **28** may be returned to the inlet of a sewage works with which the installation is associated as with the installation described with reference to Figure 3.8a. Typical operating conditions may be similar to those described above with reference to the installation of Figure 3.8a, but in this example the waste heat alone produced will be sufficient to cause the required evaporation without additional fuel due to the multieffect operation.

To ensure that the sludge is properly heat conditioned in the processing retort or first effect vessel **22**, such vessel is preferably designed as a horizontal cylindrical pressure vessel with rounded end caps as shown in Figures 3.8c and 3.8d. Within the vessel there is a spiral partition **50** to give an extended flow path for the sludge which enters through a line **51** at the periphery of the vessel and is discharged through a line **52** at the far end of the spiral path. The partition **50** extends about half way up the height of the vessel leaving a free space at the top of the vessel allowing steam to escape through the line **53**. The partition **50** is preferably comprised by heat exchanger plates which are heated by steam from the waste heat boiler.

The sludge is pumped into the vessel at a controlled rate to maintain a retention time in the retort of 15 minutes to 1 hour depending upon the heat conditioning requirements for a particular sludge. The sludge moving around the spiral path in the vessel is heat conditioned and partially evaporated, the steam bubbles produced maintaining a clean heat exchanger surface.

At the end of the spiral the sludge is pumped into the second effect vessel where it is joined by the filtrate. As far as possible the filtrate is kept out of the processing retort to avoid secondary breakdown of the organics in solution which would impair the condensate quality. Over a range of inputs, the correct retention time in the processing retort can be maintained by varying the height of the sludge within the vessel. The processing retort described with reference to Figures 3.8c and 3.8d is also suitable for use as the evaporator vessel in the installation of Figure 3.8a.

Rotary Type Filter-Dryer

N.D. Hazzard; U.S. Patent 3,287,819; November 29, 1966; assigned to The Air Preheater Company, Inc. describes a method for effectively minimizing the moisture content in a supply of sludge that is being supplied to a reactor whereby it will readily support combustion.

Figure 3.9 illustrates a system in which a quantity of sludge or waste entrained slurry is continuously supplied through an inlet duct **20** to a rotary type filter-dryer having a cylindrical drum type filter **24** in the housing **26**. The drum **24** is moved continuously about its axis **28** by drive means. The level of the slurry maintained in the housing **26** is such that it submerges only the lower portion of the filter **24**.

As the drum filter rotates through the slurry it becomes coated with a layer of the slurry which is carried with the drum **24** to the upper portion of the housing **26** and over to the outlet duct **44** for removal from the filter. An inlet **30** opening into the housing admits hot exhaust gases from duct **32** which originate in a reactor **34**. The hot gases which enter the housing **26** are supplied thereto under a slight positive pressure sufficient to force them through the filter drum and its coating of slurry to outlet openings **36** which lead to outlet duct **38**.

As the hot gases from reactor **34** pass through the coating of slurry on the filter drum **24**, the moisture in the wet slurry is evaporated and the temperature of the dried sludge is raised before it is scraped off the drum and returned via duct **44** to the reactor **34** where it is then subjected to conditions that support its combustion.

FIGURE 3.9: ROTARY TYPE FILTER-DRYER

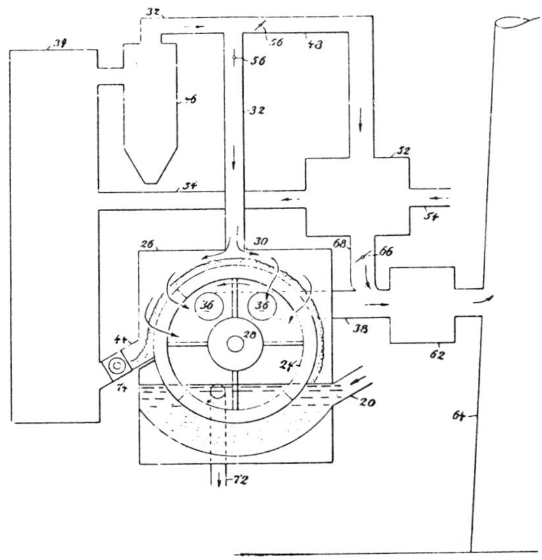

Source: U.S. Patent 3,287,819

As the liquid in the slurry is being removed by the action of the perforate filter drum **24**, the filtered solids on the drum **24** simultaneously serve as a continuously renewable filter bed that removes particulate matter from the gases exhausting from reactor **34** before the gas is exhausted to the atmosphere through duct **38** and stack **64**. If excessive amounts of particulate material are given off by the combustion process in the reactor, it may be desirable to utilize an additional gas cleaning apparatus **46** in advance of the filter-dryer to remove the bulk of the large particles from the exhaust gas in order that the particle buildup of the drum **24** will not be excessive and thus interfere with the drying of solids in the slurry.

In the event that gas being exhausted from the reactor **34** is in excess of that required to satisfactorily dry the slurry being supplied by the filter-dryer to the reactor, the excess exhaust gases may be directed through branch duct **48** and a heat exchanger **52** where such exhaust gases will give up their heat to air for combustion being supplied to the reactor **34** through duct **54**.

Fluid flow through ducts **32** and **48** may be proportioned in accordance with the demands of the system by means of the damper valves **56**. Hot exhaust gas which has traversed the filter-dryer **20** and/or the heat exchanger **52** is preferably directed through a scrubber **62** before it is exhausted to the atmosphere through a stack **64**. Valve means **66** in duct **68** that connects the heat exchanger to exhaust duct **38** is adapted to preclude back flow through the heat exchanger due to any variations in pressure. Filtered liquid that collects inside the drum **24** of the filter-dryer is drawn off through an outlet pipe **72** and discarded through any suitably disposed drain or it is otherwise utilized as deemed appropriate. Dried waste products are scraped off the rotary drum **24** and are directed through outlet duct **44** and a positive feeding apparatus **74** to the reactor **34**. The form of the feeding apparatus may vary from system to system inasmuch as variable factors such as the physical consistency of the solids, quantity of solids, distance of the filter from the reactor and their relative position are all factors that influence the final selection.

R.B. Burden, Jr. and E.J. O'Gieblyn; U.S. Patent 3,491,708; January 27, 1970; assigned to General Incinerators of California, Inc. describe a system for destroying by burning wet waste materials such as sewage sludge, garbage, wood chips, paper, and general refuse. Central to the entire system is a rotary dryer which predries the material to be burned. The dryer includes a circular horizontal conveyor table which rotates at slow speed on a central shaft and an annular gas chamber which is stationarily mounted over an outer peripheral portion of the table so that the table forms the bottom wall of the chamber.

Wet material to be burned is conveyed continuously from a dewatering station onto the table where it is conveyed at slow speed through a circular path to a discharge screw conveyor which moves the dried material continuously to a sludge burner. As the wet material rotates, warm furnace gases from the sludge burner pass through the gas chamber and thus over the surface of the material to dry the same. Plow devices within the chamber slowly stir the material as it moves to promote drying of the entire mass. After the furnace gases pass through the gas chamber, they are led into a gas scrubber and then to an afterburner, after which they are discharged to the atmosphere. Dried wastes may be led from the dryer to a hammer mill or other comminuting device before they pass to the burner.

Papermaking Machine Used for Drying Oxidized Slurry

A.S. Goldberg; U.S. Patent 3,561,598; February 9, 1971 describes an economical process for instantly disposing of sludge from sewage and industrial wastes and manufacturing a useful product. The process comprises reducing suspended solids in the sludge and wastes to a concentration of less than 1% by dilution to form a slurry, oxidizing the slurry, rapidly filtering the oxidized slurry and drying it on a papermaking machine which rolls the dried residue onto a roller. The method eliminates odors and produces a product useful as a synthetic top soil or soil conditioner or as a paperlike product, useful for packing carton liners.

In the process, grit sedimentation, screening and grinding to produce ground rag sludge, settling to produce raw primary sludge, and biological treatment of primary supernatant liquid to form activated sludge may all be accomplished by known methods. Primary sludge, having been scraped from the bottom of settling basins, is pumped to dilution tanks where it is reduced to a total solids content of 1% (10,000 mg/l) to form a sludge slurry.

The dilution may be accomplished by the addition of ordinary tap water in proper amounts to the primary sludge to be diluted. The sludge slurry is pumped into an open mixing or reacting tank through a mixing pipe wherein a suitable oxidant is added to the sludge slurry. The oxidizing material may consist of commercially available hypochlorite solutions (5 to 15% by weight) or any suitable oxidant in dissolved form which will intimately mix with the sludge. Among those oxidizing materials found to be satisfactory is sodium hypochlorite (NaOCl) which may be introduced to the sludge slurry in any suitable manner, one method being to form the oxidant in situ by introducing appropriate quantities of caustic soda (NaOH) and chlorine (Cl_2) into the mixing pipe as it is being pumped into the open mixing or reacting tank. The caustic soda and chlorine react to form sodium hypochlorite which, when added in quantities ranging from 100 to 1,000 mg/l of the chlorine equivalent of the oxidant (or 500 mg/l for each 1% of total solids), provides sufficient oxidant to completely oxidize the sludge.

In the open mixing or reacting tank, high speed mixing apparatus as well as piped recirculation equipment will ordinarily enable completion of the oxidation of the sludge in less than 10 minutes. In this regard, complete oxidation is ordinarily characterized by a change of color from black to tan or gray. After the oxidation is complete in the reacting tank, the oxidized slurry may be analyzed to determine if further dilution is required, e.g., by dilution to 0.5 to 0.9% suspended solids, and thereafter diluted as required, or if not required, then pumped as a slurry to the feed box of a papermaking machine of known type for filtration, drying and rolling. The filtration may be accomplished through a screen of virtually any mesh size that will accommodate rapid filtration. It has been found, however, that a 40 by 50 mesh screen is recommended.

Upon completion of filtration, a solid sheet of residue is formed which is paper-like and fibrous in composition. The solid sheet, which is quite thin, is rapidly fed through the heated drying rolls of the papermaking machine and wound on receiving rolls to obtain suitable units, e.g., 1 ton dry rolls. The resulting product is a stable, sterile and odor-free rolled sheet which, when viewed through the microscope, has the appearance of paper fibers.

Example 1: This example involves the treatment of sludge from purely domestic sewage. A primary sludge comprising combined raw primary sludge and activated sludge with a fairly consistent 4 to 4.5% suspended solids content was diluted to 25% by taking a 250 ml sample and diluting to 1 liter with tap water to form a sludge slurry.

The sludge slurry was then treated for 5 minutes with 10 ml of 5% sodium hypochlorite (NaOCl) (500 mg Cl_2) which turned the slurry from odorous black to a nonodorous pearl gray. The oxidized slurry was filtered through a 40 mesh screen with vacuum. An analysis of the filtrate showed that 70.3% of the solids were retained as residue by the screen, most of the passed solids being activated sludge. The residue was dried and compacted to form a paperlike product which exhibited a nitrogen content of 3.2% total nitrogen.

Example 2: This example involves the treatment of sludge from a combination of domestic sewage and industrial wastes in an unknown ratio. Primary sludge was diluted to approximately 1% suspended solids by the addition of tap water and thereafter treated with 500 mg/l chlorine (10 ml NaOCl). The oxidized slurry was filtered through a 40 mesh screen and an analysis of the filtrate showed that 81.1% of the solids were retained as residue by the screen.

The residue was dried and compacted to form a paper-like product which exhibited a nitrogen content of 2.1% total nitrogen. With papermaking machines of known types, e.g., those which roll at speeds of 1,800 to 2,500 ft/min and produce approximately 5 to 7 tons per hour depending on the sheet thickness, width of sheet and the specific gravity of the solids, it is possible to produce 120 to 160 tons of dried sludge product with one machine in a 24 hour period.

Distillation and Oxidation Using Molten Salt Bath

A process described by *J. Greenberg and D.C. Whitaker; U.S. Patent 3,642,583; February 15, 1972; assigned to Anti-Pollution Systems, Inc.* involves a one-step process for distillation of water and catalytic oxidation of residues in sewage and other contaminated liquids.

The process comprises the step of introducing wet pollutants onto the surface of a molten salt bath, which bath is characterized by a relatively low melting temperature and high stability with respect to the wet pollutants and air, and which acts both as an efficient heat transfer mechanism, thereby causing evaporation of the water contained in the wet pollutant, and as a catalyst, thereby inducing oxidation of remaining organic residues. Air is forced across the surface of the bath in order to provide oxygen for combustion, to carry away water vapor, and to promote agitation of the bath and resultant intermixture of the salt with the wet pollutants.

Referring to Figure 3.10, a sewage system is shown comprising a storage tank **50** containing the wet pollutant, in the form of sewage or other contaminated liquid, a conventional pump **51** for controlling the feed of the liquid out of the tank, and a trough **52** having a plurality of outlets through which the contaminated liquid is gravity fed onto the surface **53** of a molten salt bath **54**. The molten salt is housed in a container **55** constructed of a suitably noncorrosive material, the container having within it a main chamber **56** where evaporation and combustion of the wet pollutant take place, and an additional baffled chamber **57** where the gases of combustion produced in the main chamber are again brought into contact with the salt in order to further oxidize any incompletely oxidized products.

FIGURE 3.10: MOLTEN SALT BATH PROCESS

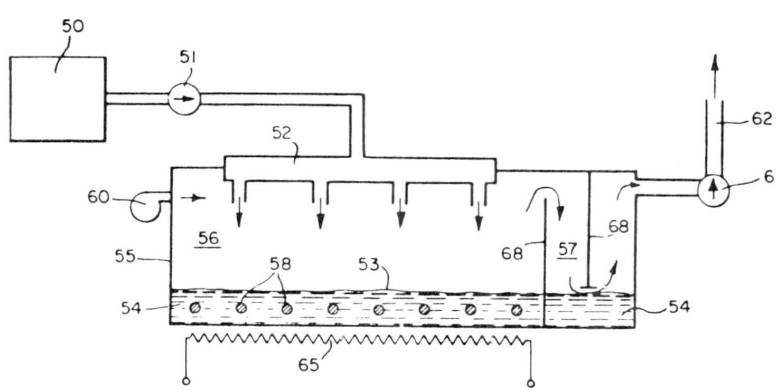

Source: U.S. Patent 3,642,583

The baffles **68** also prevent salt carryover from the lifting of the bath when the water flash evaporates in chamber **56**. A conventional blower mechanism **60** and a negative pressure fan **61** provide a continuous flow of air through the chamber for discharge of all products out of exit channel **62**. The capacities of blower **60** and fan **61** are engineering parameters which are determined in accordance with the air flow rate required, which in turn will depend upon the rate of sewage treatment for the unit.

Heater units **65** are shown positioned external to the container **55**, and are utilized to maintain the salt bath at a desired operating temperature. The heaters may be conventional open flame heaters or electrical heaters, with the heat being transferred through the container **55** to the salt. Alternately, electric or fuel fired immersion heaters, typically 6" to 12" in diameter, may be immersed within the liquid in order to provide more efficient heat transfer to the bath. The method is for the efficient treatment of sewage, wet human wastes, wet animal wastes, and other liquids containing wastes composed of acrylics, other hydrocarbons, organics and carbonaceous materials.

Example 1: In the purification of textile manufacturing wastes having a residue of acrylics, apparatus shown in Figure 3.10 is utilized. The waste, in liquid form, is fed at a rate of 500 gal/hr and gravity fed onto the surface of a bath composed of 62 mol percent KNO_2, 38 mol percent $Ca(NO_3)_2$. The bath has a melting temperature of 140°C and is maintained at an operating temperature of 450°C (850°F) to accomplish flash evaporation of the water contained in the waste; no water remains in the bath under these operating conditions.

The bath has a contact surface of about 5' x 20', sufficient to accommodate the feed rate. The bath is maintained at a 2" depth, and heated by external open gas flame. Air is provided, at a flow rate of 50 to 100 cfm, a steady flow being maintained by a blower at the input and a pump at the output. The output, in addition to air, contains water vapor and CO_2, there being no measurable hydrocarbons or CO in the output.

Example 2: Wet cow, chicken and human wastes, at dilutions from 1 to over 50% by weight, have been burned in oxidizing baths. The following baths, expressed in mol percent, have been used: 50 $Ca(NO_3)_2$-50 $NaNO_3$, 50 $Ca(NO_3)_2$-50 KNO_3, 34 $Ca(NO_3)_2$-33 KNO_3-33 $NaNO_3$, 25 $NaNO_2$-25 KNO_2-25 $NaNO_3$-25 KNO_3.

The baths have melting temperatures of about 275°F and were operated within the range of 750° to 850°F. The waste was introduced from a trough as in Figure 3.10 into a main bath chamber, and the resulting smoke and vapor passed through an end baffle chamber for further purification of the smoke, particularly to eliminate any residual odors. All other operating conditions were as in Example 1. The resulting output was comprised of water vapor and CO_2, with no measurable CO. No detectable residue remained in the bath, there being complete elimination of the wet wastes.

FLOTATION

Fuming Hydrochloric Acid

W. Thorn; U.S. Patent 3,772,191; November 13, 1973 describes a method of treating sewage where the sludge is acidified with a mineral acid to a pH value just below 5 and then the pH value is adjusted to the isoelectrical range of the high molecular sludge substances by addition of hydrochloric acid. The floated hydrophobic substances are removed from the surface and the precipitated sludge substances are separated from the supernatant liquid. The precipitated sludge is mixed with calcium hydroxide or calcium oxide and dehydrated in a filter press to a water content of 50 to 30% by weight.

It has been found that by adding mineral acids to the fresh sludges and sapropels in such an extent that the pH value is reduced to a level of below 5 the hydrophobic substances, i.e., lipids, fatty acids, higher hydrocarbons and the like may be separated from the mixture by floating quantitatively and may be gathered at the surface of the liquid charge. Among the hydrophilic components the high-molecular substances having basic groups comprising proteins, proteides such as nucleo-, glyco-, phosphoproteides, and furthermore nucleic acids etc., are precipitated simultaneously and will sink to the vessel bottom together with other sludge components. This process step is designated as digestion and thickening.

Apart from this, it has been found that especially by the addition of fuming hydrochloric acid the isoelectric range of the most protein particles is adjusted and at the same time the metabolism of the microorganisms is stopped by means of the hydrogen chloride and gaseous chlorine dissolved in this acid by killing the microorganisms. The acidification of the sludges for separating the mixture may be performed with any mineral acid. What is decisive for successful precipitation, especially for the killing of the microorganisms and thus for the digestion of the sludge is the addition of fuming hydrochloric acid.

A combination of fuming hydrochloric acid with crude sulfuric acid has proved to be most economical. Crude concentrated sulfuric acid is added to the sapropels for the purpose of eliminating the buffer capacity of the high molecular substances, predominantly the proteins, and then the fuming hydrochloric acid is added to adjust the pH value in order to obtain a complete elimination and a precipitation of the proteins.

Using the above described process all the samples of sapropel from a large variety of towns of the Federal Republic of Germany have been thickened to a solids content of 20 to 24% within a few days. It was possible to remove 66 to 83% maximum of the charged volume in the form of a relatively clear supernatant liquid from the residual thickened solids material.

Aeration-Flotation Cell

J.A. W. Kaeding and W.A. Grunert; U.S. Patent 3,525,437; August 25, 1970; assigned to Institut für Wasserwirtschaft, Germany describe a flotation process and apparatus for separating solids from liquids and for thickening sludges.

Referring to Figure 3.11, the liquid which is to be treated and which has a small amount of solids is introduced into a resting chamber before being fed into a flotation cell, so that in this way the speed of flow of the liquid is sharply reduced. In the flotation cell, the flotation process is carried out by way of vertical aeration, and the floating solids are thickened in an upper portion of the flotation cell. These solids which are thus separated and thickened are removed by way of a suitable structure which carries them away and the clear water is directed to a discharge chamber having a clear water outlet. By the addition of chemicals, for instance polyamide or polyacrylate, before or in the resting chamber, the liquid which is to be treated can be activated, and in this way the flotation in the flotation cell is improved.

The liquid which is to be treated is introduced into the flotation cell **4** beneath the surface of the body of liquid. By way of a regulated vertical aeration, which can take place in some cases over only a part of the cross section of the flotation chamber, it is possible to achieve a circulation of the liquid around a partition **5** which is built into the flotation cell.

FIGURE 3.11: AERATION-FLOTATION CELL

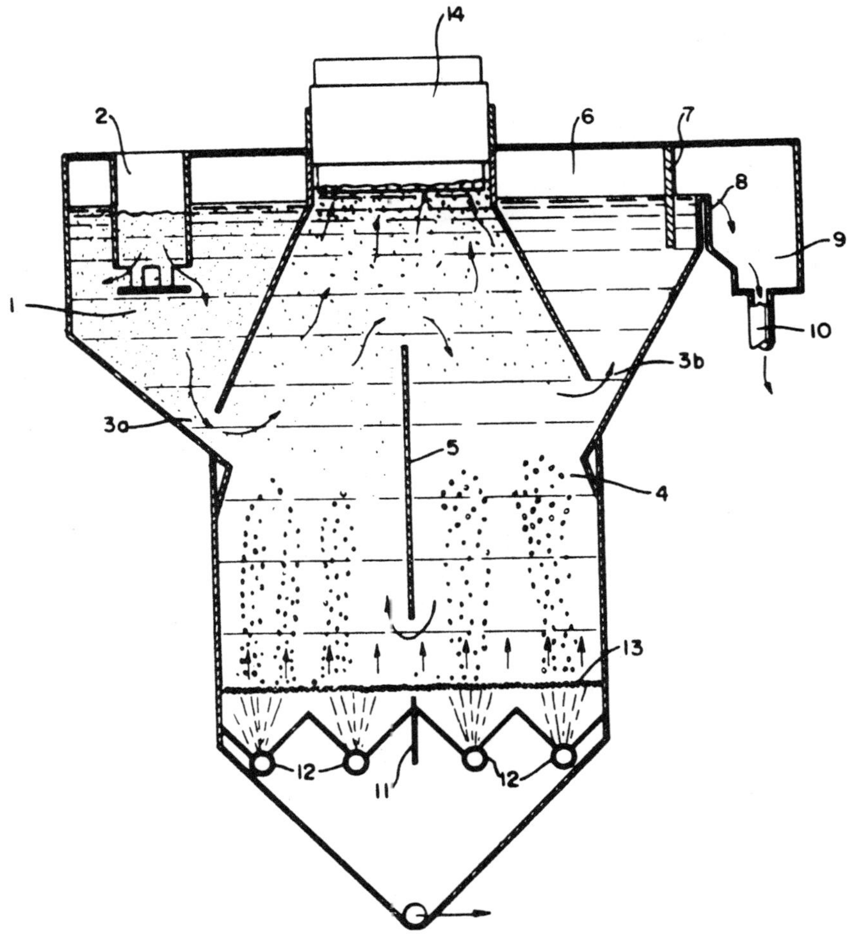

Source: U.S. Patent 3,525,437

With the structure of the process the flotation cell **4** has the resting chamber **1** situated at the inlet side of the flotation cell and at the outlet side is the discharge chamber **6** for the clear water, these chambers **1** and **6** communicating directly with the flotation cell. The flotation cell, the resting chamber, and the discharge chamber form a single constructive unit of very small size.

The flotation cell is provided in a lower region with a pressurized water inlet **12** as with a mesh structure **13** situated thereover so as to reduce turbulence resulting from the introduction of the pressurized water, this mesh structure extending over the entire cross section of the cell. In the ground of the cell is an outlet.

The flotation cell is of a tapered configuration at its upper region, preferably by having its side walls upwardly inclined toward each other, to form in this way a tapered hood which brings about a further thickening of the floating solids. These side walls of the flotation cells are respectively formed with an inlet opening **3a** for the liquid which is to be treated, as required by the operations, as well as with a discharge opening **3b** for the discharge of the clear water, and these openings **3a** and **3b** extend for the most part over the entire width of the flotation cell. Within the flotation cell there may be a partition **5** which is centrally mounted in the cell, and at the region of the pressurized water feed **12** there is a partition plate **11** for controlling the feed over the entire cross-sectional area or over only a part.

In order to remove the thickened floating solids, a removing means **14** is mounted on the flotation cell. The resting chamber and the discharge chamber can both be of the same configuration and size and are arranged in the mirror-image fashion with respect to the central axis of the flotation cell.

The resting chamber as well as the discharge chamber have in cross section a trapezoidal configuration and are elongated at their lower portions by way of inclined walls which extend into the side walls of the flotation cell through the inlet **3a** in the case of chamber **1** and the outlet **3b** in the case of chamber **6**.

In order to regulate the elevation of the surface of the liquid there is situated either within the discharge chamber or outside at the clear water outlet **9** an interchangeable and adjustable weir **8**.

The flotation cell can be formed by a pair of identical constructive bodies or of bodies which are stepped in cross section, in such a way that a return flow of the treated liquid either into the inlet or into the outlet is not possible.

With the process several advantages are achieved. Thus, there is a reliable floating separation and enriching of the suspended solids, particularly solids of flock properties accompanied by the smallest speeds of flow within the flotation cell. With the structure it is possible for larger aggregations of fine suspended solids to form. Also, a high flotation output is achieved by way of the manner in which the liquid flows through by the feeding of the pressurized water. Also, the additional thickening effect achieved for the floating solids before removal thereof is of considerable advantage.

Biological Process

J.F. Laboon; U.S. Patent 2,975,122; March 14, 1961 has found that raw sludge can be concentrated biologically by flotation without use of chemicals and the concentrate can be incinerated without further mechanical dewatering or chemical treatment.

In the method raw sewage as received at the treating plant is screened to remove inclusions and then passed through a suitable grit tank where grit and silt are removed. The sewage may be aerated in the grit tank or this step can be omitted. The sewage is then passed into settling tanks where the sludge settles and supernatant water is removed. The remaining raw or primary sludge is then transferred to concentration tanks where it is held for the required period for concentration biologically by flotation at optimum temperatures.

The flotation process is aided by heating the primary sludge to controlled temperatures. Flotation of the primary sludge is obtainable at temperatures from 20° to 55°C. The sludge at such temperatures will stratify in varying periods of time. At the lower temperatures flotation and concentration are effected in 5 days. Temperature of 35°C provides optimum results. At 35°C the formation of gases, mainly CO_2, proceeds with sufficient rapidity to effect complete flotation in less than 2 days. Flotation should not proceed too rapidly since concentration or compaction is necessary to prevent undue escape of gas through the floating sludge. Once the gas has escaped the sludge again settles to the bottom of the tank and can not rise again. The addition of heat to the upper surface of the flotation material aids in preventing escape of the gases by encrusting the surface.

Daily temperatures and seasons of the year make necessary the provision of suitable controls to maintain optimum temperatures. Heating of the raw sludge does not take place until after removal from the settling tanks. As the sludge is removed from the settling tank, it should be heated until the body of sludge in the concentrating tanks is at the desired temperature and is maintained during the flotation period. The nature of the sewage may also affect the time of flotation and the optimum temperature, thus necessitating frequent supervision of temperatures.

The time of concentration or compacting of the sludge is also of major importance. The maximum period of concentration appears to be about 5 days. After this period the sludge again begins to settle. For this reason it is preferable to separate the subnatant water as soon as desired flotation and stratification occurs and removal of the water does not carry off undue quantities of solids. The water so withdrawn may be advantageously returned to the grit tank instead of being discharged from the system.

The manner in which flotation and concentration of the sludge is obtained has not been exactly determined. After the heated raw sludge from the settling tanks is placed in the concentration tanks bacteritic and physical reactions occur in the sludge resulting in generation of gases and agglomeration of the solid components of the sludge. The gases in the agglomerate displace the water therein causing the agglomerate to float. The flotation material concentrates at the surface of the water and such concentrate has a moisture content of 80 to 90%.

After the subnatant water is removed from the concentrating tanks, the remaining flotation sludge may be disposed of in various ways. Such sludge when mixed to a uniform consistency with the water content the same throughout its mass may be economically burned in an incinerator. Where desired, the flotation sludge may be passed directly to a suitable vacuum filter and dewatered to form sludge cake. If it should be desired to make further use of the sludge by passing it through a suitable digester, the high percentage of solids in the flotation sludge permits a reduction in size of the digester and thus provides for greater economies in operation of this type of apparatus. The patent describes a complete processing plant in detail.

Agglomeration Through Air Bubble Release

In a process described by *T.R. Komline; U.S. Patent 3,622,508; November 23, 1971; assigned to Komline-Sanderson Engineering Corporation* the sewage is subjected to sequential treatments, including at least conventional primary and secondary treatments, to derive by the primary treatment a primary sludge consisting of heavier solids and to derive by the secondary treatment a secondary sludge consisting of relatively lighter solids. The primary sludge with its heavier solids is concentrated by dewatering on a filter unit to form a solid filter cake of relatively low moisture content, while the secondary sludge is concentrated by flotation, the concentrated sludges then being blended into a liquid slurry which is discharged as a fine spray for drying and incineration.

In the preferred form of the process, the concentrated secondary sludge, with its relatively high moisture content, is employed as a vehicle for the relatively drier filter cake, the proportions of filter cake and secondary sludge being chosen to produce a liquid slurry of a consistency for formation of a finely divided spray, without the addition of liquid, such as would detract from the speed of the drying and incineration.

The primary sludge is subjected to treatment by a suitable flocculating agent, while the secondary sludge, following removal of the primary sludge, is separately subjected to treatment by a different flocculating agent adapted for efficient action on the secondary sludge.

In a modification of the process, the filtering of the primary sludge is eliminated and such sludge with its comparatively heavy concentration of solids is directly blended with the sludge derived by secondary and any subsequent treatments, to form a slurry which is then subjected to spray drying and incineration as in the preferred mode of practicing the process.

Referring to Figure 3.12, the raw sewage is first subjected to a primary treatment or clarification of a conventional nature in which it passes sequentially through a grit chamber **10**, a comminuter **12** and then into a primary settling tank **14**. It will be understood that certain of the heavier solids contained within the raw sewage will have been eliminated at the grit chamber in accordance with usual practice, while the remaining solids are finely shredded by passage through the comminuter and are carried along into the settling tank, where the larger and heavier sludge particles settle to the bottom of the tank in the form of a sludge, hereinafter referred to for purposes of convenience as a primary sludge.

The primary tank effluent minus the primary sludge then is subjected to secondary treatment or clarification in which it is passed through a conventional oxygenation tank **16** in which oxygen is blown through the sewage. The sewage is then passed into a secondary settling tank **18** in which the finer, lighter particles of sludge settle to the bottom, such sludge being hereinafter referred to as secondary sludge. The effluent from the secondary settling tank is then discharged into receiving waters or if desired, may be subjected to still further treatment or clarification.

The settled primary sludge, which normally includes about 5% of solids, is subjected to a dewatering process as by subjecting it to the action of a conventional rotary drum-type vacuum filter **20** to concentrate the primary sludge into a substantially solid filter cake of approximately 65% moisture content, which filter cake is progressively discharged into a blender **22**. Before its delivery to the filter unit **20**, there may be added to the primary sludge a suitable chemical flocculating agent, as indicated at **23**, and as described in U.S. Patent 2,852,584.

The filter unit may be constructed as described in U.S. Patent 3,090,488. Preferably, the filtrate from the filter unit is returned back to the grit chamber for recycling with the inflowing raw sewage. Secondary sludge from the settling tank will consist of lighter and finer flocculent solid particles having a substantially higher water content than the primary sludge and having an appreciably higher water retention capability than the primary sludge. Because of this, the secondary sludge, if concentrated by filtering on the unit **20**, will greatly reduce the rate of filtration and capacity of that unit and of the entire process.

However, it has been found that such secondary sludge may be rapidly and efficiently concentrated by a conventional flotation process such as is described, for instance in U.S. Patent 3,175,687. Thus, the secondary sludge from the secondary settling tank is delivered into a flotation tank **24** at a location near the bottom and thus substantially beneath the surface of the body of liquid within the tank.

FIGURE 3.12: FLOW DIAGRAM FOR SEWAGE DISPOSAL PROCESS

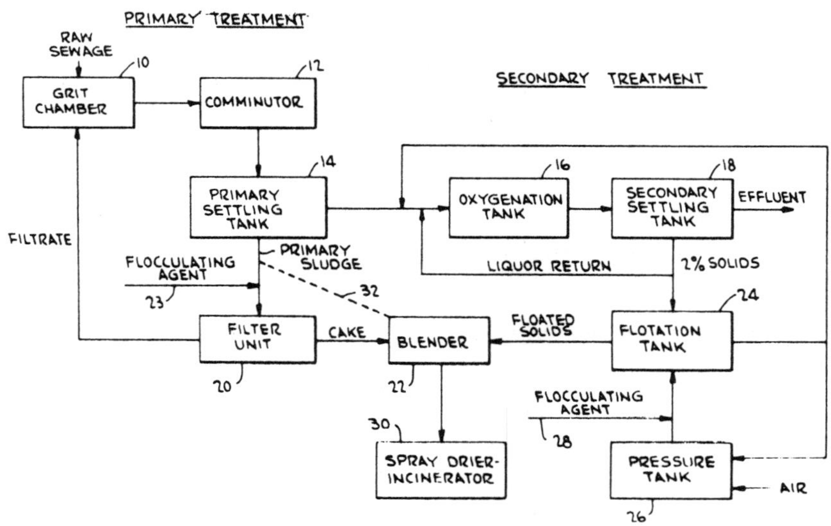

Source: U.S. Patent 3,622,508

A pressurized solution of air in water is continuously delivered from a pressure tank **26** into the flotation tank together with the secondary sludge. A flocculating agent preferably is delivered into the air-water solution at **28**, as indicated to assist in agglomerating the solid particles of secondary sludge within the flotation tank.

As the air-water solution is released from pressure incident to its introduction into the flotation tank, the air is caused to come out of the solution and form fine bubbles which attach themselves to the particles of secondary sludge to float the agglomerated particles to the surface of the body of water within the tank to form a floating blanket of sludge which is removed by usual skimming means.

These floated solids are then transferred to the blender **22** which may be of a conventional type functioning to break up the filter cake from the filter unit so that this cake may be intermixed and blended with the floated secondary sludge to form a generally homogeneous suspension or slurry of the mixed sludges; floated secondary sludge with its relatively high moisture content thus is employed as a vehicle for the relatively dry filter cake. The filter cake and the secondary sludge are delivered into the blender in such proportions as to produce a liquid slurry of a consistency for efficient spray drying and incineration as hereinafter described, without the addition of further liquid or the consequent necessity for removing the same in further processing.

From the blender the slurry is transferred to a spray dryer furnace or sludge incinerator **30** which may be of the type described in U.S. Patent 3,322,079. In such a furnace, the liquid slurry is discharged as a fine spray through a centrifugal atomizer into a current of hot gases in a spray-drying chamber, the spray-drying chamber preferably being located above an ignited bed of solid fuel onto which the heavier particles of sludge may fall to complete their oxidation.

The sizes and capacities of the filtering unit and the flotation unit will be so related as to supply filter cake and floated solids to the blender in such proportions as to provide a liquid slurry of the proper consistency.

With the various components of apparatus for carrying out the process so selected and interrelated, it will be apparent that the primary and secondary solids are separately concentrated by filtering and by flotation, respectively, which modes of concentration are especially adapted to and afford substantially maximum efficiency and capacity for the respective concentrating steps. Since such steps are carried out separately, neither will interfere with or detract from the other. Moreover, moisture contents of concentrated sludges from the respective treatments are so related as to supplement each other when delivered in proper proportions into the blender **22** to form a blended, substantially homogeneous liquid slurry adapted for treatment in a sludge incinerator such as disclosed in U.S. Patent 3,322,079.

It is thus possible to obtain a quite high capacity of both primary and secondary treatment of the raw sewage solids with substantial economies in the apparatus required. By elimination of the conventional digesting steps or stages, there are eliminated the potential deteriorating effects upon the receiving waters commonly associated with digestion.

S.J. Dea; U.S. Patent 3,397,140; August 13, 1968 describes a method for rapidly and efficiently dewatering sewage sludge by maintaining a batch of sludge in a closed space and by permitting gas to be generated by the sludge, or by introducing gas into the space, for floating solids to the top of the batch and for producing a high pressure which compresses the floated solids into a relatively compact cake.

Continuous Separation Using Liquids Discriminator Unit

A treatment process described by *M.G. Talon and M.R. Stasse; U.S. Patent 3,224,963; December 21, 1965; assigned to Equipements Techniques Industriels, France* does not make use of the common step of allowing sludges to settle, but is characterized by a continuous method of operation which does not comprise at any moment the total or partial settlement of all or part of the constituent substances of the sludges. The stages constituting the method according to the process are as follows.

(1) A sieving operation to eliminate large particles from the sludge, whether these are floating or not, such particles being collected to determine whether they will be rejected or recovered. The filtrate resulting from the sieving operations is collected in a reservoir or appropriate receptacle suitably designated as a primary tank.

(2) A dilution of the filtrate collected in the primary tank with both pure water and a fluid substance derived from a subsequent stage, wherein the latter fluid substance has acquired at that subsequent stage appropriate physico-chemical properties and/or biological properties.

(3) Transfer by an appropriate means of the diluted filtrate containing both floating and nonfloating portions, to a discriminator which is an apparatus which delivers two fluid substances essentially discriminated, one comprising water and the floating portion and the other comprising water and the nonfloating portion.

(4) Simultaneous transfer, both of the aqueous fluid substance containing the floating portion to a precipitation and decantation device, and the aqueous fluid substance containing only the nonfloating portion to a separator. The separator separates the latter fluid material into an aqueous magma having a consistency such that it can then be handled by simple shovelling and collected in an appropriate means with a view to its rejection or retention, and an aqueous filtrate which is sent into the precipitation and decantation device at the same time as is the aqueous fluid material which contains the floating portion.

(5) The stay, due to slow continuous circulation of the fluid material containing a mixture of the floating portion and the aqueous filtrate, in the precipitation and decantation device for a predetermined period having a duration such that the phenomena of precipitation and decantation have sufficient time to take place with a satisfactory efficiency, the latter phenomena terminating first in a yield of purified water and secondly in the production of an aqueous precipitate which has acquired appropriate physico-chemical properties, and in some cases biological properties during the course of its formation. The smaller portion of the aqueous precipitate is then directed into the primary tank, in which it is

mixed with the first filtrate while the major portion of the precipitate is fed into the upstream side of the precipitation and decantation device, the latter device thus operating simultaneously in a continuous closed loop cycle, the cycle including the preceding stages starting from the primary tank, and as a continuous delivery device for pure water, the purified water being discharged into an appropriate means.

Immersion Electrodes

J.W. Abson, E.I. Clark and I.M. Gray; U.S. Patent 3,543,936; December 1, 1970; assigned to Simon-Carves Limited, England describe a method of treating sludge which comprises steps of immersing electrodes in effluent sludge, applying a potential difference across the electrodes at a low current density sufficient to cause gassing, maintaining the current flow so that a layer of thickened sludge is formed, and removing the layer of thickened sludge from the remainder of the liquor.

Advantageously, the effluent sludge is treated for an average period of from 1 to 2½ hours, preferably 1½ to 2 hours and more preferably 1.85 hours. The current density during treatment may be between 0.003 amp/in^2 of cathode (0.00046 amp/cm^2 of cathode) and 0.02 amp/in^2 of cathode (0.0031 amp/cm^2 of cathode).

Referring to Figure 3.13 the apparatus comprises a tank **12** in which is mounted along a base **14** thereof a horizontal planar aluminum cathode **16**. A plurality of vertical, parallel, planar, evenly spaced, shallow, mild steel anodes **18** are also mounted in the tank. The anodes are suspended above the cathode by electrically conductive rods **20** which are mounted in two spaced insulated support members **22**.

FIGURE 3.13: IMMERSION ELECTRODE PROCESS

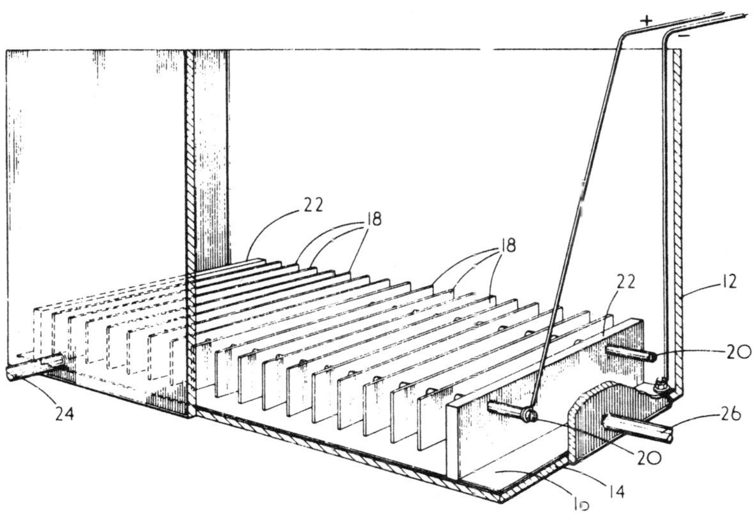

Source: U.S. Patent 3,543,936

In the operation of the apparatus an effluent sludge of high organic content is pumped continuously into the tank along a line **24** and a potential difference at a low current density applied between the anodes and the cathode causes separation of sludge and aqueous liquor; the sludge rises to the liquid surface and forms a layer which is removed by removal means (not shown); the clear liquor is pumped out of the tank along a line **26**. The liquid level in the tank is near the top of the tank well above the anodes.

The apparatus may also be used under batch conditions instead of continuous flow conditions. The space between adjacent anodes is preferably between 0.5" (1.27 cm) and 2" (5.08 cm) and the distance between the cathode and the bottom edges of the anodes is preferably between 0.25" (0.635 cm) and 1" (2.54 cm).

It is believed that the action of the potential difference involves flocculation promoted by hydroxyl ions, the flocs being carried to the liquid surface by gas bubbles produced by electrolysis.

Example 1: The apparatus described with reference to Figure 3.13 was used with the following parameters:

Cathode: 22.5 inches (57.2 cm) x 11.38 inches (28.9 cm) = 256 in^2 (1,653 cm^2)

Anodes: each 11.375 inches x 1.5 inches = 17 in^2 (110 cm^2)

Distance between cathode and bottom edges of anodes: 0.75 inch (1.9 cm)

Voltage: 8.6 volts

Current: 4 amp

Current density: 0.016 amp per square inch of cathode (0.0024 amp/cm^2 cathode)

Capacity of tank: 8 gallons (36 liters).

An activated sludge derived from the treatment of municipal sewage was treated in the apparatus under batch conditions for 2 hours. The initial solids content of the sludge before treatment in the apparatus was 0.81% by weight and the final solids content after treatment was 4.23% by weight.

Example 2: The apparatus described with reference to Figure 3.13 was used with the same parameters as in Example 1. An activated sludge similar to that of Example 1 was treated in the apparatus under continuous flow conditions with an average residence time of 2 hours. Clear liquor was pumped from the tank at ⅘ of the flow rate of sludge into the tank. The initial solids content of the sludge was 1.45% by weight and the final solids content was 5.25% by weight.

Bubble Flotation

According to a process described by *E.H. Brink and W.G. Palmer; U.S. Patent 3,787,316; January 22, 1974; assigned to FMC Corporation* a suspension of activated sludge from a secondary settling tank is concentrated by a bubble flotation process. Air is forced through a mixture of water and a surfactant in a shear type foamer at low pressure to produce a stable foam of small air bubbles in water. The foam is pumped into the suspension entering the concentrator and the bubbles attach to the solids in the suspension. The mixture of foam and suspension is released within a baffle in the concentrator tank, causing the surfactant shielded air bubbles to lift the solids to which they have become attached and form a blanket of concentrated sludge. A rotating skimmer head removes an upper layer of the sludge blanket whereas the clarified effluent flows under the baffle and out over an annular weir for discharge. The concentrated sludge removed from the concentrator is introduced into an anaerobic digester, along with primary sludge from the primary settling tank of the system. One overall effect is to reduce the size required for the digester.

OTHER PROCESSES

Cryogenic Freezing

D.W. Stout; U.S. Patent 3,892,662; July 1, 1975 describes a process and apparatus for water purification, recycling and conservation combined with waste reduction and disposal. Household sewage or commercial and industrial waste is collected, homogenized and solid waste particle size reduced, following which the sewage is atomized into a cryogenic environment where it is instantly frozen into crystalline form. The crystals are homogenized and further reduced in size until the mass of crystals takes on the characteristics of a liquid.

Separation of the water crystals from the light and heavy solid waste crystals is accomplished in a centrifugal separator and the water crystals are converted to steam. Alternatively, water vapor is removed from the atomized sewage prior to freezing whereupon only the solid waste freezes into crystalline form. The water vapor is converted to steam. The steam is all condensed and recycled to the household, commercial or industrial plant and in small part utilized as a heat source for the condensed steam to furnish hot water. The solid waste crystals are incinerated and the combustion gas and ash separated and disposed of by venting and collection, respectively.

Turning to Figure 3.14 the sewage purification, waste reduction and water purification and recycling system is shown generally at **10**. **A** represents the influent to the system which consists generally of all household waste products, including waste water and recharging water, human waste, detergents, soap, oils, household chemicals, and the like. All influent is collected in storage influent tank **12** where it is continually agitated by a screw type circulator **14** operated from a central shaft **16** suitably journaled on bearings and powered by a conventional electric motor **18**.

The constant agitation imparted by the screw type circulator is necessary to prevent the settling out of heavy and light solids in the influent. At the same time the screw serves to reduce the bulk size of solids in the influent to a smaller and more uniform size.

Influent **A** is fed from the influent tank to a unit **20** which further reduces influent particle size and which injects the influent under pressure into the influent homogenizer, size reducer and pressure injection portions of the system. Preferably, unit **20** is a screw-type cutter-injector which operates from shaft **16**. A screw-type unit is particularly suited for use herein because, at equal speeds, two such cutter-injectors of different diameters will produce a differential pressure which promotes efficiency of operation. Moreover, this type cutter-injector provides automatic overpressure relief. The primary function of unit **20** is to produce solid particles of uniform size which, upon proceeding through the system, will promote smooth and steady operation of subsequent units. Of course, the smaller the particles expelled by unit **20** the less work demand on subsequent size reduction units.

Influent A under pressure from the unit is forced through the primary particle size reduction and homogenization unit **22** wherein the solid particles in the waste are reduced to atomizable size. The waste particles must be as small as possible at this time in order that they can pass freely through the spray-fog nozzle or atomizer **26** without risk of plugging or malfunctioning the atomizer. Moreover for subsequent solid waste water separation to be most effective, the solid waste particles must be very small.

FIGURE 3.14: CLOSED RECYCLING SYSTEM EMPLOYING CRYOGENIC FREEZING

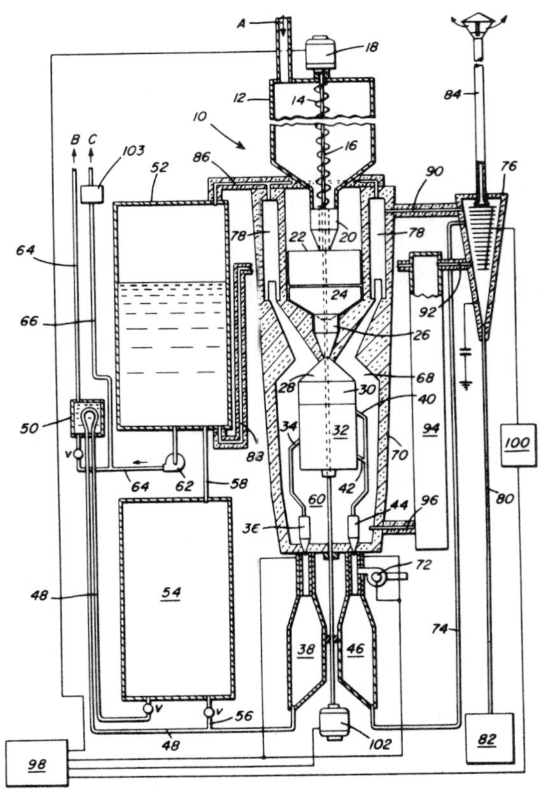

Source: U.S. Patent 3,892,662

Homogenizer **22** should preferably be a multipass unit which is capable of producing a uniform mixture of the influent (small waste particles and water) intermixed with air bubbles to aid the homogenization. It has been noted that single pass units are of greatly reduced effectiveness for this purpose and are not recommended.

The homogenized influent is next forced into an injection unit **24** which injects it through nozzle unit **26**. As previously indicated nozzle unit **26** may be an atomizing spray fog nozzle. The fogged or atomized influent upon leaving the nozzle is sprayed into a cryogenic crystallization chamber **28** where it is exposed to temperatures of about -120°C. The particulate influent is instantly frozen in crystalline form. Due to the differences in freezing points, at least three identifiable categories of crystals appear to form, water, light crystals of solid waste and heavy crystals of solid waste.

Instantaneous freezing as occurs at cryogenic temperatures produces a number of desired and necessary results. First, the cryogenic temperature maintains the influent in permanent solid state which is required for subsequent separation. Higher temperature freezing would permit particles to return to the liquid state when subsequently subjected to mechanical or electronic energy incident to the production of extremely fine particle crystals and during centrifugal separation.

Also, the influent must remain in the solid state in order that the purity of the crystals be maintained and to prevent undesirable mixing of water and other portions of the influent. Further, cryogenic freezing creates individual crystals as it acts upon the fogged influent material and thus becomes a first stage in the waste separation process.

The cryogenically frozen crystalline particles are next processed to further homogenize the crystal mixture and to finally reduce the particle size to the critical size for effective centrifugal separation. In homogenizer-reducer **30** the crystals are subjected to a blending to prevent the collection of multiple crystals into individual large crystals which may trap impurities and create nonuniform crystal sizes.

In addition, the individual crystal size is reduced to produce solid material that is of extremely fine texture, approximating the particle size of fine flour particles or smaller, which material takes on and exhibits the flow characteristics of a liquid. While it is not desired to limit the process to a specific crystalline size, since the precise size will depend upon the design of the centrifugal separator and associated equipment, it is necessary that the frozen crystalline particles be sufficiently small that the mass takes on the characteristics of a liquid.

The frozen, properly sized crystals are introduced into the cryogenic centrifugal gravity separator **32**, preferably a multiple stage separator, which separates water in crystal form from the light and heavy solids of the influent. Separator **32** acts upon the crystals in much the same manner as conventional centrifugal separators effect liquid-liquid density separations. The water crystals are metered from separator **32** via line **34** through a crystal injector **36** into a high temperature steam chamber **38**. The light and heavy solid waste crystals are directed from separator **32** via lines **40** and **42** through crystal injector **44** into an incinerator or combustion chamber **46**.

Freeze-Thaw Process

C.J. Raineri, Z.P. Mandy, M. Lesczynski, J.S. Styron and G. Akerhielm; U.S. Patent 3,880,756; April 29, 1975; assigned to Carrier Corporation describe a system for sequentially freezing and thawing raw sludge to effect the separation of the sludge into concentrated and dilute parts. The system comprises a reversible refrigeration circuit including a pair of heat exchangers adapted to receive and discharge sludge, a compressor, and appropriate valves and pumps for controlling the circulation of refrigerant through the circuit.

The heat exchangers serve alternate and opposite functions as evaporators and as condensers. The evaporator receives raw sludge which is cooled and frozen by virtue of the heat transferred from the sludge to the evaporating refrigerant. The condenser operates to thaw previously frozen sludge by virtue of the transfer of heat from the condensing refrigerant to the sludge. Melted sludge or cool water can also be circulated through the condenser to aid in the thawing process.

The system further includes a raw sludge tank for supplying raw sludge to the evaporator, and for receiving for recirculation liquid raw sludge discharged from the evaporator. A melted sludge tank receives sludge discharged by the condenser, and acts as a supply for recirculated melted sludge. Dampers associated with each of the heat exchangers direct sludge discharged by the exchangers to the appropriate tanks.

The direction of refrigerant flow is regulated by a series of valves disposed in the various refrigerant lines, these valves and other mechanical devices in the system being automatically controlled by various electrical switches and timers whose functioning can be regulated by the levels of sludge in the respective tanks. The system and the method for treating the sludge have been found to be very effective and efficient, and are quite capable of operating on a continuous basis to treat large quantities of sludge.

Freezing Technique

In a process described by *C.J. Regan, R.J. Stephenson and G.S. Clements; U.S. Patent 2,703,782; March 8, 1955; assigned to London County Council, England* sewage sludges are completely frozen and then permitted to thaw and are then settled and/or filtered. The apparatus for carrying out the process utilizes the latent heat of fusion of ice to effect the alternate refrigeration and thawing of the sludge under treatment.

Experiments show that the filterability of the sludges without the addition of chemicals, is as a result of the process, greatly enhanced, filtration occupying only 1/5 of the time generally required for untreated sludges or sludges treated according to well known processes. With the addition of suitable amounts of selected chemicals, singly or in a mixture, to the sludge before freezing, the rate of vacuum filtration may be very greatly increased as compared with the rates obtained by normal good practice in the use of sludges treated according to processes generally used. Examples of suitable chemicals for this purpose are ferric compounds, chlorinated ferrous sulfate, aluminum compounds and chlorine.

It has been found that filtration of digested sludge, chemically treated and frozen according to the process, consistently gave, on vacuum filtration through a flannel filter cloth, filter cakes containing well over 20% solids, and in some cases, at rates higher than 300 pounds per square foot per hour dry solids.

Experiments were carried out on the filterability of digested sludges treated according to the process with chemicals on a gravity sand filter of ¼ sq ft area and made up of 1" of builders' sand over 1" of granite road chippings. In one experiment the filter cake produced could easily have been removed after 20 minutes representing an output of 9 lb/sq ft/hr, but it was actually removed as a cake of 24.8% solids after 30 minutes, that is, at 6 pounds per square foot per hour.

Similarly it has been found that activated sludge after chemical treatment and freezing according to the process filters more easily. Experiments carried out with the activated sludge treated with 1,000 parts of ferric iron per million parts of wet sludge in the process resulted in the production of dry solids at rates up to 52 lb/sq ft/hr by vacuum filtration. In a typical case using vacuum filtration 80% of the liquid was extracted from a sludge previously settled to 40% by freezing, in one minute.

Conical Screw Compaction Device

E. Condolios and M. Merle; U.S. Patent 3,923,652; December 2, 1975; assigned to Societe Generale de Constructions Electriques et Mecaniques (ALSTHOM), France describe a method for compacting fluid sludges which involves slowly stirring the sludges by means of perforated blades rotating at low speed in a tank and extracting them from the tank by pressing by means of a perforated conical screw rotating slowly in a conical housing.

The controlled stirring of the sludge is effected mechanically, slowly, at all points, by inclined blades placed judiciously in the tank containing the sludges and formed with holes which are well determined for obtaining the settling of the flakes of sludge, the sludges thus settled being extracted through the bottom of the tank, by a perforated conical screw rotating slowly in a conical recess whose point, constituting the sludge outlet orifice, is directed downwards, so that the discharge of sludge decreases as the extraction orifice of the device is approached, so as to send the interstitial liquid which is between the flakes back towards the top and to make the latter come out in a compact mass towards the orifice.

The tank may comprise one or several conical screws for extracting and the drive of the conical screws for extracting may be common to that of the mechanical stirring blades or distinct therefrom.

With reference to Figure 3.15, the sludges, after a preliminary primary and secondary flocculation treatment by the known methods, are inserted into a device according to the process by a tube 1.

The large flakes which have been formed during primary and secondary flocculation operations are partly destroyed by the passing in the tube 1 due to the speed required for the transfer of the sludge into the device. To enable the reconstituting of the flakes and to speed up the settling thereof, a flocculation agent which is generally a synthesizing polymer, called a settling flocculation agent, is inserted at a point 2 on the tube.

FIGURE 3.15: SLUDGE COMPACTION DEVICE

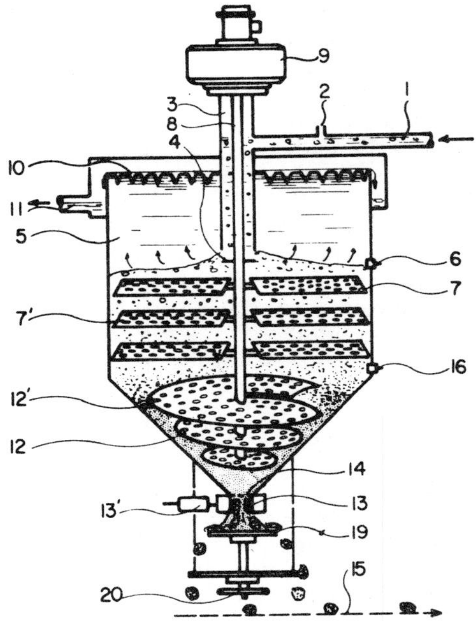

Source: U.S. Patent 3,923,652

At the inlet of the tube **3**, the settling flocculation agent gathers together the flakes which are more or less broken by their hydraulic transfer in the tube **1** to form again large flakes which go down the tube **3** towards a distribution deflector **4** placed at the bottom of that tube. The deflector separates the clear liquid from the large flakes which it contains in a tank **5**. The large flakes are then deposited on the bottom of the tank and the liquid and the water which are practically clear rise towards spillway **10** of the tank **5** and are decanted in a regular way by V-shaped channels placed on the spillway to be discharged by the scavenging spout **11**.

The flakes which have been deposited on the bottom of the tank are then pushed by a series of blades **7** rotating slowly in the tank. These blades are driven by a shaft **8** and a speed reducing motor **9** enabling the required speed to be obtained.

The flakes which have thus been enlarged and have settled in the central zone of the tank **5** move down towards the cylindro-conical zone placed at the bottom of that tank. The flakes are then taken up again by a conical screw **12** comprising holes **12'** which push them slowly towards the orifice **13** placed at the bottom of the conical part of the tank **5**. That conical screw **12** may be controlled by the same shaft **8** and the same speed reducing motor unit **9** such as shown in Figure 3.15, either by another shaft concentric with the first, driven by the same gear speed reducing motor **9**, or by another drive unit. The latter arrangement enables a rotation speed of the conical screw different from that of the presettling blades **7** to be obtained.

The arrangement is one of the important points of the process, for it is necessary to give the conical screw a rate such that the discharge of the sludge discharged by the screw decreases as progress is made towards the orifice **13**. This particular arrangement enables the flakes to settle completely and the water or liquid still contained between the flakes to be discharged towards the top of the conical screw. It is also necessary to place on the conical screw holes having dimensions close to those of the blades **7** so as to enable the water or liquid to rise more easily towards the top of the tank **5** and to obtain a low conical screw speed which never exceeds that of the presettling blades **7**.

The flakes which have completely settled at the bottom of the screw obstruct completely the orifice **13** and prevent any pouring out of liquid towards the bottom. Under the mechanical action of the screw **12**, these settled flakes have gathered together to form a very compact mass which pours out through the orifice **13** to be removed by classical means **15** (conveyor belt, lorry, wagon).

The orifice **13** is provided with a grill **14** having hole dimensions chosen according to the sludges and comprises a variable cross section adjusted by the height of the sludge in the tank **5**. A plate **19** which may be adjusted for height by means of a control device **20** is placed beneath the orifice **13** in order to maintain a continuous mass of sludge at the outlet of the extraction orifice to avoid the driving out of water through cracks which may exist in the sludges.

To obtain a normal operation of the device described, taking into account the fact that the discharge of the sludge inserted in the tube **1** is generally variable, a minimum sludge level detector **16** and a maximum sludge level detector **6** will be placed on the tank **5**. If the level of sludge is found to be below the detector **16**, it is necessary either to reduce the speed of the conical screw **12** or to even stop that screw for a more or less long time so that the level may rise above the detector, or to reduce the cross section of the orifice **13** by suitable means, such as closing device **13'** controlled by the detector.

If the sludge level increases too much and exceeds the detector **6** which is generally placed at the height of the deflector **4**, it is necessary to increase either the speed of the conical screw **12** without ever exceeding the maximum permissible speeds, or to increase the dimension of the orifice **13** or both. If, at the limit of the speed possibilities of the conical screw **12** and of the diameter of the orifice **13** the level of the sludge exceeds the level **6**, it is then necessary to reduce the discharge of the sludge to be treated which is admitted in the tube **1**.

Conversion to Carbonate Form

P.J. Cardinal, Jr.; U.S. Patent 3,947,350; March 30, 1976; assigned to Envirotech Corporation describes a process for preparing sludge for dewatering in a tertiary line precipitation treating system in which raw sewage has been dosed with calcium hydroxide to remove phosphates. A source of carbon dioxide is introduced into the treating system to convert calcium hydroxide salts and other salts to the more readily dewatered carbonate form, and to convert magnesium hydroxide salts to the bicarbonate form to facilitate removal of magnesium from the system. Carbon dioxide may be introduced in gaseous form, an available source of which is the exhaust gases from sludge incineration following dewatering. Alternatively, sulfuric acid or hydrochloric acid or other suitable chemical reactant may be introduced into the system to neutralize the higher pH of the sludge.

Conversion of the sludge to the carbonate form is effected in conjunction with the production of carbon dioxide during the neutralization reaction effected by the chemical reactants introduced into the system. Desirably, the carbon dioxide is utilized in the system in a molar ratio of approximately 1 to 1 with the calcium hydroxide dosed into the system.

Referring to Figure 3.16, a system utilizing the treating process is shown. The process is illustrated in conjunction with a standard raw sewage line precipitation tertiary treating procedure with the improved concepts of this process incorporated. In such a procedure, raw sewage influent is introduced through conduit **1** into a primary clarifier **2** for clarification in known fashion. A predetermined quantity of calcium hydroxide is introduced through conduit **3** into the primary clarifier for phosphorous removal from the sewage in the primary clarifier in known fashion.

After primary clarification, the clarified mixture is subjected to a secondary treatment in an activated sludge system, summarized by reference numeral **4**, in which aeration or like treatment of the sludge is effected. Sludge from the primary clarifier is introduced through conduit **6** in the activated sludge system.

The sewage effluent from the activated sludge system passes through conduit **7** into a secondary clarifier **8**. Following secondary clarification, in known fashion, the effluent passes through conduit **9** into a flash mixer **11** from which the effluent passes through conduit **12** into a chemical clarifier **13**. From the chemical clarifier a suitable conduit **14** carries the clarified sewage to subsequent treatment, in separation beds and the like. Procedures effected within the activated sludge system, secondary clarifier, flash mixer and chemical clarifer are conventional and known in the tertiary treating art.

Waste activated sludge from the sludge system is transferred through conduit **16** into a sludge thickener **17** of known construction. In known fashion, a predetermined portion of the activated sludge is returned through conduit **18** into the tertiary system in advance of the activated sludge system for retreatment for most effective results.

Sludge from primary clarifier **2** is transferred through conduit **19** into a treating tank **21** in which the primary sludge is subjected to the recarbonation effects of carbon dioxide which transforms the hydroxide salts and the other salts in the sludge to the carbonate or bicarbonate form, in which form they may more readily be dewatered.

Following carbon dioxide treatment in tank **21**, the recarbonated sludge is introduced through conduit **22** into the aforementioned sludge thickener **17** in which settling is allowed to take place. The fluid overflow from the thickener passes from the thickener in known fashion and the solids are transferred through conduit **23** into a dewatering apparatus **24**. Such dewaterer may be of any type known and used in tertiary systems, such as a centrifuge, or vacuum filter.

FIGURE 3.16: SLUDGE TREATING PROCESS WITH TERTIARY SEWAGE TREATMENT SYSTEM

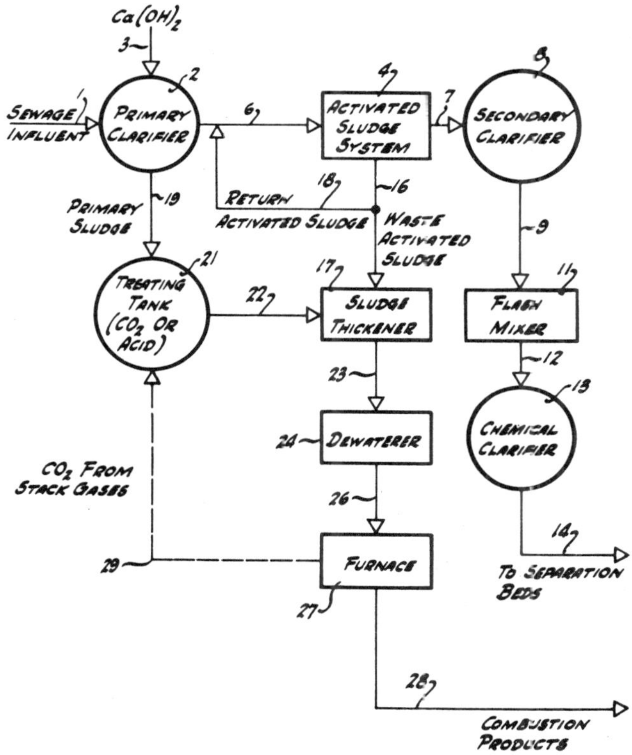

Source: U.S. Patent 3,947,350

The dewatered sludge thereafter preferably is transferred through conduit **26** into a furnace apparatus **27**, which may be of any known type, such as a multiple hearth incinerator, a lime recalcining kiln, or a combination incinerator-kiln, as described in U.S. Patent 3,623,975. The dried solid products of combustion are removed from the furnace through suitable conduit means **28** for subsequent treatment and handling in known fashion.

In one form of this process, the gaseous products of combustion of furnace **27** are utilized as a readily available and relatively inexpensive source of carbon dioxide which can be employed to recarbonate the sludge in the treating tank **21**. Such exhaust gases resulting from incineration or recalcination of the sludge in the furnace are discharged from the furnace and preferably are passed through a gas scrubber of known construction to cleanse and concentrate the carbon dioxide therein for subsequent reuse. As thus concentrated, the carbon dioxide may be selectively reintroduced through suitable conduit means **29** into the aforementioned treating tank **21**. Such carbon dioxide gas thus bubbles through the mixture in tank **21** to produce the recarbonating results desired.

As an alternative to utilization of gaseous carbon dioxide from the stack gases of furnace **27**, suitable chemical reactants or compositions which provide a source of carbon dioxide may be introduced into the treating tank **21** to effect transformation of the hydroxide salts therein to the carbonate or bicarbonate form. That is, by introducing a suitable chemical reactant or reactants into the system, the pH of the lime dosed sludge is reduced from its normal range of approximately 8.0 to 11.0 to approximately 7.0 pH, thereby converting the calcium compounds in the sludge to calcium carbonate and the magnesium compounds in the sludge to magnesium bicarbonate.

Various known chemical reactants may be employed for that purpose, the criterion for their use being their ability to produce carbon dioxide when mixed with the salts contained in the sludge introduced into treating tank **21**. In that connection, various acids known to be reactive with the sludge to produce or liberate carbon dioxide therefrom during the chemical reaction therebetween may be employed. Sulfuric acid and hydrochloric acid are typical and exemplary of the type of chemicals which may be employed to reduce the solubility of the hydroxide salts to convert the same to the carbonate or bicarbonate form.

The amount of carbon dioxide introduced into the various systems described herein to react with the calcium hydroxide of the system will be selected in accordance with the requirements of particular treating installations. Generally, effective recarbonation requires approximately one mol of carbon dioxide for each mol of calcium hydroxide introduced into the system. That is, a molar ratio of approximately 1:1 CO_2:$Ca(OH)_2$ has been found to produce effective results. However, in those systems in which magnesium hydroxide is formed in appreciable amounts, the molar ratio noted should be modified to increase the amount of carbon dioxide, for example to 1.5 or 2:1, to insure desired conversion of the magnesium hydroxide to magnesium bicarbonate.

By way of example, a tertiary sewage treating plant having 1 million gallons per day capacity, using 1 ton (240 ppm) of calcium hydroxide per day requires about 0.8 tons of carbon dioxide for complete and effective conversion of the calcium hydroxide salts to the calcium carbonate form.

In such an arrangement the sludge constituency is altered so that the calcium hydroxide is converted to the carbonate form. The other materials in the sludge, such as primary solids and waste secondary sludge solids, remain essentially the same as in a conventional tertiary treating procedure. The introduction of carbon dioxide into the system subsequent to lime dosing produces the carbonic acid employed to produce recarbonation in the manner described.

It should be understood that the calcium hydroxide quantities dosed into the system to attain a minimum acceptable phosphorus removal level of approximately 70% are a direct function of the alkalinity of the sewage and its phosphorus content. Thus, alkalinity and phosphorus content should be carefully monitored in the system.

Therefore, it should also be noted that, for effective results, the raw sewage should be dosed with calcium hydroxide in amounts sufficient to insure that the pH of the sewage is increased to or maintained at 7.0 or above. In preferred operating systems, such pH is increased to within the range of approximately 8.5 to 12.0, to effect minimum acceptable phosphorus removal of approximately 70%.

However, if the pH level is controlled to fall within the range of 9.5 and above, magnesium hydroxide precipitation results. Such precipitation, unless counteracted, produces two undesirable results: (1) dewatering with mechanical devices is rendered more difficult, and (2) particulate buildup of magnesium hydroxide within the system increases. Thus to counteract such precipitation, carbon dioxide is utilized in the system to convert the magnesium hydroxide to magnesium bicarbonate, as noted, which is discharged from the system with other soluble effluents.

USE OF FLOCCULANTS, PHOSPHATE AND NITRATE REMOVAL

FLOCCULATING AGENTS

Condensation Products of Polyalkylene Polyamines, Ketones and Aldehydes

W.T. Grundner; U.S. Patent 3,849,307; November 19, 1974; assigned to Sybron Corporation has found that reacting a polyalkylene polyamine, an aldehyde and a ketone in molar ratios of about 0.9 to 1.1 polyalkylene polyamine to about 0.9 to 1.1 ketone to about 1.8 to 3.0 aldehyde produces a viscous, water-soluble, condensation product which has utility as a flocculant, a dewatering aid and a retention aid. It is essential to avoid water-insoluble resin in preparing the condensation product, and therefore, it is critical to add the polyalkylene polyamine, the aldehyde and the ketone within the specified ranges.

In general, an aqueous solution of a polyalkylene polyamine cooled to ambient temperature is mixed with a ketone and an aldehyde and heated at 75° to 100°C until the viscous product forms. The aldehyde and ketone are added to the polyalkylene polyamine solution over a period of time to prevent excessive heating due to the reaction exotherm. Subsequent to the final addition of ketone and aldehyde heat is externally applied. The following examples illustrate the process.

Example 1: Preparation of Condensate — Into a sealed 10-gallon stainless steel reactor equipped with stirrer, thermometer, reflux condenser and addition port were placed 11.93 kg of water and 5.805 kg (30.66 mols) of tetraethylene pentamine (TEP). The resulting solution was cooled to 25°C and a mixture of 1.78 kilograms (30.65 mols) of acetone and 5.971 kg (73.49 mols) of a 37% formaldehyde solution was added over a period of 1.75 hours.

The first two-thirds of the acetone-formaldehyde mixture was added over 0.75 hour. The temperature reached 65°C as a result of reaction exotherm and externally applied heat. The balance of the mixture was added over a period of 1 hour while maintaining a temperature in the range of 65° to 70°C. The reaction was allowed to continue for 16 hours at 93° to 95°C after the addition was complete. The heating was discontinued and 6.812 kg of cold water was added to the viscous material. The resulting mixture was filtered through a single layer of cheesecloth and cooled. The product, 32.75 kg, contained 24.8% solids. Upon dilution to a solids concentration of 20%, the viscosity of the resulting solution was found to be 33 cp at 22°C. The reactants were added at a molar ratio of 2.4 formaldehyde to 1.0 of acetone to 1.0 of tetraethylene pentamine (2.4:1:1). Test results are shown in the table below.

Example 2: Several reactions using various polyalkylene polyamines were carried out in accordance with the method disclosed in Example 1. The polyalkylene polyamines utilized in various experiments were ethylene diamine (ED), iminobispropylamine (IBP) and diethylene triamine (DET). Molar ratios, temperature and other reaction conditions were identical with Example 1 above, unless shown otherwise in the table. The viscous condensate products obtained from each reaction were diluted to working concentration of 0.1% for use in testing and are recorded in the table.

Another set of reactions, in which reaction temperature and formaldehyde, acetone, and tetraethylene pentamine concentrations were varied, were conducted in accordance with the method described in Example 1. Test results and variables are shown in the table.

Example 3: Measurement of Flocculation — Into a 1,000-ml graduated cylinder was placed 1,000 ml of tap water. A sample of each condensate prepared in Examples 1 and 2 was diluted to a working concentration of 0.1%. A sufficient amount of the 0.1% condensate solution was added to the 1,000 ml of water in order to provide a concentration of 1 part per million of condensate in the water. The pH was adjusted to about 6.7. Fifteen grams of a finely divided kaolinite was added to the graduate and dispersed through the liquid by vigorously shaking and inverting the covered graduate.

The graduate was placed on a flat surface and illuminated from behind. The fine clay (kaolinite) particles were found to agglomerate and settle to the bottom of the graduate. The rate at which the particles settled was determined by observing the length of time required for the interface between the relatively clear supernatant liquid and the zone of settling particles to fall to a level representing 20% of the original liquid depth.

The test was repeated without the addition of the condensate, and the ratio of settling times with and without polymer addition was determined. This value of time with polymer addition divided by time without polymer addition is designated as the settling ratio. Settling ratios using 1.0 ppm of the condensate are shown in the table. The settling ratio is calculated from the following formula:

$$\frac{\text{Settling Time With Polymer Addition}}{\text{Settling Time Without Polymer Addition}}$$

The table shows the settling ratio (settling improvement factor) for the condensates tested. The lower settling ratios are an indication of superior flocculation. The condensates having the lowest settling ratios, and therefore, the ones which demonstrate superior flocculation, are those made with tetraethylene pentamine (TEP). The table points out various other polyalkylene polyamines (polyamines) which produce condensates having flocculating characteristics, but they are inferior to condensates made with TEP. Condensates made without acetone have very high (0.90 and 0.75) settling ratios, and are therefore, not suited to flocculation. Condensates having formaldehyde molar ratios of 3.5 or above are insoluble in water, and therefore, are not suitable as flocculants.

Ratio of Formaldehyde: Acetone:Polyamine	Polyamine	Reaction Temp. °C	Water Solubility	Settling Ratio 1.0 ppm Condensate
1:0:1	*TEP	75	Soluble	0.90
2.2:0:1	TEP	100	Soluble	0.75
2.2:1:1	TEP	100	Soluble	0.18
2.3:1:1	TEP	100	Soluble	0.19
2.4:1:1	TEP	95	Soluble	0.15
2.5:1:1	TEP	75	Soluble	0.32
3:1:1	TEP	75	Soluble	0.23
3:1:1	TEP	100	Soluble	0.28
3.5:1:1	TEP	75	Insoluble	—
4:1:1	TEP	75	Insoluble	—
2.3:1:1	*DET	100	Soluble	0.41
2.3:1:1	*IBP	100	Soluble	0.51
2.3:1:1	*ED	100	Soluble	0.50

*TEP = tetraethylene pentamine
*DET = Diethylene triamine
*IBP = iminobispropylamine
*ED = ethylene diamine

Acid Treated Cationic Starch

W.G. Hunt and R.J. Belz; U.S. Patent 3,901,878; August 26, 1975; assigned to Anheuser-Busch, Incorporated describe the use of an acid treated, high fluidity, high degree of substitution, waxy maize cationic starch in the flocculation and dewatering of municipal raw primary sludge. From 0.007 to 0.200% (based on the dry weight of the sludge) of the treated starch of about 30 to 50 fluidity is added to raw primary municipal sludge and contacted with the sewage before the mixture is dewatered to a moisture content below about 70% moisture, which is the moisture level at which the solids will adequately sustain combustion. The starch has from about 0.15 to 0.25 degree of substitution of an amine butene halide, specifically, 4-chloro-2-butenyl-trimethylammonium chloride.

Oxidized Cationic Starch

W.G. Hunt and R.J. Belz; U.S. Patent 3,875,054; April 1, 1975; assigned to Anheuser-Busch Incorporated describe the use of a cationic oxidized starch in the flocculation and dewatering of municipal raw primary sludge. From 0.016 to 0.1895% (based on the dry weight of the sludge) of the treated starch of 35 to 60 fluidity and 1.75 to 2.25% oxidation is added to the municipal sludge and contacted with the sewage for a period of 3 to 5 minutes before being dewatered to a moisture content below about 70% moisture, which is the moisture level at which the solids will adequately sustain combustion. The starch has from 0.15 to 0.25 degree substitution of an amine butene halide, specifically, 1-chloro-4-butenyl-trimethylammonium chloride.

The dent corn starch is oxidized to an available chlorine content of about 1.75

to 2.25%. This oxidation treatment can be carried on by any suitable method, but should be carried to the extent that it is equivalent to a sodium hypochlorite treated corn starch of this available chlorine content. Preferably, the available chlorine content is about 2.2%. The fluidity of the oxidized starch is about 35 to 60.

In reacting sodium hypochlorite with dent corn starch, from about 2.0 to 2.25% by weight (based on the dry weight of the starch) available chlorine is reacted at 80° to 85°F with about 42 to 50% by weight starch (dry basis) based on the combined water and starch which has been adjusted to pH 9.0 with a 2.0% NaOH solution. The sodium hypochlorite is added dropwise so as not to allow the pH to exceed 11.0. The reaction is then allowed to proceed for 1½ to 2½ hours after addition is complete or until the desired fluidity is achieved.

The excess chlorine is removed with minimal amounts of solid sodium bisulfite by a spot plate test using o-tolidine (complete removal of excess chlorine indicated by no color change). The starch is then adjusted to pH 6.5 to 5.0 with 1:1 HCl and diluted with excess water to 15°Be, filtered on a Buchner funnel, reslurried to 15°Be, filtered again and dried.

In reacting the amine butene halide (particularly 1-chloro-4-butenyltrimethylammonium chloride) with starch, from about 50 to 55% by weight (based on the dry weight of the starch) of the amine alkene halide is reacted with oxidized gelatinized starch for 4 to 5 hours at a temperature of about 65° to 75°C using a technique of Example 1. The final product has a carboxyl number of about 0.1 to 0.3% which corresponds to a degree of substitution (DS) of about 0.0036 to 0.0108. The average molecular weight is about 89,000.

Example 1: In preparing the cationic starch, sodium hypochlorite treated dent starch of 53 fluidity and 2.2% chlorine is added to water in the proportion of 108 g of starch (dry basis) to 400 ml of water. From about 20 to 25% by weight starch (dry basis) based on the combined water and starch can be used. The oxidized starch is prepared as described above. This mixture is gelatinized with live steam while stirring. Specifically, the starch is treated with steam at a temperature of 212° to 220°F for a period of 5 to 10 minutes while being stirred.

The temperature is adjusted to 70°C and the stirring is continued. At this point 32 g of a 50% NaOH solution is added to the mixture of gelatinized starch and water with continued stirring. The NaOH is added as a catalyst and not as a gelatinizing agent. From about 12 to 18% by weight NaOH (based on the dry weight of the starch) can be added. The temperature can be 65° to 75°C.

At this point, 111 g of a 50% TAC solution (1-chloro-4-butenyltrimethylammonium chloride) is added. This reagent is prepared as hereinbefore described. This solution is allowed to react for 4 hours with stirring at 70°C before it is diluted to 25% solids by weight with water to product a gelatinized oxidized cationic starch. From about 50 to 55% by weight (based on the dry weight of starch) of the amine butene halide can be used. This reaction between the amine butene halide and the starch can continue for 4 to 5 hours at a temperature of 65° to 75°C. Following are examples of treating raw primary sewage with the cationic oxidized starch of this process.

Example 2: A mixture of 108 g (dry basis) 7303 (lightly oxidized dent starch) in 300 ml water is gelatinized with steam and diluted with 300 ml water. The dent starch had a chlorine content of 2.20%. The gelatinized starch is transferred to a 70°C water bath and 32 g of 50% NaOH solution and 111 g 50% TAC solution are added. This mixture is reacted for 4 hours at 70°C. The resultant product is 21% solids by weight and has a DS of 0.25.

Composite samples of primary sludge were picked up from a municipal sewage treatment plant. Cationic starch at a concentration of 0.5% was examined using the following method: A predetermined amount (based on the dry weight of the sludge) of a 0.05% by weight aqueous solution of an anionic polyelectrolyte is added to 200 ml of raw primary sludge and mixed back and forth 10 times between two beakers. This anionic polyelectrolyte is a very high molecular weight synthetic water-soluble hydrolyzed polyacrylamide (Purifloc A-23).

A specific weight (based on the dry weight of the sludge) of a 0.5% by weight aqueous solution of cationic polyelectrolyte prepared according to Example 3 is then added, mixed back and forth at 76°F ten times between two beakers, and transferred to a Buchner funnel fitted with Whatman No. 1 filter paper. The pressure in the filtrate receiver was reduced to 13 inches Hg and the filtrate collected after 30, 60, 90 and 120 seconds determined. Results are as follows:

Run	Sludge (ml)	% Solids	Amount of Anionic Polyelectrolyte Added (%)*	Amount of Cationic Oxidized Starch Added (%)*	Amount of Filtrate per Unit Time (ml/sec) 30	60	90	120	% Solids After Filtration
1	200	5.0	0.025	0.100	130	167	173	175	40.0
2	200	6.4	0.035	0.097	95	142	160	168	40.0
3	200	5.4	0.037	0.097	112	150	160	167	32.7
4	200	11.4	0.022	0.098	60	97	120	130	32.6
5	200	3.5	0.011	0.100	70	125	170	183	41.2
6	200	9.6	0.017	0.047	83	123	147	158	45.7
7	200	2.0	0.025	0.100	85	125	155	172	14.3
8	200	9.5	0.025	0.092	137	160	163	165	55.9**
9	200	5.0	0.015	0.100	112	158	172	177	43.5
10	200	5.2	0.029	0.096	150	175	178	178	44.3**
11	200	9.9	0.025	0.101	115	130	135	137	31.4
12	200	5.2	0.012	0.096	115	160	172	175	41.6
13	200	5.2	0.024	0.096	123	168	174	176	43.3
14	200	3.6	0.024	0.097	92	137	158	170	24.0
15	200	9.4	0.027	0.098	73	100	114	123	24.4
16	200	6.0	0.025	0.100	102	140	153	160	30.0

*Based on dry weight of sludge. **Average.

In all tests, the minimal quantities of polyelectrolytes required to increase solids to approximately 30% in 120 seconds was used as criteria for adequate flocculation and dewatering. The 30% solids level after filtration is an adequate solids content for sustaining combustion when the filter cake is later incinerated.

Example 3: A predetermined amount (based on the dry weight of the sludge) of a 0.05% by weight aqueous solution of an anionic polyelectrolyte (Purifloc A-23) is added to 2000 ml of raw primary sludge and mixed. Then a specific amount of a 0.5% by weight aqueous solution of cationic polyelectrolyte (based on the dry weight of the sludge) prepared according to Example 1 is added and mixed, being sure not to deflocculate the sludge with the agitation. A filter leaf with a reduced pressure of 13 inches Hg is then lowered into the conditioned sludge. Filtrate is collected for 60 seconds and the filter leaf is removed and allowed to dry under reduced pressure for another 180 seconds. The filter cake is then removed and percent solids and total dry solids are determined. Results are as follows:

Run	Sludge (ml.)	% Solids	Amount of Cationic Oxidized Starch Added (%)[1]	Amount of Anionic Polymer Added (%)[1]	% Solids After Filtration
1	2000	6.5	.045	.032	23.8
2	2000	7.0	.045	.032	27.8
3	2000	6.0	.045	.032	29.4
4	2000	7.4	.045	.032	24.9

[1]Based on dry weight of sludge.

In this example, the test procedure for determining adequate flocculation and dewatering by the polyelectrolyte has been changed from the Buchner funnel test that was used in Example 2. This is due primarily to the small percentage polyelectrolyte (dry basis) by weight used based on the dry weight of the sludge. In all tests, approximately 30% solids after 4 minutes vacuum filtration on a filter leaf was used as criteria for adequate flocculation and dewatering.

Example 4: Following is a plant-size run using oxidized dent cationic starch in which the starch is 2.2% oxidized and has a degree of substitution of 0.25 for the amine butene halide. The sewage treatment plant is a commercial operation and the sludge had an initial solids content of 10.8% by weight. The starch is added to the sewage as a 0.5% by weight solution at a 0.04785% by weight basis (based on the dry weight of the sewage).

The amine polyelectrolyte (Purifloc A-23) is added as a 0.05% by weight aqueous solution at a 0.0275% by weight basis (based on the dry weight of the sludge). The treated sludge is filtered on a rotary filter at a rate of 277.5 wet pounds per minute. The treated sludge has a solids content of 31.95%. On a dry solids basis, 0.975 pound of starch reactant is added per ton of dry sludge recovered. This is a very satisfactory performance both from a cost and a technical viewpoint.

Cationic Polyelectrolyte Polymers

R.D. Lees; U.S. Patent 3,472,767; October 14, 1969; assigned to Hercules Incorporated describes a process for facilitating the dewatering of waste having suspended solids by contacting the waste with a cationic polyelectrolyte polymer in the presence of an acid salt containing polyvalent metal ions. The following examples illustrate the process.

Examples 1 and 2: Sewage Plant Runs — Runs were carried out to determine the effect on the filtration characteristics of unctuous sewage slurry of polymer alone as contrasted with the same polymer together with acid salt containing polyvalent metal ions as dewatering aid. The dewatering aid was added to the sewage treatment plant by mixing the dewatering aid with sewage sludge after the digestion step and pumping the resulting mixture directly into the filter pan of the rotary vacuum filter employed in the vacuum filtration step, i.e., the plant was not employing an elutriation step. Further details appear below in the table. As will be seen from this table, the process increased by more than 40% the efficiency of a sewage treatment plant, efficiency being reported as customary in dry solids yields in a given operating period.

Sewage Plant Runs

Example No.	Dewatering Aid: Name	Amount[1]: Polymer	Amount[1]: Salt	Percent Solids in Influent Sewage Sludge	Dry Solids Yields, lb/hr	Time, hr
1	Polymer A only	131	None	14	1,960	19.0[2]
2	Polymer A plus Salt A	64	6,000	14	2,920	20.5[3]

[1]Ppm dry weight basis by weight of total dry solids in sewage treated. Polymer added as 0.1% aqueous solution and salt added as 1.0% aqueous solution.
[2]Had to be shut down 42% of this time for unclogging the filter.
[3]No shut-down was necessary—gave continuous good filtration.
Polymer A = 83% acrylamide, 17% MTMMS copolymer.
Salt A - $AlCl_3$.
MTMMS = Beta-methacrylyloxyethyltrimethylammonium methyl sulfate.

Preferred polymers include the copolymers of acrylamide monomers ("acrylamide" being used to designate acrylamide per se as well as acrylamide substituted on the alpha carbon atom as well as on the nitrogen atom) with other ethylenically unsaturated monomers such as for example (1) vinyl quaternary ammonium salts including those derived from dialkylaminoalkyl methacrylate (e.g., beta-methacrylyloxyethyltrimethylammonium methyl sulfate), dialkylaminoalkyl acrylate, dialkylaminoalkyl vinyl ether, (2) vinyl pyridinium salts including those derived from 2-vinyl pyridine, 4-vinyl pyridine, 2-methyl-5-vinyl pyridine (e.g., 1,2-dimethyl-5-vinyl pyridinium methylsulfate), 2-vinyl-5-ethyl pyridine.

Salts applicable herein, in combination with the polyelectrolytes, as dewatering aids comprise substantially water-soluble inorganic acid salts containing polyvalent metal ions and preferably trivalent metal ions such as Al^{+++} or Fe^{+++} or both. Particularly good results have been obtained with, for instance, $AlCl_3$ and $Al_2(SO_4)_3$.

F.C. Buhl and R.D. Lees; U.S. Patent 3,414,513; December 3, 1968; assigned to Hercules Incorporated found that carrying out the elutriation step in the presence of a small amount of a certain acrylamide-beta-methacrylyloxyethyltrimethylammonium methyl sulfate (acrylamide-MTMMS) copolymer provides a highly effective and economical means of increasing the solids recovery and decreasing the "hold" time of this step. The particular copolymer effects these substantial improvements by increasing the flocculation rate, floc size and floc density which in turn gives not only a substantially faster sedimentation rate but also gives an appreciable increase in percent total solids recovered.

F.C. Buhl; U.S. Patent 3,414,514; December 3, 1968; assigned to Hercules Incorporated has found that contacting sewage with a small amount of a certain acrylamide-beta-methacrylyloxyethyltrimethylammonium methyl sulfate (acrylamide-MTMMS) copolymer and vacuum filtering provides a highly effective and economical means of dewatering sewage. The copolymer flocculates the sewage particles and substantially facilitates dewatering by vacuum filtration to give the desired relatively thick and dry sludge cakes in a short period of time as contrasted with much thinner and wetter sludge cakes requiring substantially longer periods of time with prior art flocculants. In general, somewhat smaller amounts are required of the copolymer flocculant of the process than with prior art flocculants. The following examples illustrate the process.

Examples 1 through 4: These sewage plant runs were carried out in order to determine the effect of various flocculating agents on the filtration characteristics of sewage sludge. The flocculating agent was added to the sewage treatment plant stream by mixing the flocculating agent with sewage sludge after the elutriation step and pumping the resulting mixture into the filter pan of the rotary vacuum filter employed in the vacuum filtration step. As noted below, the particular copolymer of the process gave substantially higher production rates at far lower concentrations than the conventional $FeCl_3$-CaO flocculating agent.

Sewage Plant Runs—Filtration
Effect of Various Flocculating Agents on the Filtration Characteristics of Sewage Sludge

	Flocculating Agent			Percent Solids in Influent Sludge	Dry Solids Yield lbs./sq. ft. filter/hr.	Total Dry Solids Produced lbs.
		Amount*				
Example	Name	Lbs.	Percent			
1	$FeCl_3$-CaO	3,189 to 18,336	1.60 to 9.2	10.8	3.80	199,300
2	$FeCl_3$-CaO	2,892 to 17,930	1.50 to 9.3	11.0	3.71	192,800
3	Acrylamide-MTMMS	13.9	0.11	10.7	5.08	12,700
4	Acrylamide-MTMMS	28.9	0.11	11.7	4.79	26,300

* Dry weight basis by weight total dry solids in sewage treated, added as 1.0% aqueous solution. Examples 3 and 4 used 83% acrylamide-17% MTMMS copolymer.

The copolymer flocculating agent of the process consists by weight essentially of 99 to 20% acrylamide and 1 to 80% MTMMS, preferably 97 to 50% acrylamide and 3 to 50% MTMMS, 95 to 60% acrylamide and 5 to 40% MTMMS being specifically preferred.

Sodium Polystyrene Sulfonate

A method described by *C.P. Priesing and S. Mogelnicki; U.S. Patent 3,247,102; April 19, 1966; assigned to The Dow Chemical Company* involves the elutriation process. This process comprises the steps of feeding a digested sewage sludge to a liquid mixing zone concomitantly with an aqueous washing medium and thereafter settling the washed sewage solids and withdrawing the wash water elutriate as an overhead stream.

The process particularly involves the improvement in the foregoing process which consists in conjunctively (1) feeding a controlled amount of an additive nondigested sewage sludge into the mixing zone and (2) incorporating a sufficient amount of an anionic organic polymeric flocculant into the mixing zone to enhance the rate of solids-liquid separation. The adjective "additive" is applied to the specified nondigested sewage sludges inasmuch as such sludges are not normally in the feed to the mixing zone.

The amounts of the additive sewage sludge and anionic polymeric flocculant blended with the digested sludge and aqueous washing medium will vary for optimum results between relatively narrow limits. To enhance the settling rate of solids in the resulting blend, the amount of the additive sewage sludge used should be controlled to provide from about 5 to 50% by weight additional solids based on the weight of the digested sewage sludge solids, with 10 to 30% additional solids being preferred. The amount of the organic polymer flocculant utilized is within the range from about 0.01 to 10 lb/ton of the total sewage sludge solids fed to the mixing zone, with from about 0.1 to 1 lb being preferred. The term "solids" refers to the mass of solid materials determined by filtering a given aliquot of the sludge and drying the filter cake at 110°C for 2 hours. The weight of such dry solids is taken as the solids content of the given aliquot.

Specific polymeric flocculants for use in the process include the high molecular weight, anionic, water-soluble polyalkanes obtained by the vinyl polymerization of one or more monoethylenically unsaturated monomers bearing a carboxylic or sulfonic acid group, including alkali metal and ammonium salts with the optional inclusion of other monoethylenically unsaturated monomers to provide a water-soluble polymer. A preferred class of anionic polymeric flocculants is the polymers of an alkali metal or ammonium salt of vinyl aromatic sulfonic acid having a weight average molecular weight of about 6 million or more.

Such polymers correspond to water-soluble homopolymers of the sulfonate monomers and copolymers prepared by the vinyl polymerization of a major proportion of the vinyl aromatic sulfonate with a minor proportion of a monomer such as acrylamide, methacrylamide, acrylonitrile, styrene, vinyltoluene, and methyl acrylate. Such polymers are also obtained by the direct sulfonation of a poly(vinyl aromatic) backbone polymer. A suitable method is described in U.S. Patent 3,172,618. In general, the vinyl aromatic nuclei of the sulfonated polymers are derived from styrene, vinyltoluene and α-methyl styrene. The following examples illustrate the process.

The following operations were carried out on a laboratory scale. As a consequence, the elutriation process was conducted as a batch process. The digested sewage sludge to be elutriated was obtained from the sewage treatment plant of a small Midwestern city. The solids content of this sludge was 117,698 parts by weight per million parts by weight of sludge. The solids contents of individual sludges and blends are determined for the purposes herein by filtering a known aliquot of the sludge or blend of interest, in a Büchner funnel vacuum filter and weighing the filter cake after drying it at 110°C for 2 hours.

The additive nondigested sewage sludge to be mixed with the digested sewage sludge was obtained from a second sewage processing plant in the form of an activated sludge containing 8,300 parts of solids per million parts by weight of sludge. Normally, large sewage processing plants have both types of sewage sludges available at a given location.

The aqueous washing medium was obtained from the overhead effluent stream of a primary settling clarifier for raw sewage. The polymeric flocculant utilized was a high molecular weight, water-soluble anionic polymer in the form of a sodium polystyrene sulfonate having a weight average molecular weight of about 6.2 million. The polymer was applied to the system to be flocculated as a 0.05% by weight solution in water.

To carry out the elutriation of the digested sludge, the aqueous washing medium was premixed with the additive nondigested sewage sludge. Particularly, this blend was prepared by mixing together 2,118 ml of the activated sludge with 1,382 ml of the primary effluent. The solids content of the blended mixture was measured and found to be 5,763 parts of solids per million parts by weight of the mixture. 3,500 ml of the above mixture were added to 700 ml of the aforementioned digested sewage sludge. This gave a liquid dilution ratio of 5 to 1. The additive nondigested sewage sludge solids comprised 20% of the total sludge solids. The resulting mixture was mildly agitated by stirring to accomplish blending of the suspended sewage solids.

Subsequently, the blend was divided into four 1-liter aliquots. Each of three aliquots was treated with a different dosage of the aforementioned polymer solution. Each polymer treated mixture was then stirred with a paddle stirrer at 50 rpm for one minute to accomplish complete dispersion of the polymer throughout the suspension. Thereafter the paddle was rotated at 25 rpm for 30 minutes to promote efficient flocculation of the suspended solids.

At this point the stirring was stopped and settling allowed to occur for 30 minutes after which a 100-ml sample was taken from the supernatant (overhead elutriate) to ascertain residual suspended solids. It is to be noted that in conventional elutriation operations, which are operated on a continuous process basis, there would be need but for one mixing step to achieve both blending of the sludges and distribution of the polymer.

The results of the batch elutriation operations with regard to the purity of the overhead elutriate are reported in the following table as Run 1 along with results of a second run involving a similar operation, except that the dilution ratio was doubled. Also included for the purposes of comparison are two runs according to the above general procedure utilizing the same feed streams except that no additive nondigested sludge solids were used.

These operations designated Runs 3 and 4 show that, at roughly comparable dilution ratios, the residual suspended solids, both with and without the anionic flocculant treatment, is substantially greater than in the instance when additional nondigested sludge solids were incorporated into the digested sludge to be elutriated. In addition, it is apparent that the polymer dosage to achieve effective results is substantially reduced when the additive nondigested sludge is used conjunctively with the anionic polymeric flocculant.

Run Number	Blended Sludge Solids Conc, ppm[1]	Dilution Ratio[2]	Polymer Dose, lb/ton[3]	Overhead Solids Conc, ppm[1]
1	21,252	5/1	0.05	603
	21,252	5/1	0.1	399
	21,252	5/1	0.25	197
2	4,500	10/1	0.25	214
	4,500	10/1	0.5	191
	4,500	10/1	1.25	234
3 (compare with Run 1)	15,000	4/1	0	2,607
	15,000	4/1	0.5	795
	15,000	4/1	1.0	840
4 (compare with Run 2)	6,100	12/1	1.37	663
	6,100	12/1	3.30	700

[1]Parts by weight of solids per million parts by weight of blend.

[2]This equals the total volume of aqueous liquid represented by the aqueous washing medium and activated sewage sludge divided by the volume of digested sewage sludge. The amount of water added with the polymer was omitted from this ratio. Due to the small amount of polymer used, its effect on the dilution ratio is negligible.

[3]The polymer dosage is specified in terms of pounds per unit weight (ton) of suspended solids.

J.G. Sak; U.S. Patent 3,397,139; August 13, 1968; assigned to The Dow Chemical Co. describes an improved secondary waste treatment process. Such processes normally comprise the unit operations of primary settling, biological oxidation, secondary settling and dewatering of a mixed sludge from the primary and secondary settling operations. These operations are modified with the use of a high molecular weight organic anionic polymer flocculant in the primary settling operation and the use of cationic reagents, especially cationic organic polymers, to condition the mixed sludge for dewatering.

Preferably, the cationic polymeric conditioning agent is a cationic, water-soluble organic polymer containing nitrogen in and along the polymer chain in the reduced form. A large number of examples of such polymers are described in U.S. Patents 3,014,896 and 3,259,570.

The polymers used in the primary settling operation are high molecular weight, synthetic polymers obtained by the polymerization of ethylenically unsaturated monomers. Preferably such polymer chains are characterized by substitution with a plurality of carboxylate and/or sulfonate groups. Illustratively, such polymers can be prepared from sodium acrylate, potassium styrene sulfonate, potassium vinylbenzyl sulfonate, sodium ethylene sulfonate, sodium maleate, and ammonium methacrylate.

Usually, a minimum of at least about 50 mol percent of the combined monomer moieties should bear water-solubilizing groups to ensure that the resulting polymer is properly water-soluble. It is also essential that a minimum of about 4 mol percent of the monomer moieties combined in the finished polymer bear an anionic substituent.

C.P. Priesing and S. Mogelnicki; U.S. Patent 3,259,570; July 5, 1966 and U.S. Patent 3,300,407; January 24, 1967; both assigned to The Dow Chemical Company also describe the use of anionic polyelectrolytes such as sodium polystyrene sulfonate in combination with cationic polyelectrolytes as well as with primary inorganic coagulants such as ferric chloride.

Modified Acrylamide-Diallylamine Copolymers

R.J. Schaper; U.S. Patent 3,514,398; May 26, 1970; assigned to Calgon Corporation describes a method of flocculating suspended matter in an aqueous medium by adding a small amount of a copolymer of acrylamide and diallylamine or quaternary ammonium compound having condensed on the nitrogen through a Michael addition reaction a compound having an activated double bond and an electron-withdrawing (activating) group. A copolymer of acrylamide and diallyl methyl (β-propionamido) ammonium halide is preferred for sewage flocculation and filtration and general use in solids-liquid separation processes.

A municipal sewage plant was filtering an 8½% solids content sewage sludge over a coil spring filter under a 13 vacuum. The 2% ferric chloride-7% lime filter aid being used produced a ⅛-inch to ½-inch filter cake. Use of this filter aid was terminated and 335 to 350 ppm of a copolymer of 50% acrylamide and 50% methyl (β-propionamido) diallyl ammonium chloride was maintained in the filter sludge. A ½-inch to ¾-inch filter cake having 17% solids content was produced which was dry enough to handle. Only a very small percentage of the solids content of this cake consisted of copolymer; under previous treatment about 5%

of the solids content consisted of ferric chloride and lime. Most important, the amount of sewage sludge which could be filtered in a given period of time was double that filtered using the ferric chloride-lime filter a d.

While a minute amount of copolymer will flocculate some suspended particulate matter, the amount of polymer needed to clarify a specific water sample depends on many factors, such as particle concentration, particle size, particle charge, the presence of other interfering chemicals, etc. Without these factors, it is difficult to predict the exact amount of copolymer necessary to clarify a given water system without adding excess polymer. Ordinary river water would usually not require more than about 10 ppm, while sewage water might require as much as 500 ppm and industrial waste as much as 2,000 ppm.

The most one can say is that sufficient copolymer should be added to clarify the water, without adding so much as to waste copolymer, this amount being dependent upon the various characteristics of the medium being clarified. However, it may be said that sewage flocculation usually requires a much lower concentration than does sewage filtration; a range of about 0.1 to 2,000 ppm is generally suitable for sewage applications, and about 100 to 500 ppm is the preferred range.

Calcium Chloride and Sodium Hydroxide Added Sequentially

N.S. Davis and O.J. Foust; U.S. Patent 3,431,200; March 4, 1969; assigned to North American Rockwell Corporation describe a process for treating waste material for reducing the organic content and recovering reusable water. Algae are photosynthetically grown in the wastes and the algae is recovered from the slurry as an economically valuable product. In order to remove the algae from the slurry it is flocculated by sequential additions of calcium chloride and sodium hydroxide for producing calcium hydroxide in the suspension in a concentration below the solubility limit of calcium hydroxide. Such sequential additions induce instantaneous flocculation of the algae and permit ready settling and filtering without introducing potentially toxic flocculating agents in the resultant product.

Example: Flocculation and settling tests were made on aqueous suspensions of algae grown in liquid raw sewage. The growing conditions were such that the algae comprises principally *Scenedesmus dimorphous* with a somewhat smaller proportion of Chlorella and other species of algae. A typical suspension comprised about 0.2% by weight of algae. In each of the tests set forth below, 20 liters of algae suspension was flocculated by adding a weighed quantity of calcium chloride dissolved in water to the algae suspension and the mixture was stirred for about 30 seconds.

Thereafter, a measured quantity of 12 normal sodium hydroxide was added and the mixture stirred for an additional 30 seconds. At this point, 250 ml of slurry was placed in a graduated cylinder for determination of the settling rate as indicated by the volume of settled algae after a selected time interval. The balance of the 20-liter algae suspension was permitted to settle from 2 to 4 hours after which time about 16 liters of the supernatant liquid was removed by suction leaving 4 liters of concentrated algae.

A conventional 0.1-square-foot filter leaf was then placed face downward in the suspension with about one-inch clearance between the bottom of the leaf and the bottom of the container. Suction at about three inches of mercury was applied to the filter leaf with the rate of vacuum buildup controlled by a relatively small orifice. The slow buildup of suction permits an algae cake to form on the filter cloth with minimized plugging of the filter. After about 45 seconds, suction of about 17 inches of mercury is applied for a total filtering time of exactly three minutes. At the end of three minutes the filter leaf was carefully lifted vertically out of the container and held for exactly 30 seconds, with the lower face of the filter horizontal to permit most of the water in the wet cake to be sucked through the filter cloth. The filter leaf was then inverted and held in a horizontal position for a total vacuum drying time of exactly three minutes.

The settling rate was determined by noting the depth of settled algae in the bottom of the 250-ml graduated cylinder. In most instances a clear interface could be seen between the settled algae and the supernatant. The settling time was noted at about 20 minutes and after 24 hours. The data obtained in a series of such tests with various quantities of addition agents are set forth in the following table.

--Additives--		Filter Cake			Settled Volume, ml/250 ml	
$CaCl_2$, g/l	NaOH, g/l	Yield, g/ft²	Solids, %	Filtrate Volume, ml	20 min	24 hr
2.0	0.09	41.8	24.2	-	-	13
1.5	0.18	31.0	24.0	-	(*)	19
0.5	0.18	37.5	25.5	-	(*)	-
1.0	0.36	37.5	24.4	-	(*)	-
1.5	0.36	67.6	20.2	-	50*	19
2.0	0.36	25.0	21.9	1,950	50	21
0.5	0.54	31.4	24.9	600	(*)	-
1.0	0.54	47.3	22.9	-	50*	18
1.5	0.54	35.7	22.3	1,310	50	18
2.0	0.54	35.6	20.6	875	50	23
1.0	0.72	76.1	20.6	-	50	19
2.0	0.72	42.8	20.4	2,450	62	26
0	0	7.6	21.1	1,500	-	-

*Secondary flock of small algae remained in suspension.

Carbon Dioxide and Hydrated Calcium Oxide Added Sequentially

N.S. Davis and O.J. Foust; U.S. Patent 3,440,166; April 22, 1969; assigned to North American Rockwell Corporation describe a process for treatment of sewage where activated sludge is produced which may be treated directly or may be digested for further treatment. The sludge is flocculated by sequential additions of carbon dioxide and hydrated calcium oxide. The flocculated sludge is filtered to remove water from the solids and the filter cake is incinerated. Flue gas from the incinerator is recycled to the sludge for providing carbon dioxide and calcium oxide in the solid product of the incinerator is ground and recycled to provide hydrated calcium oxide for inducing flocculation.

In related work, *N.S. Davis, O.J. Foust and T.W. Withers; U.S. Patent 3,440,165; April 22, 1969; assigned to North American Rockwell Corporation* describe a process for treatment of sewage wherein activated sludge is produced which may be treated directly or may be digested before further treatment. The sludge is flocculated by sequential additions of calcium chloride and hydrated calcium oxide. The flocculated sludge is filtered to remove water from the solids and the filter cake is incinerated at a temperature sufficient to calcine the calcium compounds therein to calcium oxide. Calcium oxide in the solid product of the incinerator is ground and recycled to provide hydrated calcium oxide for inducing flocculation thereby substantially reducing the cost of operation.

Magnesium and Zinc Hydroxide

In a process described by *E.D. Cann; U.S. Patent 3,377,271; April 9, 1968* waste-containing waters are treated for purification, disposal, or reuse with a regenerable floc of magnesium or zinc hydroxide. Procedures are described for recovering and recycling the magnesium and/or zinc. Thus, the effectiveness of waste-containing water treatment techniques is enhanced by introducing, as a flocculating agent, a soluble sulfur oxide salt of magnesium and/or zinc (e.g., zinc or magnesium sulfate, zinc or magnesium bisulfite, or zinc or magnesium bisulfate), which is alkalized to form a magnesium or zinc floc.

The resultant floc coalesces suspended solid wastes which are then collected and contacted with an aqueous sulfur oxide (SO_2 and/or SO_3) to dissolve and thereby recover the magnesium and/or zinc as a soluble sulfur oxide salt. The resultant salt is then available for cycling back to the initial flocculation step or steps.

In one form of the process, a stream of the magnesium or zinc floc and coalesced wastes is burned under combustion conditions (dry combustion) to burn off the waste and to form magnesium or zinc oxide as part of the furnace ash. The ash is then contacted with water and with sulfur oxide to dissolve or extract the magnesium or zinc as the soluble sulfur oxide salt, which is then in a form suitable for recycling.

In a second form, the stream of magnesium or zinc floc and coalesced wastes is first contacted with sulfur oxide (or a sulfur oxide affording substance such as elemental sulfur) to form the sulfur oxide salt, and the wastes and salt burned under combustion conditions (dry combustion) to burn off the wastes and provide a furnace ash containing the magnesium or zinc as a soluble sulfur oxide salt. This salt is extracted with water and is used to provide recycle flocculating agent.

FLOCCULATING PROCEDURES

Recovery of Precipitating Agent in Protein Removal Process

A process described by *P. Nettli; U.S. Patent 3,936,375; February 3, 1976; assigned to A/S Apothekernes Laboratorium for Specialpraeparater, Norway* provides a method for treating a proteinaceous sludge and comprises a chemical-thermal combination method yielding a concentrated proteinaceous sludge with a high content of solids simultaneously with the utilized precipitating agent going into solution in the separated aqueous phase and thus renders it possible to recover the precipitating agents.

In the process, an alkaline earth compound is added to the protein sludge and, if necessary, an alkali metal hydroxide or alkaline earth metal hydroxide is added until a pH of about 7 to 9 is obtained, followed by heating, during which the protein complex formed from the precipitating agents and proteins splits, the proteins coagulate as an alkaline earth proteinate and the precipitating agent dissolves in the separated hydrous phase. By washing the coagulated material, it is moreover possible to recover the precipitating agents almost quantitatively. If alkaline earth hydroxide, e.g., $Ca(OH)_2$, is used alone before the protein sludge is heated, a smaller proportion of the precipitating agent is dissolved and consequently less precipitating agent is recovered, probably because the calcium salts of the precipitating agents are less soluble. One advantage of calcium alone is, however, that the coagulated material has a firmer consistency and separates better; it is, for example, easier to filter.

By using an alkaline earth salt in a quantity sufficient to form an alkaline earth proteinate and alkali hydroxide adequate for neutralization, the precipitating agent will dissolve as an alkali salt that is highly soluble and thus makes it possible to recover the precipitating agent almost quantitatively. In the process, the proteinaceous material is also supplied with heat-coagulable protein, whereafter the mixture is agitated and heated to a temperature of at least 60°C, preferably by the aid of steam supplied directly to the pretreated sludge, and the separated water is removed, preferably in a centrifuge or a desludger.

Thus blood, which is a heat-coagulable protein, is added to this proteinaceous sludge. The mixture is slowly heated with thorough agitation until coagulation begins. A continuous flow of the mixture is then rapidly heated to 70° to 100°C, whereupon the separated water is removed in a manner known per se. In this manner a good separation of concentrated proteinaceous material is achieved with a content of solids of approximately 40%. The addition of a heat-coagulable protein also contributes to a better coagulation and dewatering but best results are achieved with a combination of heat-coagulable protein and alkaline earth compounds at a pH of about 7 to 9.

Flocculant and Sewage Mixing Module

According to a process described by *R.P. Briltz; U.S. Patent 3,886,073; May 27, 1975* effluent is pumped from the primary sewage treatment location through a module at which time a flocculating agent is introduced into the sewage. The sewage and flocculent are agitated and thoroughly mixed in this module and it then passes to a retention of flocculent module at which time the majority of the flocculated material is separated out by baffles. The deflocculated effluent then passes to conventional first and second stage filtering tanks. Means are provided to back-flow the modules to remove the flocculated material and cleanse the modules when necessary.

Figure 4.1a is a schematic view of the treatment system in the form of a process and flow diagram. Figure 4.1b is an enlarged fragmentary top view of one of the conduits of the agitating module with the casing broken away in part to show the interior. Figure 4.1c is a longitudinal section substantially along the line **3–3** of Figure 4.1b. Figure 4.1d is a cross-sectional view substantially along the line **4–4** of Figure 4.1c. Figure 4.1e is an end view of Figure 4.1c along the line **5–5** of Figure 4.1c.

Figure 4.1f is a longitudinal section of one of the conduits of the flocculent module substantially along the line **6–6** of Figure 4.1g. Figure 4.1g is a cross-sectional view substantially along the line **7–7** of Figure 4.1f. Figure 4.1h is an enlarged end view of the agitating module per se along the line **8–8** of Figure 4.1a.

FIGURE 4.1: SEWAGE AND EFFLUENT TREATING SYSTEM

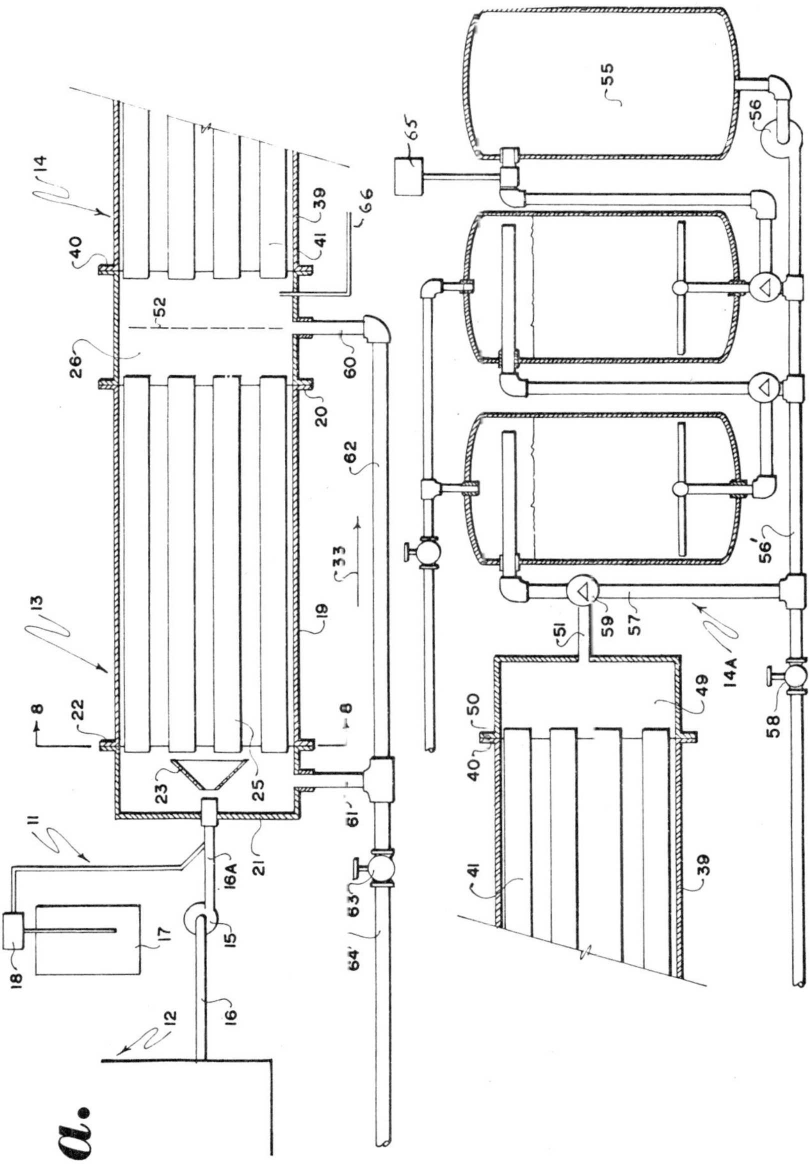

(continued)

FIGURE 4.1: (continued)

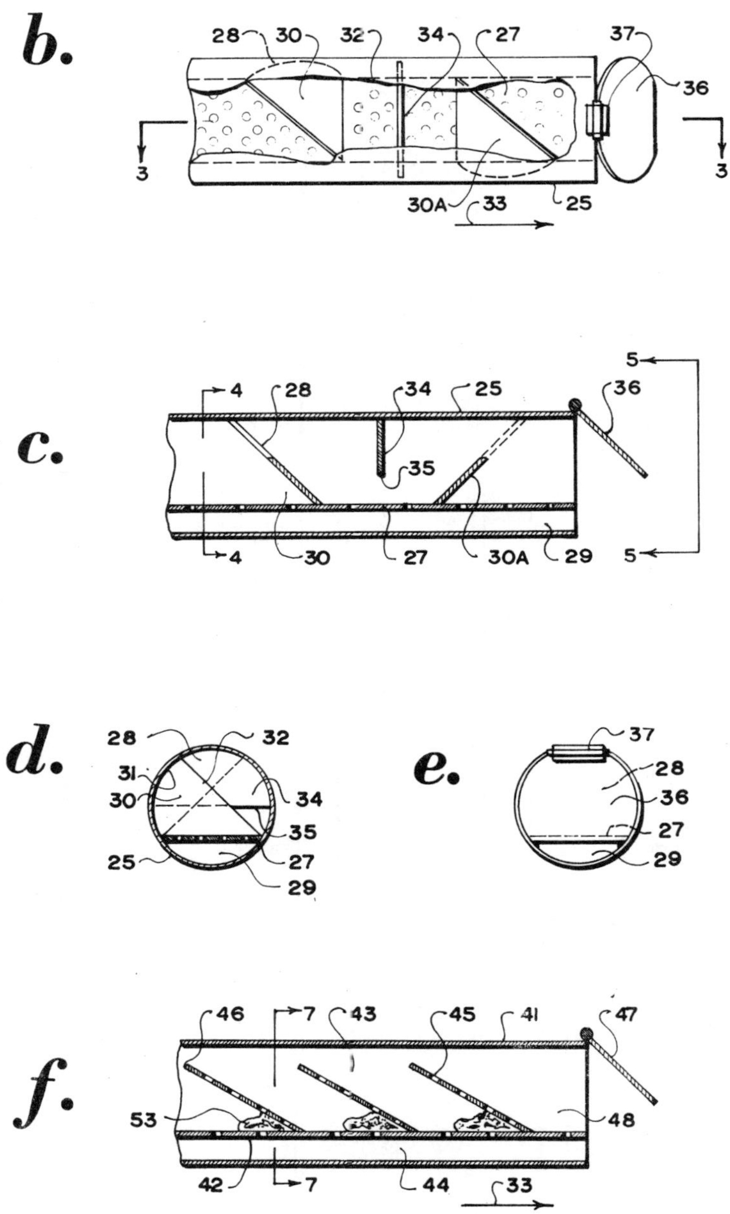

(continued)

FIGURE 4.1: (continued)

g.

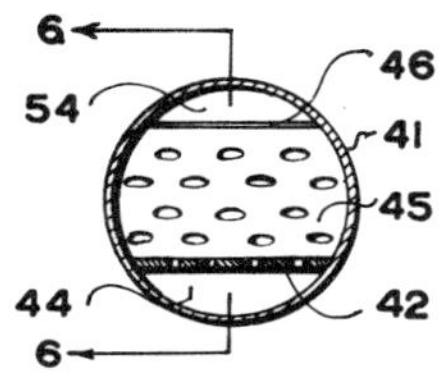

h.

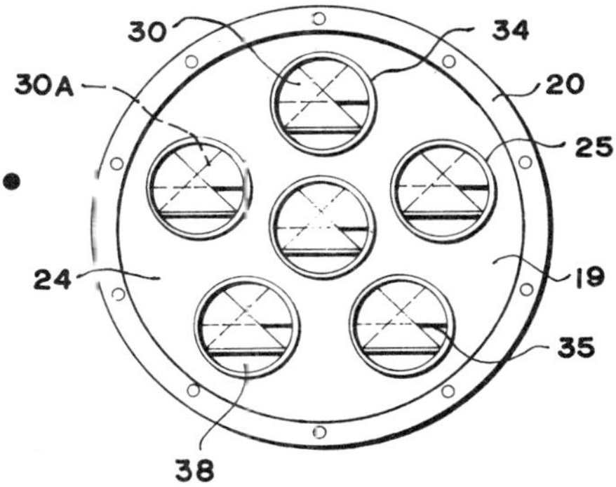

Source: U.S. Patent 3,886,073

Referring to the figure, a secondary sewage treatment apparatus includes means **11** to add a flocculation chemical to sewage after it has been treated in the primary sewage area **12**. Means **13** are provided to agitate the sewage and the flocculating chemicals and further means **14** are provided to separate the flocculated material from the liquid effluent which then passes to conventional filtering tanks **14A**. In Figure 4.1a, **12** illustrates schematically a holding unit for primary treated sewage. This holding unit may be a reservoir or tanks as desired. A pump **15** pumps sewage from the holding area **12** via conduit **16** and a conduit **16A** extends between the pump and the means to agitate collectively designated **13**.

Reference character **17** illustrates a container which holds flocculating chemicals and a pump **18** injects metered amounts of these chemicals into the line or conduit **16A**. The flocculating chemical utilized will of course depend upon the contaminants held in solution within this sewage being treated. However, there are many well-known flocculents which could be used such as soda ash, silicates, polyelectrolytes and liquid alum.

The means to agitate the chemical and sewage indicated by reference character **13** consists of a module which in turn is connected to a further module comprising the means to separate the flocculated material indicated by reference character **14**. The module **13** consists of a cylindrical casing **19** having a flange **20** at either end and an intake header **21** is secured by a similar flange to the intake end **22** of the casing. Conduit **16A** connects centrally into this intake header and a perforated dispersal cone **23** is situated immediately in front of the connection of the conduit **16A** with the header so that the mixture of sewage and flocculating chemicals are evenly distributed through the agitating module **13**.

End plates **24** are provided at each end of the casing and a plurality of conduits **25** extend through the casing between end plates **24** and communicate with the intake header **21** at one end of the casing and an outlet baffle chamber **26** secured to the other end of the casing by similar flanges **20**. Each of the conduits

25 is shown in detail in Figure 4.1b to Figure 4.1e inclusive. The conduits are preferably cylindrical and made of a noncorrosive material such as stainless steel. Each conduit is provided with a horizontally situated perforated partition baffle **27** which spans the conduit and extends from end to end. This perforated partition baffle separates the conduit into a normal flow agitation portion **28** above the partition baffle **27** and a reverse flow cleaning portion **29** below the partition baffle **27**, it being understood that portion **28** is considerably larger in volume than portion **29** due to the positioning of the partition baffle **27**.

Situated within the normal flow agitating portion **28** is a plurality of baffles which include a first baffle **30** which is substantially triangular in front elevation and is secured by the lower edge thereof to the partition baffle **27** and by the curved side **31** to the wall of the portion **28**. The diagonal edge **32** extends between one side of the partition baffle **27** diagonally upwardly to the upper end of the curved edge **31**. This baffle is inclined towards the intake header end of the conduits **25**. That is to say they are inclined at an angle towards the normal direction of flow of sewage therethrough which is indicated by arrow **33**.

A similar baffle **30A** is also situated within the portion **28** of the conduit which is similar in configuration to baffle **30** but is oppositely handed or reversed with respect to baffle **30**. However, this baffle also inclines in the same direction as baffle **30**, namely, at an angle towards the normal sewage flow. Situated between baffles **30** and **30A** is a vertical baffle **34** which forms part of a segment of the cylindrical conduit **25**. This baffle is secured by the curved edge thereof to the upper wall of a portion **28** of the conduit and terminates spaced from the partition baffle **27** as indicated by reference character **35**.

The provision of these baffles causes severe agitation or turbulence of the sewage passing therethrough thus thoroughly mixing the flocculating chemicals and ensuring or assisting in the flocculating action thereof. The length of the conduits and the volume thereof is such that when the sewage reaches the outlet baffle chamber **26**, flocculation has commenced. A flap valve **36** is hinged as at **37** to the upper edge of the portion **28** of the conduit at the discharge end thereof and this flap valve normally rests against the separating or dividing partition baffle **27** as shown in Figure 4.1e. However, the hinging is such that sewage passing in the normal direction, namely in the direction of arrow **33**, opens this flap valve whereas any back pressure or movement of effluent in the opposite direction closes off the portion **28**.

Finally, it should be observed that the intake end of the reverse flow cleaning portion **29** of the conduits is closed as illustrated in Figure 4.1h by reference character **38**. The conduits and various baffles are preferably designed so that they can be readily dismantled for replacement if desired.

Dealing next with the flocculent module **14**, this also consists of a cylindrical casing **39** having a flange **40** at either end thereof and a baffle chamber **26**, being similarly flanged, bolts to this flange as clearly shown in Figure 4.1a, it being understood that the baffle chamber **26** acts as a connecting portion between modules **13** and **14**. Module **14** is also provided with a plurality of conduits **41** extending therethrough between end plates which are similar to end plates **19**. Details of these conduits **41** are shown in Figure 4.1f and Figure 4.1g. A perforated horizontally situated partition baffle **42** spans the conduits in a position similar to the position of the partition baffle **27** and separates the conduits into a normal flow flocculent-trapping portion **43** above the partition baffle **42** and a reverse flow cleaning portion **44** below the partition baffle **42**. Each conduit includes a plurality of baffles **45** which are perforated and which are secured by the lower edges thereof to the partition baffle **42** and extend into the portion **43** of the conduits. These baffles extend or span the conduits but terminate spaced from the upper wall as indicated by reference character **46** and shown in Figure 4.1g. The baffles **45** are in spaced and parallel relationship one with the other and are inclined at an angle in a direction against the normal flow of effluent therethrough indicated by reference character **33**.

A hinged effluent-flow-operated flap valve **47** is provided at the outlet end **48** of each of the conduits **41**, and this covers the normal flow flocculent trapping portion **43** resting against the end of the partition baffle **42** by gravity. The construction is similar to flap valves **36** and they operate in a similar manner. An outlet header **49** is secured to flange **40** by a corresponding flange **50** and an outlet conduit **51** connects to the settling or filtering tanks collectively designated **14A** which are conventional in construction and operation. A perforated baffle **52** spans the baffle chamber **26** and the arrangement of apertures in the baffle assures an even flow of the effluent from the agitating module **13** to the flocculent module **14**.

In operation, the treated sewage is pumped by pump **15** to the inlet header **21** and the necessary flocculent chemical is added in a metered amount just prior to entry to the header **21**. The cone **23** distributes the sewage evenly to the conduits **25** and as it passes through these conduits it is completely agitated thus initiating the flocculating action. It then passes to the baffle chamber **26** and through the perforated screen **52** into the module **14**. At this point, due to the time factor which can be controlled by dimensions and by the capacity of pump **15**, the flocculating action starts and the flocculated material is trapped by the baffles **45** within the angle defined by these baffles and the partition baffle **42**. Such trapped flocculent is illustrated by **53** in Figure 4.1f.

It should be understood that the perforations within the partition baffle **42** are small enough to prevent the flocculated material from passing through. The upper edges **46** of baffles **45** act as a dam or weir and permit clear effluent to pass over the upper edges thereof within the space indicated in Figure 4.1g and identified by reference character **54**. This clear effluent which may contain some small particles of flocculent, then passes to the filter tanks shown in **14A** which eliminates any solids which may be held in suspension within the effluent. The clear effluent is then held within reservoir **55** for disposal into the aquifers as necessary.

When it is desired to clean the modules **13** and **14**, a pump **56** pumps clear effluent from the reservoir **55** through a conduit **56′** and into the module **14** via a conduit **57**. In this connection a gate valve **58**, normally used for the distribution of the clear effluent from reservoir **55**, is closed, and a gate valve **59** normally routing the effluent from module **14** to the tanks **14A** is also set so that clear effluent passing through conduit **57** enters the outlet header **49** and does not enter tanks **14A**.

As soon as this back flow commences, the flap valves **36** and **47** close so that the effluent under pressure has to pass through the conduits **41** and **25**, via the reverse flow cleaning portions **44** and **29** respectively. As both of these portions are closed at the one end as shown at **38**, the clear effluent under pressure is forced upwardly through the perforated separator baffles **42** and **27** thus removing any flocculated material **53** which has been collected by baffles **45** and any flocculated material which may have lodged within the agitating baffles **30** and **30A**. This washed away flocculated material discharges via conduits **60** and **61** to a common conduit **62**. Conduit **60** is connected to the baffle chamber **26** and conduit **61** to the inlet header **21**.

A gate valve **63** is open under these conditions so that this effluent carrying the flocculated material flows via conduit **64** back to the primary treatment area **12** where it settles out by gravity. This means that the modules **13** and **14** can be cleaned on a regular basis depending upon the amount of flocculated material collected thereby. The collection of solids, of course, is conventional and these are more readily disposed of than highly contaminated sewage effluent normally remaining after conventional treatments.

Reference character **65**, in Figure 4.1a, illustrates schematically a chlorination unit for post-treatment of effluent prior to passing to the reservoir **55**. Also shown in Figure 4.1a is an air line **66** to aerate the chamber **26** if required, line **66** being connected to a source of compressed air which, of course, is conventional. Finally it should be noted that the system will reduce lagoon holding requirements by one-half or better by having the system on a continuous flow, due to the fact that over 80% of sewage is liquid. With the present system it would not be necessary to hold this liquid for long periods as is the case with present lagoons thus giving a reduction in land costs, reduction in lagoon construction costs and, of course, a reduction of unsightly ponds and odors.

NITRATE REMOVAL

Using Nitrosomonas Bacteria

G.V. Levin and G.J. Topol; U.S. Patent 3,654,147; April 4, 1972; assigned to Biospherics Incorporated describe a process for treating raw or treated sewage to obtain an effluent substantially free of nitrogen-containing compounds which is returned to natural water resources. The process comprises mixing influent sewage material with activated sludge to provide a mixed liquor. The mixed liquor is passed to an aeration zone where it is aerated at a rate sufficient to reduce the BOD content and to convert ammonia present in the sewage to nitrate. Under conditions of sufficiently high aeration, e.g., at least about 2 cubic feet of air per gallon of mixed liquor, Nitrosomonas bacteria present in the mixed liquor convert ammonia in the raw sewage to nitrite and Nitrobacter bacteria convert nitrite to nitrate.

The mixed liquor is then passed to a zone where it is maintained under conditions in which there is insufficient oxygen present to satisfy the metabolic needs of the microorganisms in the mixed liquor, i.e., under anaerobic or semiaerobic conditions. This induces denitrifying microorganisms present to break down the nitrate content. These microorganisms present in the sludge obtain oxygen by the reduction of the nitrate content. Nitrogen gas is formed in the process and is evolved from the system. Nitrate-depleted sludge is separated from this zone to provide a substantially nitrate-free effluent. The sludge is then recycled for mixing with influent sewage material.

Referring to Figure 4.2, a raw sewage influent stream **1** is passed through conventional screening and grit-removing units and is optionally subjected to primary settling in a tank **2** from which primary sludge is removed in line **3**. The primary settled sewage is mixed with recycled activated sludge to form a mixed liquor and is passed by line **4** to the aeration tank **5**. In the aeration tank **5**, the mixed liquor is aerated at a rate sufficient to convert ammonia present in the sewage to nitrate. During aeration, the bacteria present take up phosphate and consume organic matter present in the sewage. A high degree of BOD removal is obtained during aeration.

FIGURE 4.2: NITRATE REMOVAL FROM SEWAGE

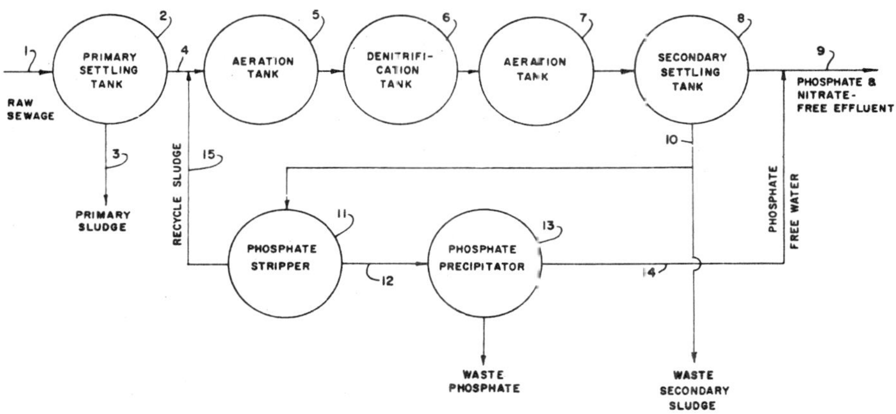

Source: U.S. Patent 3,654,147

After the aeration, the mixed liquor is fed into a tank **6** wherein it is maintained under conditions in which there is insufficient oxygen present to satisfy the needs of the microorganisms in the mixed liquor. This induces the microorganisms to consume the nitrate content of the sewage. After depletion of the nitrates and release of the nitrogen as nitrogen gas, the mixed liquor is passed to the aeration tank **7** in which it is again aerated. In this tank, the microorganisms in the sludge take up any phosphate which has leaked out during the period the mixed liquor was in the tank **6**. This step, and the phosphate stripping operation hereinafter described, may be omitted if it is only desired to remove nitrogen in the sewage, i.e., wherein phosphate removal is not required.

After aeration in the tank **7**, the mixed liquor is fed into a secondary settling tank **8**. In the secondary settling tank **8**, phosphate-enriched sludge settles and thereby separates from the mixed liquor. The sludge contains a substantial portion of the phosphate present in the sewage. The substantially phosphate-free and nitrate-free effluent is discharged for disposal in a conventional manner by line **9**. The phosphate-enriched sludge is removed from the settling tank by line **10**. A portion of the sludge may be delivered to waste and the remainder is passed to the phosphate stripper **11**. In the phosphate stripper **11**, the phosphate-enriched sludge is treated to cause the microorganisms in the sludge to release phosphate. This treatment may be accomplished by holding the mixture under anaerobic conditions as described in U.S. Patent 3,236,766; by aerating as described in U.S. Patent 3,654,146; or by appropriate pH adjustment, i.e., adjusting the pH to less than 6.5 and maintaining it at this pH for at least 10 minutes. This treatment causes the organisms in the sludge to release the phosphate which they have taken up in the aeration tank **5**. The phosphate leaks out of the sludge into the liquid phase.

A phosphate-enriched supernatant liquor is produced upon settling of the sludge. After settling, the sludge is passed by line **15** for mixing with the raw sewage which is being fed to the aeration tank **5**. A phosphate-enriched supernatant liquor is produced by the phosphate stripper **11** and is passed by line **12** to the phosphate precipitator **13**. A phosphate precipitant, such as aluminum or iron salts or lime, is mixed with the phosphate-enriched supernatant liquor in the phosphate precipitator **13** to precipitate phosphate.

The phosphate precipitate may be combined with any waste phosphate-enriched sludge removed from the secondary settling tank **8** and converted into a fertilizer or otherwise disposed of by conventional methods. A phosphate-free supernatant liquor is withdrawn from the phosphate precipitator **13** and passed by line **14** to line **9** where it is combined with the phosphate-free effluent from the secondary settling tank **8**. The following example illustrates the process.

Example: The pH of influent raw sewage is adjusted to 7 to 8 and is passed through conventional screening and grit removal units. The raw sewage is mixed with recycled activated sludge having a low phosphate content in an amount sufficient to provide about 15% by volume of return sludge in the mixed liquor. The mixed liquor is then fed at the rate of 15 gallons per hour to an aeration zone wherein it is aerated at a rate of 0.1 liter of air per minute per liter of mixed liquor for 6 hours.

The mixed liquor is then passed to a denitrification tank wherein it is held under aerobic conditions for 4 hours. During this time, the microorganisms break down the soluble nitrate content of the mixed liquor formed in the aeration tank and nitrogen gas is evolved. The mixed liquor is then again aerated at a rate of 0.1 liter of air per minute per liter of mixed liquor for 2 hours to cause the sludge to take up any phosphate which leaks out during the anaerobic treatment.

The aerated mixed liquor is passed to a secondary settling tank. Clarified effluent which is substantially free of phosphate, nitrate and ammonia is discharged to the effluent outflow after chlorination. The phosphate-enriched sludge is passed to a phosphate stripper and sludge thickener where it is held under anaerobic conditions for several hours. The conditions existing in the stripper induce considerable quantities of intracellular phosphate to leak out into the liquid phase. The phosphate-depleted sludge is recycled for mixing with incoming raw sewage.

The phosphate-enriched supernatant liquid is withdrawn from the settling tank and is fed into a chemical precipitation tank where alum is added and mixed to form a phosphate precipitate. The phosphate precipitate is wasted and the phosphate-depleted effluent is discharged to the effluent outflow along with the clarified effluent from the secondary settling tank. This process removes about 95% of the phosphate and 90% of the ammonia contained in the original raw sewage.

PHOSPHATE REMOVAL

Internal Precipitation of Phosphate

G.J. Topol; U.S. Patent 3,681,235; August 1, 1972; assigned to Biospherics Incorporated describes an activated sludge sewage treatment process in which influent sewage material is mixed with activated sludge containing precipitated phosphate particles to provide a mixed liquor. The mixed liquor is aerated to cause the microorganisms present in the sludge to take up phosphate. Phosphate-enriched sludge is separated from the mixed liquor to provide a substantially phosphate-free effluent. The phosphate-enriched sludge is treated to cause the microorganisms in the sludge to release phosphate to the liquid phase and a phosphate precipitant is added to precipitate the water-soluble phosphate content. The sludge containing the precipitated phosphate particles is recycled for mixing with influent sewage material.

Example: The pH of influent raw sewage is adjusted to 7 to 8 and is passed through conventional screening and grit-removing units. The raw sewage (about 1,000,000 gallons per day—gpd) containing about 100 parts per million (ppm) of solids is mixed with recycled activated sludge (about 190,000 gpd) containing about 500 ppm of phosphate precipitate. The mixed liquor is fed to an aeration zone and is aerated at a rate of 5 cubic feet of air per gallon of sewage for six hours. The effluent mixed liquor from the aeration zone is fed to a secondary settling tank. Clarified waste liquid which is substantially free of phosphate is discharged to the effluent outflow after chlorination at a rate of about 1,000,000 gallons per day.

The settled mixture of phosphate-enriched sludge and particles of precipitated phosphate is withdrawn from the secondary settling tank at a rate of about 200,000 gpd. A portion (about 10,000 gpd) is passed to waste sludge and the remainder is passed to an anaerobic phosphate stripper wherein it is held under anaerobic conditions for several hours. The conditions existing in the stripper induce considerable quantities of intracellular phosphate to leak out into the liquid phase. An approximate stoichiometric amount of alum required to precipitate the soluble phosphate present is added to the sludge and the mixture of sludge and precipitated phosphate is recycled for mixing with incoming raw sewage.

In related work, *G.V. Levin and G.J. Topol; U.S. Patents 3,730,882; May 1, 1973 and 3,756,946; September 4, 1973; both assigned to Biospherics Incorporated* describe an activated sludge sewage treatment process which comprises mixing influent sewage material with activated sludge to provide a mixed liquor. The mixed liquor is passed to an aeration zone wherein it is aerated to reduce the BOD content thereof and to cause the microorganisms present to take up phosphate. The mixed liquor is then passed to a settling zone containing an anaerobic layer of sludge. In the settling zone, phosphate-enriched sludge, in which the phosphate is present in the cells of the organisms, settles into the anaerobic layer of sludge and a substantially phosphate-free effluent is removed from above the layer of sludge.

The sludge is maintained under anaerobic conditions in the sludge layer of the settling zone for a time sufficient to cause the organisms in the sludge to release phosphate to the liquid phase of the sludge. A phosphate precipitant is added to the sludge to precipitate the soluble phosphate in the phase thereof, resulting in a mixture of phosphate-depleted sludge and precipitated phosphate particles. This mixture is recycled for mixing with influent sewage material in the activated sludge sewage treatment process.

In related work, *G.V. Levin and G.J. Topol; U.S. Patent 3,654,146; April 4, 1972; assigned to Biospherics Incorporated* describe an activated sludge sewage treatment process in which phosphates are removed from phosphate-enriched sludge by aerating the phosphate-enriched sludge with an oxygen-containing gas. During aeration, the organisms in the sludge after consuming the available food substrate, go into endogenous respiration, consuming much of their own cellular material. Thus, the aeration serves to reduce the volume of sludge as well as to cause the organisms in the sludge to release phosphate.

A phosphate-enriched supernatant liquor is formed on settling. The sludge, having a reduced phosphate content, is separated from the phosphate-enriched supernatant liquor and at least a portion is recycled for mixing with influent sewage material in an activated sludge sewage treatment process. The mixed liquor is aerated to reduce the BOD content and to cause the organisms present to take up phosphate and phosphate-enriched sludge is separated from the mixed liquor to provide a substantially phosphate-free effluent.

By Controlled Additions of Sodium Aluminate and a Flocculant

According to a process described by *R.D. Sawyer and J.D. Tinsley; U.S. Patent 3,673,083; June 27, 1972; assigned to Nalco Chemical Company* soluble phosphates are removed from a sewage effluent by controlled additions of sodium aluminate, $NaAlO_2$, and a flocculant introduced into the sewage effluent at some point after secondary (aeration) treatment and prior to discharging the effluent to a receiving body of water, thereby reducing considerably the amount of phosphates responsible for proliferation of primitive aquatic plant life deemed ecologically inimical.

The preferred flocculants are anionic water-soluble polymeric flocculants. The most readily prepared polymers that give the best results are the copolymers and homopolymers of acrylic acid which contain at least 10% by weight of acrylic acid or acrylic acid salts. A particularly useful group are those which contain 20 to 60 weight percent acrylic acid and 40 to 80 weight percent acrylamide with such polymers having a preferred molecular weight range of 1,000,000 to 3,000,000.

By Acidification

A process described by *M. Spiegel and M.F. Hobbs; U.S. Patent 3,522,171; July 28, 1970; assigned to FMC Corporation* involves a method of treating sludge to reduce the phosphate content prior to recycling the sludge as seed material to the aeration zone of an activated sludge sewage treatment system. The method involves subjecting the first sludge concentrate produced in the separator successively to acidification followed by separation of a second sludge concentrate and then to dilution with a low phosphate content aqueous medium followed by separation of the third sludge concentrate which is the reduced phosphate content concentrate of microorganisms to be recycled.

In one mode of operation, the process of removing, for example, orthophosphates from sludge to be recycled as seed material comprises removing mixed liquor from the aeration tank to sludge separation apparatus, removing a sludge concentrate of a solids content of about 4% from the settling tank in a liquid volume of about 10 to 15% of the volume of influent sewage, i.e., 100 gpm influent sewage, and 14 gpm of sludge concentrate, and having a phosphate content of, for example, 175 to 250 mg/liter.

The process continues by acidifying with H_2SO_4 to a pH of about 5, passing the acidified sludge through a second separation step where the sludge is removed from the liquor as a concentrate at a rate of about 5 gallons per minute having about 8 to 10% solids content and a phosphate content of about 175 to 250 mg per liter, mixing the flotation concentrate with an approximately equal volume of effluent from the settling tank and passing the diluted sludge through a third separation unit to produce a sludge recycle medium of 10% solids concentration, containing approximately 110 mg/liter phosphate content, at a rate of 5 gallons per minute. This concentration amounts to about 1.1 mg of desorbable phosphate per gram of activated sludge compared to the mixed liquor content of about 6 mg of desorbable phosphate per gram of activated sludge.

Example: An aeration tank which may be 22-foot square and having a liquid operating depth of 14 feet 6 inches, may be arranged with an influent channel along the top of one wall of the tank with a weir arranged to provide relatively uniform distribution of influent flow along the wall. The opposite wall may be provided with five equally spaced ports for discharge of mixed liquor. Air dispersion equipment consists of one header positioned 2 feet from the tank floor and 3 feet from the wall topped by the influent channel and a second header similarly positioned adjacent the wall provided with spaced ports. The header may support a series of horizontally projecting cord wound air disperser units each capable of introducing 5 to 8 cubic feet of air per minute.

The aeration tank is provided with about 53,000 gallons of mixed liquor having a mixed liquor sludge solids content of 5,000 mg/liter. Influent sewage having a phosphate content of approximately 30 mg/liter is introduced into the tank at a rate of about 295 gpm into which is also introduced about 15 gpm of recycle sludge. After about 3 hours' detention, mixed liquor is delivered to a settling tank which produces approximately 272 gpm of effluent having a phosphate content of about 3.2 mg/liter and 38 gpm of sludge concentrate having a phosphate content of approximately 220 mg/liter.

The 38 gpm of sludge concentrate is mixed with approximately 33 cc per minute of 66°Bé sulfuric acid and delivered to a sludge flotation tank. In this tank, the acidified liquor, after an appropriate holding time, is separated into approximately 23 gpm of effluent containing approximately 220 mg/liter of phosphate and 15 gpm of approximately 10% solids content sludge containing approximately 220 mg/liter of phosphate. The 10% solids content sludge is mixed with 30 gpm of effluent from the settling tank containing 3.2 mg/liter of phosphate and the diluted sludge concentrate is subjected to a second flotation operation.

The products of the second flotation can be 30 gpm of a clarified liquor containing approximately 75 mg/liter of phosphate and 15 gpm of approximately 10% solids content recycle sludge of approximately 75 mg/liter. This content of phosphorus amounts to about 0.8 mg of desorbable phosphate per gram of return activated sludge whereas the mixed liquor activated sludge contains about

6 mg of desorbable phosphate per gram of sludge. From the above example, it will be apparent that the volume of desorbed phosphate-containing aqueous medium which must be treated to precipitate phosphates before such an aqueous medium is discharged from the system, is 53 gpm having an average phosphate content of about 140 mg/liter, which is to be compared to other systems where sludge washing operations will produce a representative dilute solution totaling 100 to 200 gpm of liquor having a phosphate content of only 10 to 15 mg/liter.

O.E. Albertson; U.S. Patent 3,409,545; November 5, 1968; assigned to Dorr-Oliver Incorporated describes a process for decreasing the phosphorus and nitrogen content of organic waste material which comprises mixing the waste with a chemical precipitant; settling and removing precipitated phosphorus-bearing solids; stripping the ammonia; aerating the effluent to biologically remove phosphorus and settling and removing precipitated phosphorus-bearing solids. The solids from the first sedimentation are then thermally oxidized in a reactor and the solids product of the reactor is run through a recovery system to recover reusable chemical precipitant for the first step.

Dispersed Culture Reactor with Carbon Enrichment

J.H. Bruemmer; U.S. Patent 3,356,609; December 5, 1967; assigned to Union Carbide Corporation describes a sewage treatment process which is capable of returning a water effluent to resources substantially free of carbon, nitrogen and phosphorus. In essence, the process comprises subjecting a sewage stream to a primary clarification to remove as much as practical of the solids content, delivering the clarified effluent to a dispersed culture aerobic reactor, and the solids, i.e., the primary or raw sludge, to a sludge aerobic reactor.

The effluent from the primary clarifier is enriched with a suitable source of carbon and is subjected to adequate aerobic biological treatment in the dispersed culture aerobic reactor, after which the entire mass is transferred to a flocculation tank and chemically and/or mechanically treated to cause an agglomeration of the cell mass grown in the dispersed culture aerobic reactor, and from this flocculation tank is sent to a secondary clarifier where the agglomerated cell masses are mechanically removed.

The effluent from the secondary clarifier being free of at least 99% of the original N and P content can safely be returned to the water resources after chlorination or similar disinfection. The raw sludge from the primary clarifier is fed to a sludge aerobic reactor in which the oxygen tension is maintained at a high level by circulating gas containing more than 30% oxygen and the carbon content is enriched as in the dispersed culture aerobic reactor. In addition, any part or all of the bacterial sludge from the secondary clarifier can be combined with the sludge from the primary clarifier in the sludge aerobic reactor. Or if desired, the secondary sludge can by-pass the aerobic reactor by going directly into centrifuge or filter.

This solid product may be further processed as supplement for animal feed. Moreover, any part or all of the sludge from the secondary clarifier can be fed back to the stream going into the primary clarifier to aid in clarification of incoming sewage, and/or to the dispersed culture aerobic reactor where it provides inoculum for the incoming settled sewage from the primary clarifier. After suitable treatment in the sludge aerobic reactor, the combined sludges from the primary and secondary clarifiers are centrifuged or filtered or otherwise separated into liquid and solid fractions and the solids disposed of. The liquid fraction can be recycled if desired, or returned to the natural water resources.

The dispersed nature of the microorganisms in the dispersed culture aerobic reactor renders the operation of the process step carried out in this reactor quite different from that which occurs in an activated sludge operation of conventional treatment procedures. In the latter operation, the bacterial culture present is in its declining phase of growth in which the bacteria use much of the available carbon to produce a flocculating substance, and therefore, has less carbon available for incorporation of nitrogen and phosphorus into cell mass, and in which their facility for consuming organic material is relatively low. Also, the types of bacteria which are capable of producing the flocculating substance are limited. Thus, the number of different kinds of organic pollutants which can suitably be treated by the activated sludge process is also limited.

In contrast, the dispersed culture aerobic reactor of this process contains the microorganisms in a highly dispersed state, existing as individuals or as small clusters of no more than about twenty or so bacteria. In this dispersed condition, the bacteria are in their most active state of growth and utilize nutrient material for cell mass at a much more rapid rate than do the microorganisms of a conventional activated sludge system.

The dispersed condition of the bacteria is maintained in the dispersed culture aerobic reactor by rapid mixing and by insuring that the nutrients in the sewage stream are introduced at a sufficiently high rate and that the balance of carbon, nitrogen, and phosphorus in nutrient form is appropriate for protoplasm production. Maintenance of the proper CNP ratio is vital if the phosphorus and nitrogen values of the raw sewage are to be removed chemically combined in cell protoplasm.

Accordingly and in view of the general deficiency of usable carbon in the sewage streams treated in the process, a carbon source is added to the sewage in the dispersed culture aerobic reactor to maintain a weight ratio of carbon to phosphorus in the sewage of from about 150:1 to 100:1. The carbon source can be any of a vast variety of substances, including organic compounds (preferably free of nitrogen and/or phosphorus) such as monosaccharides and industrial wastes such as molasses, food processing, pulp, pharmaceutical and fermentation by-products, and inorganic carbonates such as sodium carbonate.

Example: Domestic raw sewage containing 250 parts per million by weight (anhydrous) suspended solids, 540 ppm Biological Oxygen Demand (BOD), 180 ppm nitrogen, 10 ppm phosphorus and having a CNP ratio of 54:18:1 is fed continuously into a 50,000-gallon primary clarifier at the rate of 600,000 gallons per day. To the primary clarifier is added daily 60 lb lime to adjust the pH to 7.4, 10 lb of asbestos to aid clarification and 1,800 gallons bacterial sludge from the secondary clarifier containing inoculum and polyelectrolyte coagulant from the flocculator.

Residence time in the primary clarifier is about 2 hours. The effluent from the primary clarifier (essentially free of organic nonbacterial suspended solids) is fed continuously to the dispersed culture aerobic reactor, a 100,000-gallon closed cylinder at a rate of 25,000 gallons per hour, where gas containing 90% O_2 and 10% CO_2 is circulated through the liquid at a rate of 0.1 volume of gas per volume of liquid per minute.

A temperature of 30°C is maintained and the pH is adjusted to 7.2 with 5 lb lime per day. Asbestos is added at the rate of 10 lb per day to aid bacterial growth. The contents of the reactor is vigorously mixed by conventional fermentation impellers to achieve a "complete mixed" condition. A daily inoculum of 1,200 lb bacterial sludge from the secondary clarifier (20% of secondary sludge production) maintains the required bacteria concentration.

The daily throughput of the reactor, on the 4-hour detention time basis, is 600 lb N, 45 lb P and 6,000 lb suspended solids plus 1,200 lb bacterial sludge from the secondary clarifier. The contents of the dispersed culture reactor flows out at a rate of 25,000 gallons per hour to the 2,500-gallon flocculator tank where a solution of UCC Polymer X150 is added at the rate of 5 lb per day and alum at the rate of 20 lb per day.

Residence time in the flocculator is 6 minutes. The flow from the flocculator goes to the secondary clarifier, a 25,000-gallon cylinder, where in 1-hour residence time, the bacterial sludge settles out at the rate of 600 lb of N, 45 lb P, and 6,000 lb suspended solids per day. The bacterial sludge settles to bottom where it is distributed intermittently as follows: 300 lb return per day to the primary clarifier to aid coagulation of incoming sewage; 1,200 lb go to the dispersed culture aerobic reactor as inoculum and to maintain high cell concentration, 10^{10} to 10^{12} cells/ml; 4,500 lb are dewatered by centrifugation and processed for animal feed. The sedimented sludge from the primary clarifier is transferred to the sludge aerobic reactor, having a 20,000-gallon capacity, twice a day at the rate of 3,000 gallons per day of 5% solids.

The daily inflow is 290 lb nitrogen, 7.2 lb phosphorus, and 1,250 lb suspended solids. The contents of the sludge aerobic reactor are stirred slowly (impellers 150 rpm) so as not to disturb flocculated growth. A gas mixture of 90% O_2 and 10% CO_2 is circulated through the mixture by bubbles at a rate of 0.1 volume of gas per volume of liquid. Temperature is maintained at 30°C and the pH remained at 7.2. The separation of the liquid from the treated sludge by centrifugation provides a filtrate that is less than 7 ppm with respect to N, less than 2 ppm with respect to P and BOD of less than 25 ppm O_2.

This liquid volume of 3,000 gallons was recirculated through the liquid stream to effectively remove all of the N and P. The packed solids from the treated sludge is dried for high nitrogen-containing fertilizer. Daily yields of 2,000 lb of dried solids contained 290 lb of nitrogen and 7.2 lb of phosphorus. The treated water returned to the water resources contains less than 1 ppm nitrogen and phosphorus, and less than 10 ppm BOD and suspended solids.

FERTILIZERS FROM SLUDGE

COMPOSTING

Aerobic Composting

J.R. Snell; U.S. Patent 3,357,812; December 12, 1967 describes a process for composting organic waste materials to yield a product suitable for use as a fertilizer or soil conditioner. In the process oxygen-containing gas is uniformly directed through the waste materials while agitating certain zones of the material without lifting the material.

The apparatus comprises a generally rectangular open-top tank having a bottom and upstanding side walls, means for introducing organic waste material at one end of the tank and for discharging finished compost from the opposite end, means communicating with the interior of the tank for selectively effecting air circulation upwardly and downwardly, and means for agitating and loosening successive increments of the mass during treatment without substantial vertical displacement of material from any portion of the mass.

Continuous Drying Through Aerobic Precomposting

H. Gujer; U.S. Patent 3,553,844; January 12, 1971 describes a method for drying wet organic waste matter, particularly fresh sludge, through aerobic precomposting. In the method wet organic waste matter is continuously added to a bed of precomposting decaying matter in such proportions that the moisture balance is not effected to the point of halting the process.

Constant recirculation of the material in the bed is conducted in order to keep the bed relatively loosely arranged so that sufficient air is available throughout the bed. The preferred apparatus includes an annular shaped receptacle with a work platform suspended above on which are mounted a circulating device, a feeding device, and a removal device. In the preferred forms, either the annular receptacle revolves or, alternatively, the annular receptacle remains stationary while the work platform revolves around the same. The drying device is depicted in Figures 5.1a through 5.1c.

FIGURE 5.1: CONTINUOUS DRYING ORGANIC WASTE THROUGH AEROBIC PRECOMPOSTING

a.

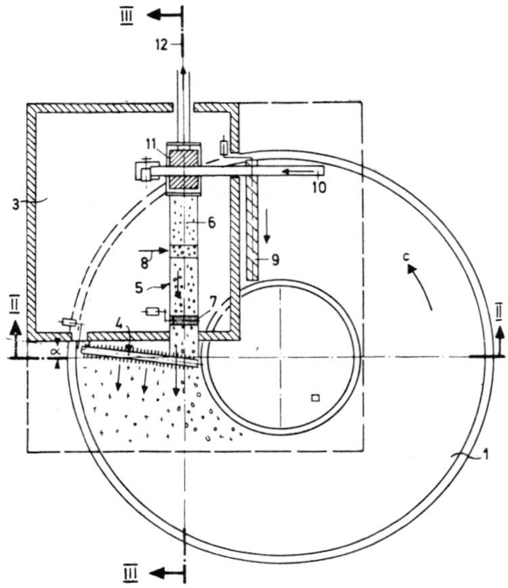

Top View of Drying Device

b.

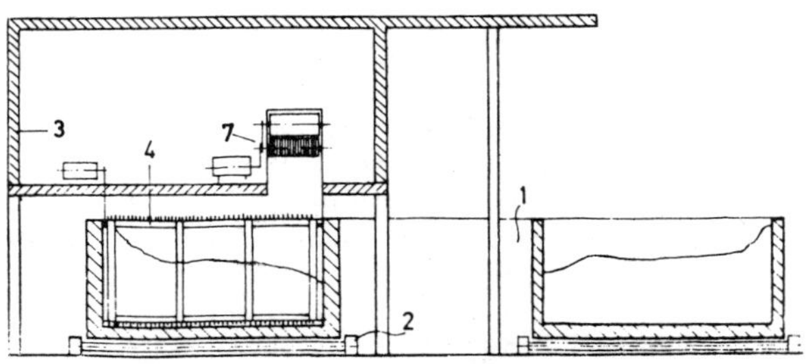

Vertical View Along Line II-II of Figure 5.1a

c.

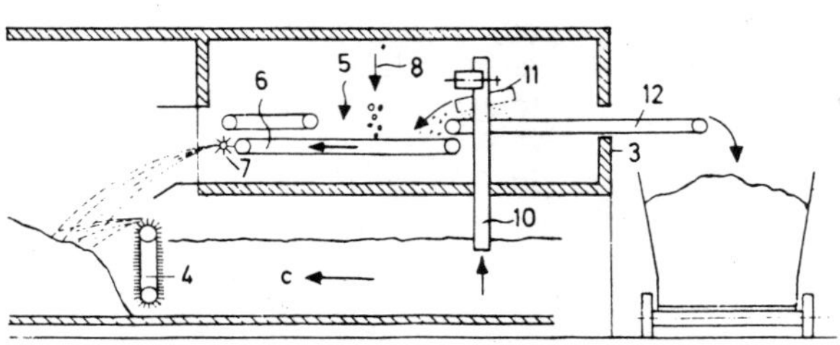

Sectional View Along Line III-III of Figure 5.1a

Source: U.S. Patent 3,553,844

The drying device for performing the precomposting process includes an annular receptacle **1** which is supported on two rollers **2** and rotates about its axis in the direction represented by the arrow **C**. The floor and side walls forming the precompost chamber of the annular receptacle are insulated so that the entire self-induced temperature which develops during the aerobic decomposition, can be utilized to the greatest extent possible, for the evaporation of liquid.

A working platform **3**, preferably having a roof, is provided for on a stationary support above the annular receptacle, and serves to carry certain other components of the precomposting device. The working platform covers at least a portion of the area of the annular receptacle and could, if desired, cover the entire receptacle.

A turning or circulating device **4** is provided at the working platform, such turning device extending over the entire width of the annular cross section and, preferably, being mounted at an adjustable angle α to the radius of the annular receptacle so that the turned, or circulated, matter is conveyed by a certain amount in the direction of the exterior edge of the receptacle. Thus, as the decaying matter is worked over each time, it is being conveyed, or circulated, from the inside toward the outside by means of the inclined position of the turning, or circulating, device **4** relative to the tangent of the conveying direction.

By adjusting the angle, the transverse conveying can be increased or reduced. The circulating device may consist of a vertically mounted conveyor belt equipped with teeth, although other suitable circulating devices can be used.

A depositing device **5** is mounted on the work platform and is used to distribute the fresh waste matter at the inner annular region in the direction of the circulating device, as viewed in the direction of advance of the circulating device. The depositing device is equipped with a conveyor belt **6** which is preferably adjustable in its speed. The fresh waste matter is finely fragmented and evenly distributed, and is deposited at **8** onto the conveyor belt.

The drying device is further equipped with a conveyor device **9**, preferably a worm conveyor, which is also secured to the work platform support and runs essentially horizontally. The conveyor device **9** extends over the width of the annular receptacle and is mounted such that it conveys back to the interior annular area the decaying matter accumulating in the exterior area of the receptacle.

The conveyor device **9** provides an intensive transverse circulating effect and at the same time a constant feedback of the decaying matter into the mixing area of the raw matter. The inclined position of the circulating device and the feedback through the conveyor device **9** causes an intensive mixing and homogenization of the entire contents of the receptacle so that, in spite of local addition of the raw matter, the same is continuously mixed into the entire contents of the decaying matter.

The removal device **10**, which is advantageously also mounted at the working platform, has a sifting device **11** connected thereto which has associated therewith a discharge device **12** for the precomposted decaying matter and is also coupled with the depositing device for disposal of the sifted material. With this arrangement it is possible to free the precomposted decaying matter of coarse components and at the same time add those coarse components back to the bed via the depositing device **5**. This helps to loosen the decaying matter further. The coarse matter which has been sifted from the precomposted matter in sifting device **11** is preferably dried prior to adding it to the bed. This drying permits the coarse matter to absorb a part of the moisture as well as to loosen the decaying matter.

The drying devices are suitable for all organic waste matter having a water content exceeding that suitable for precomposting using ordinary methods. In particular, fresh sludges belong in this category. If it is desired that the drying devices also process refuse, it is necessary to preliminarily prepare the same according to conventional methods, that is, segregating coarse components from the raw refuse and crushing the same if necessary. Further, the raw refuse must be freed of metallic particles.

Peat Moss Used as Filter Medium

In a process described by *C.J. Geraghty and R.R. Kennedy; U.S. Patent 2,861,877; November 25, 1958; assigned to Modoc Peat Moss Company* sludge is filtered into a layer of loose porous (and preferably fibrous) cellulosic material (preferably peat moss) and thus partially dewaters the sludge. When the moisture content of the resulting mass of sludge and porous cellulosic material is in the range of 50 to 70% the mass is subjected to a tilling operation which breaks it up into a product which is ready for market.

Fertilizer Mixture

A process described by *F.T. Nielsson and A.J. Soday; U.S. Patent 3,073,693; January 15, 1963; assigned to International Minerals & Chemical Corporation* involves the composition and preparation of a nitrogenous fertilizer material comprised of sewage sludge, peat moss, urea-formaldehyde, and an ammonium salt of a strong mineral acid.

In the preparation of the fertilizer, sewage sludge, peat moss, and solid urea are mixed with an aqueous solution of urea-formaldehyde and an aqueous solution of a strong mineral acid. The reactants are mixed for between ½ and 2 minutes, whereby polymerization and condensation of the urea-formaldehyde is effected. The resulting mixture is then mixed with an aqueous solution of ammonia whereby further condensation and polymerization of the urea-formaldehyde is inhibited.

The reactants are mixed for between ½ and 2 minutes to produce a substantially homogenous mixture. The resulting solids mixture, after curing, is a substantially dust-free nitrogenous fertilizer which will not burn the plants and which does not have an offensive odor when contacted with water. The following example illustrates the process. Raw materials used to prepare nitrogenous fertilizer materials in the following example are as follows:

(1) Activated sewage sludge containing 6.2% P_2O_5, 8.2% water, and 5% nitrogen, 90% of the nitrogen being insoluble in cold water.

(2) Peat moss, a commercially available product of the partial natural carbonization of vegetable and wood, the average nitrogen content being 0.5% and the average water content being 50%.

(3) Solid urea, fertilizer grade, containing 45% nitrogen.

(4) Aqueous urea-formaldehyde solution containing 26% urea, 59% formaldehyde, and 15% water.

(5) Sulfuric acid, 60°Bé.

(6) Aqueous ammonia solution containing 25% ammonia, 35.3% urea, 11.9% formaldehyde, 3.0% carbon dioxide, and 24.8% water.

Example: Activated sewage sludge (520 parts), peat moss (520 parts), and solid urea (553 parts) were mixed in a paddle mixer for 1 minute. Aqueous urea-formaldehyde solution (368 parts) and sulfuric acid (139 parts) were mixed with the solid materials for 30 seconds. Aqueous ammonia-containing solution (150 parts) was then mixed with the reagents for 30 seconds. The resulting mixture was divided into three equal portions designated as Portions A, B and C, respectively.

Portion A was stored in a closed container at room temperature for one week. Portion B was stored in a closed container at a temperature of 150°F for one week in order to simulate the temperature conditions which prevail in storage of large quantities of fertilizer materials. Portion C was heated in a rotary dryer having an inlet gas temperature of 500°F, the temperature of the solids being maintained below 160°F. The ratio of various components in the fertilizer and chemical analysis of the fertilizer are set forth in the table below.

Component	Portion A	Portion B	Portion C
Urea, formaldehyde molar ratio	1.46	1.46	1.46
Ammonia, sulfuric acid molar ratio	2.0	2.0	2.0
Synthetic organic N, natural organic N molar ratio	11.1	11.1	11.1
Solid urea, urea in solution weight ratio	3.6	3.6	3.6
Urea, ammonia weight ratio	18.2	18.2	18.2
Organic nitrogen, inorganic nitrogen weight ratio	11.3	11.3	11.3
Total nitrogen, percent	19.0	20.14	21.73
Cold-water-insoluble N, percent	9.0	12.0	9.73
Hot-water-insoluble N, percent	5.38	6.66	6.58
Activity index	40.2	44.5	32.37
Moisture, percent	13.14	5.14	9.15

For purposes of comparison, activated sewage sludge (1,040 parts) and solid urea (150 parts) were mixed in a paddle mixer for one minute, no peat moss being added to the mixture. The resulting mixture was treated with an aqueous solution of urea-formaldehyde, sulfuric acid, and an aqueous ammonia solution in the same proportions and in accordance with the same procedure as set forth above. Upon curing this mixture at either room temperature or at a temperature of 150°F for 1 week in accordance with the above procedure, the mixture in each instance, set up into a hard, solid mass which required grinding before the material could be used as a fertilizer. Eliminating peat moss as one of the components of the fertilizer material, adversely affected both the process and the product. Portions A, B and C, which contained peat moss, were friable mixtures and easy to handle.

Separation of Inorganic Impurities

J.R. Liljegren; U.S. Patent 3,476,683; November 4, 1969 describes a method of separating inorganic impurities from sewage and of making a soil-improving composition from the separated impurities. The process comprises adding to the sewage, on the basis of the measured quantity of the sewage and the degree of pollution, a precipitating agent in the form of lime, hydrate of lime or slaked lime, leading the water and lime into pools in which the heavy precipitates sink to the bottom as sludge and the lightweight precipitates assisted by the gas formation resulting from the reaction in the water rise to the surface, and removing the sludge and floating precipitates from the water and dewatering the sludge and the precipitates.

For the preparation of the soil-improving composition the dewatered sludge is mixed with sludge obtained from a purification plant operating biologically with active sludge, treated in the same way and preferably dewatered together with the first mentioned sludge, and during mixing a suitable quantity of unslaked lime, calcium oxide, is added to the dewatered sludge mixture to effect both sterilization and drying.

Treatment with Halogenated Hydrocarbon Solvent

A process described by *D. McDonald; U.S. Patent 3,365,395; January 23, 1968; assigned to Science Progress, Incorporated* relates to the solvent treatment of sewage and waste material. The term "solvent" is meant to include halogenated hydrocarbon solvents of which the following class of solvents is representative: perchloroethylene (tetrachloroethylene), trichloroethylene, tetrachloroethane, chloroform and carbon tetrachloride.

The preferred solvent, however, is perchloroethylene, which has a boiling point of 250°F, an azeotropic boiling point with water of between 190° and 195°F, and a specific gravity of 1.62. The solvent in the method serves a dual purpose in that it not only dissolves the solvent solubles (fats and the like), but also serves as an effective dehydrant.

In the process, it is the objective to so treat sewage that all products or by-products will be free of contaminating properties while simultaneously providing a self-sustaining sewage treatment. When the end product is intended to be a dry sterile fertilizer, the sewage is preferably ground in the presence of an excess of solvent so as to comminute the larger particles of sewage material. The resulting comminuted material (sludge) is allowed to settle for a predetermined period of time (relatively short, that is, about 15 to 60 minutes, or shorter). Due to the presence of the solvent, the sludge will settle rather rapidly and separate itself from a majority of the effluent water present.

The effluent water is continuously removed and further treated, in numerous ways, to be thoroughly decontaminated and then discarded. The sludge, meanwhile, will also be continually removed and passed to a surge tank where it is further exposed to solvent, thereby allowing the solvent to thoroughly remove from the sludge, all solvent-soluble materials.

There will be formed in the surge tank two major layers, the lower or heavier layer containing the heavier materials like solvent and heavy inorganic material, while the light layer will contain sludge, solvent containing solvent-soluble material and entrained effluent which is still present from the preceding step. The lower or heavier layer will be withdrawn and passed through suitable apparatus to separate the inorganic material from the solvent.

The lighter layer will be passed through a centrifuge which will remove therefrom any remaining effluent and the residual solvent and sludge will then pass to the extraction chamber. The sludge will separate into meal and solvent, and the meal having a specific gravity less than the solvent will float on the surface thereof and be continuously removed and dried and readied for consumer use.

High Paper Content Waste

In a process described by *V. Brown; U.S. Patent 3,533,775; October 13, 1970; assigned to Metropolitan Waste Conversion Corporation* sewage sludge and municipal waste, the latter comprising at least 50% paper, are processed by removing the salvageable scrap and inert materials from the municipal waste, comminuting the remaining waste, removing a part of the combustible materials, comprising a portion of paper content, from the comminuted waste, mixing dewatered sewage sludge with the remaining waste and composting the mixture by aerobic digestion. The composted material is dried by hot air obtained by burning the removed combustible materials.

In the process, the salvageable materials are manually, pneumatically, mechanically or electromagnetically removed and sold as scrap and the inert materials are removed and sent to landfill disposal tracts. The remaining waste is comminuted and a substantial quantity of the paper and film plastic is removed and burned to supply heat for evaporating water from the sewage sludge, for drying the composted organic waste, or for other energy needs.

In a city of approximately one million population, the town and house waste amounts to about 2,000 tons per day, about 75% of which is paper. One ton of paper provides the same quantity of heat as one barrel of oil or approximately 5,000 Btu per cubic foot. This feature gives flexibility to the composting process. In seasons when compost demand is light, a proportionately larger quantity of paper may be removed and burned. When compost demand is heavy, the minimum amount of paper necessary to provide fuel may be removed and the remainder composted.

It has also been found that aerobic digestion of the comminuted waste and sludge is accelerated by enriching with oxygen the air fed to the composting tank. A preferred apparatus for carrying out the composting with agitation and in the presence of air is shown and described in U.S. Patent 3,294,491 and includes a false bottom or the like through which air may be intimately diffused into the organic mass being composted.

The air surrounding the organic waste particles must have a minimum of 6% oxygen for efficient composting and preferably the oxygen content should be maintained at 20 to 35%. Below the 6% minimum, the digestion is so slow as to increase the cost of the process beyond economic feasibility. Atmospheric air, of course, contains approximately 20% oxygen, but this oxygen is quickly utilized in the aerobic digestion process and the proportion of oxygen in the air within the mass is likely to fall below 6% if the incoming air is not enriched.

The presence of raw primary or activated sewage solids in the digesting mass greatly increases the biochemical oxygen demand and makes the addition of oxygen of greater value and importance than in composting town waste only. Figure 5.2 is a flow sheet showing the essential apparatus and the flow of materials.

FIGURE 5.2: PROCESS UTILIZING COMPOSTING STEP TO PROVIDE FERTILIZER

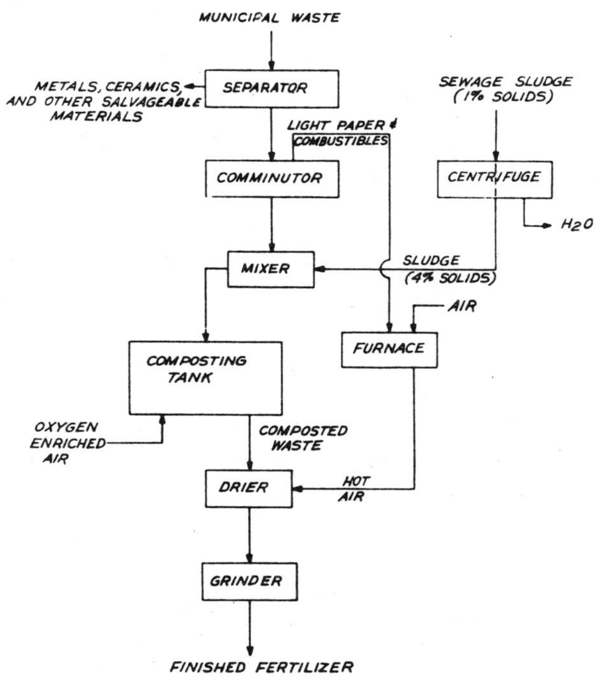

Source: U.S. Patent 3,533,775

Granular Sludge for Fertilizer

J.W. Hudson and G.D. Ferguson; U.S. Patent 3,451,800; June 24, 1969; assigned to USS Agri-Chemicals, Incorporated have found that the woolly fiber problem in dried sewage sludge can be solved by grinding all of the sludge in a hammer mill or similar equipment until approximately 95% will pass a 50 mesh screen, and then by mixing with the ground material a dry blend of about 0.5% starch and heating the same to a temperature at least as high as 185°F, the blended material can be granulated in a fertilizer granulator to obtain the maximum number of particles in the desired -8+20 mesh screen range. The granulated products may then be dried in any suitable apparatus, such as, for example, a rotary gas fired dryer.

Apparently, by grinding all of the sludge until 95% will pass a 50 mesh screen, the fiber difficulty is overcome, and with the high temperatures of heat, a relatively small amount of starch is effective in producing the uniform, strong, nondusting, granular product. The dried granulated particles in the -8+20 mesh screen range are smooth granules with well-rounded exteriors and are uniformly sized, strong and nondusting granules. Such a product permits evenly controlled application to lawns and other uses, employing the usual fertilizer applicators.

Example 1: A dried sewage sludge sample was first ground in a hammer mill until 95% passed a 50 mesh screen. The ground material was then dry blended with 0.5% cornstarch (No. 3005 starch, Corn Products Refining Co.). The blended material was then fed to a TVA-type granulator, water and steam being added to give a moisture content of 10 to 14% of dry weight of sludge and to maintain a temperature in the range of 185° to 195°F. At this temperature, the cooked starch produced the maximum number of particles in the desired -8+20 mesh screen range.

The granulated particles were dried in a rotary gas fired dryer and the material in the -8+20 mesh range was screened out as product. The oversize was finely ground and mixed with the fines and recycled to the granulator. The product consisted of relatively uniform granules which were strong and nondusting and which had smooth and well rounded exteriors.

Example 2: The process was carried out as described in Example 1 except that the dry mixture of ground sludge and starch was heated in the granulator to the temperature range of 190° to 200°F. The cornstarch in the proportion of slightly less than 0.5% was found to be effective in forming the granulated particles in the desired range of -8+20 mesh.

In a related process by *J.W. Hudson and G.D. Ferguson; U.S. Patent 3,442,637; May 6, 1969; assigned to USS Agri-Chemicals, Inc.* fertilizer ingredients including sewage and bran fibers are mixed with nitrogen solution, ammonia and steam, and the blended mixture granulated to form hot granules The granules while hot and wet are mixed with cottonseed meal and starch whereby the oil-rich particles of the meal coat the granules to provide an attrition-resistant fertilizer.

COMPACTION

Surface Soil Manufacture

K.H.R. Wikstrom and P.E. Ericson; U.S. Patent 3,824,092; July 16, 1974 describe a method of manufacturing a surface soil where sludge from sedimentation tanks having a dry solids content of approximately 10 to 90% is mixed with a sandy material and then compacted by pressing

Alternatively this surface soil may contain peat soil in addition to the sandy material. The surface soil is produced in two stages, the purpose of the first stage being to produce a premix, in which the starting materials are present in the intended proportions, while in the second stage the premix formed in the first stage is homogenized so as to transform the same to a product in which the particles of sand are enveloped by an organic substance.

This effect is obtained by subjecting the premix from the first stage to a pressing operation, it having been discovered that the dry product enters into intimate contact with the hydrophilic sandy material in spite of its hydrophobic nature. It would appear that one necessary condition is that the sludge has a dry solids content within the given limits and that the formed premix is subjected to a pressing operation which is suitably extended to comprise a series of such pressing operations at a progressively increasing pressure, similar to the case in conventional metal rolling operations.

The sandy material may be chosen from a large number of different materials of an inorganic character, containing no, or preferably minor quantities of organic substances; the essential thing being that the material is not of uniform particle size. Thus, the size of the various particles may vary from close to zero to 20 mm, preferably from 0.5 to 8 mm. The material should have a density of 1,000 to 2,500 kg/m^3. Mo-clay, common sand and gravel are examples of the types of material envisaged as the sandy material used in the method.

The pressing operation to which the premix is subjected is of particular importance, since it is this which causes a satisfactory, homogeneous mixture of the sludge with the sandy material. The material can be pressed together, for example, in a tapering annular slot, in which it is progressively compressed while being subjected to a series of blows, which causes the coarser sand particles to be crushed to smaller sizes, these particles then being forced into the liquid phase of the sludge having a relatively high dry solids content.

Subsequent to being pressed together, the premix can be suitably subjected to a loosening or expanding process, wherein the material is caused to fall freely onto a receiving surface preferably comprising a treated material, where it is broken up into a porous, very loose product suitable for use as a surface soil and which can be readily spread on grassland or similar surfaces such as lawns for example.

Because of the nature of the starting materials, one great advantage afforded by the product produced in accordance with the process is that the product is free of seeds liable to produce weeds, thereby reducing to a very large extent the task of removing weeds from ground employing the surface soil of the process. It has also been found that a lawn built on a surface comprising a top soil produced according to this process is surprisingly durable. The reason for this is that the roots of the grass become embedded in the surface of the surface soil in a manner whereby the surface is reinforced with the roots in a very effective manner. Furthermore, the ability of the surface soil to retain water is extremely high.

Example 1: Sedimentation sludge having dry solids content of approximately 10% was dewatered to a dry solids content of 25%. One hundred kilograms of the dewatered sedimentation sludge were mixed with 100 kg of sand, having a specific gravity of 1.4 and a screen analysis of 0 to 6 mm, to form a premix.

The premix was then subjected to a series of rapid pressing operations. The resulting product was then lifted and allowed to fall freely onto a receiving surface and then placed in storage.

Example 2: Sedimentation sludge having a dry solids content of approximately 10% was dewatered to a dry solids content of 30%. Thirty-five kilograms of the dewatered sludge were mixed with 35 kg of a clay material (fraction 0 to 2 mm), 28 kg of peat soil and 2 kg of a fertilizer, containing nitrogen, phosphorus and potassium, calcium and magnesium in the form of compounds, to form a premix. The premix was stored for two months and then subjected to a series of rapid press operations.

The resultant product was found to be an excellent surface soil for lawn laying purposes. A better result was obtained with a product produced according to the process above but which had been caused to fall freely onto a receiving surface and then stored for a short period. In both cases, the soil was found to be extremely durable and the surface layer, subsequent to being sown with grass seed, was found to be well reinforced with the grass roots.

Using Agglomerants

In a process described by *J. Schick; U.S. Patent 3,226,319; December 28, 1965* weighting agents are added to the sludge before or during or after the addition of the agglomerants, become associated with the solids, increase their specific weight and cause them to drop, and the resulting solid flakes are given a further addition of agglomerant to become a sediment which has a high concentration of solids and which can be separated from the purified water merely by decanting without needing filtration.

The solids contained in the voluminous sludges, which often have water contents of 98%, are therefore, combined by adsorption of the agglomerants, because of electrostatic interaction, to form a solid sediment from which so much clarified water has been separated that such water can be decanted in a very simple way, for instance, by overspill. The resulting inspissated sludge has a pudding-like consistency, is compact, water-repellent and water-insoluble; it can therefore be easily handled.

The precipitants of natural origin are extracts from plants, fruits or animal products, such as starch flour, ground carob bean, resins or the equivalent; albuminous substances made particularly suitable natural agglomerants. Synthetic high molecular carboxylic-group-containing polymers are employed as synthetic agglomerants.

The synthetic agglomerants are commercially available substances including those having the commercial names Separan NP10 (a high molecular weight synthetic polymer containing both amide and carboxylic groupings), Surosolan (an agent on the basis of polyacrylamide), Sedipur (high molecular organic substances derived from polyacrylic acid), and others. A distinction is made with these synthetic agents between mainly anionic and mainly cationic agglomerants. Preferably, anionic precipitants are used for the alkaline range and cationic precipitants for the acid range. When adding precipitants as well as when adding weighting agents an intensive oxidation of the sludge takes place simultaneously by aeration or chlorination since this aids also in reducing the sludge volume.

Advantageously, the agglomerant is used in the form of an aqueous solution, in which event the solution can have a concentration of from 0.5 to 5 g/l, preferably from 1 to 2 g/l. The concentration of the agglomerants can readily be determined individually for each case and depends upon the nature, origin and proportion of solids of the sludges. Conveniently, the agglomerant solution is used in proportion to the quantity and nature of the solid to be precipitated, preferably from 1 to 200 grams of agglomerant being used per metric ton of dry solid substance. Usually, 2 to 5 grams of agglomerant per metric ton of dry solid substance are sufficient.

In Situ Polymerization of Acrylic Monomers

A process described by *J. Schick; U.S. Patent 3,226,318; December 28, 1965* relates to a process for consolidating aqueous sludge having high solids contents and low water contents. According to the process, water-soluble polymerizable or condensable monomers are added to the sludge, then polymerized or condensed, the solids and water constituents of the sludge being locked into the polymerization or condensation product and the sludge being permanently consolidated.

This process takes much less time than the conventional drying processes, with the result that the area required for final preparation of the sludge yielded is reduced correspondingly. Another advantage of the process is that the consolidated sludges have a number of possible uses, for instance, as an insulating material or sealant or as a building material.

In a preferred form of the process, the timing of the polymerization or condensation is controlled in a manner known per se by an addition of accelerating or retarding compounds. Advantageously, water-soluble derivatives of acrylic acid or low molecular weight aldehydes together with benzene derivatives, for instance, phenol, are used as polymerizable or condensable monomers.

The polymerizable or condensable water-soluble or water-dispersable compounds, preferably while in the monomeric state, are added to the aqueous sludge in solutions or dispersions of varying concentration, with or without foaming agents. The mixture is allowed to freeze, possibly in molds, with an addition of substances which initiate the consolidation process and which are known as starters, and with an addition of substances (usually accelerators) which control the timing of the process; the organic and inorganic solids contained in the sludge and the water contained therein are locked into the polymerization or condensation product.

Shaped or unshaped compositions in the form of frozen structures or foams are yielded, the cavities of the structures or foams being full of water while the cavity or cell walls have the solids polymerized or condensed into them. These solid compositions are reversible systems, yielding water by evaporation and absorbing water.

The cell walls take the form of synthetic resin gels if a two-component aqueous solution of monomeric derivatives of acrylic acid, for instance, a mixture of acrylamide, and N,N'-methylenebisacrylamide, the commercial product known as AM-9 (American Cyanamid Co.) is used. To start polymerization, there are added to these solutions or dispersions in known manner catalysts, such as β-dimethyl amino propionitrile which forms free radicals, accelerators such as peroxides or per-salts, and inhibitors, such as ferricyano compounds, which control the timing of polymerization. Of course, only monomers which lead to nonadhesive polymers are used for polymerization. The polymers used, together with the polymerized sludge ingredients and the water present in the sludge, should lead to at least compact compositions which can be transported without returning to a thixotropic condition.

To produce consolidation by condensation, for instance, a phenol-formaldehyde solution can be added to the aqueous sludges, and condensation can be started by a further addition of formaldehyde or of some suitable curing agent. Preferably, the resins used for these condensations are of the kind which can receive a very wide variety of fillers.

Foaming agents or foamable resins can also be added as in the preparation of foamed substances; the solids contents of the sludge are, in this application locked into the foam upon the completion of condensation. However, these condensation processes require longer reaction times than do polymerization processes. Also, heat may occasionally be required to start the condensation process. It is therefore more economical to use polymerizable substances for this application instead of expensive chemicals for the consolidation of sludges in respect of which the only need is, as a rule, removal with maximum efficiency.

Example 1: Fifteen kilograms of 20% aqueous solution of AM-9, to which 120 grams of catalyst and 40 cc of a 10% ferricyano-potassium solution had been mixed, were added to 100 kg of a sludge containing 62% water. Also, a solution of 150 grams of ammonium persulfate dissolved in 1 liter of water was added to the sludge with thorough mixing. At the end of 1 hour the mixture had frozen to a compact composition which could readily be transported.

If instead of 15 kg of a 20% solution, 15 kg of a 25% solution of AM-9 containing 150 grams of catalyst and 10 cc of a 10% ferricyano-potassium solution were added and a solution of 188 grams of ammonium persulfate dissolved in 1 liter of water was simultaneously added, the resulting composition was already solid.

This example shows that a quantity of at least 4% of AM-9 is necessary to consolidate a sludge having 62% aqueous content. If the quantity of AM-9 added is increased to above 5%, for instance, to 15%, the resulting products are solid but, although they are very hard at the completion of polymerization they are shattered by slight vibrations and then look like powdered glass. The sludges may be of organic or inorganic origin.

Example 2: Instead of the polymerizable AM-9 used in Example 1, 20 kg of a 40% phenol-formaldehyde solution were added to 100 kg of the same sludge, whereafter 2 kg of a 40% formalin solution were added. At the end of 8 hours a condensation product was yielded of a consistency adequate for transportation to a dump.

Compacting and Granulating

In a process described by *J.H. McLellan; U.S. Patent 2,977,214; March 28, 1961; assigned to H.J. Baker & Bro.*, a system is provided for the compacting and granulating of highly pulverous dried sewage sludge whereby the powdered material can be compacted into useable granules of increased density for economic handling and shipment, with decreased loss due to the pulverous nature of the starting material, and in a manner which both enhances the utility of the dried sludge as by-product fertilizer or soil conditioner material and reduces the dangers incident to spontaneous combustion and continued putrefaction during bulk storage or handling.

Referring to Figure 5.3, a source of pulverous dried sewage sludge is indicated at **10**, purely diagrammatically indicating a source of pulverous dried sludge material of, perhaps 5 to 15% moisture content, more or less, conventionally produced by heat drying, as in rotary or flash dryers. The source of dried pulverous sewage sludge is shown as discharging a mass **11** of such dried sludge into a hopper **12** of the illustrated apparatus, with the outlet to the hopper controlled by a gate valve **13** with associated operating mechanism **14** for introduction of the mass of pulverous material **11** to the compacting rolls to be described, preferably sufficiently to flood the roll nip.

Compacting apparatus **15** is illustrated as including a main frame member **20** mounted on a base or floor **21** and supporting compacting rolls **22, 23** in suitable bearings and journals not shown. Compacting rolls **21, 23** are driven for rotation as indicated by the arrows thereon by conventional motor means, not shown, and are urged toward each other to provide a high nip pressure therebetween as by a hydraulic mechanism indicated at **24** as including a hydraulic cylinder and piston arrangement including bearing piston **25** actuated by hydraulic lines **26**, all in known manner.

Upon passage of the highly pulverous dried organic and fibrous or cellular sewage sludge material through the high pressure nip of the compacting rolls **22, 23**, the pulverous material is compacted into a more or less continuous sheet **30** to assume a rigid and highly frangible character. As sheet **30** emerges in compacted form from the nip between compacting rolls **22, 23**, it is subjected to a breaking or fracturing action, as by rotating breaker **35** having breaking fingers **36** therearound which break or fractionate compacted sheet **30** into rather gross flakes or chunks or fragments which flow by gravity out of the outlet **37** from the compacting and breaking apparatus and into the inlet **38** of granulating apparatus **40**.

The granulating apparatus is illustrated as having a main frame **45** supported on a base or floor **46** and comprising two sets of corrugated granulating rolls **47, 48** and **49, 50**. Nip or contact or meshing adjustment means are indicated, generally, by the handle and screw arrangements **51, 52**, respectively, and, as will be understood, as the broken flakes or chunks of the sheet pass into inlet **38** of the granulating mechanism and through the nips between rows **47, 48** and **49, 50**, the flakes are granulated or further fractionated or broken up into small particles or chunks of the desired size.

As illustrated in the figure, the corrugations of rows **47, 48** are substantially axial, although it may be preferred that the corrugations be at a slight angle to the roll axis, whereas the corrugations on rolls **49, 50** conform to the so-called Lepage arrangement for roller mills in which substantially axial corrugations are arranged on roll **50** while roll **49** has substantially circumferential corrugations. In any case, as will be understood, the several rolls are driven by motor means and, indeed, preferably at differential speeds as between the two rolls of each pair, and the large fractions or chips broken by breaker **35** from the sheet are further granulated by the interaction of the corrugations on rolls **47, 48** and **49, 50** to produce granulations or particles of the desired compacted form. The granulated particles leave the final pair of rolls **49, 50** and emerge from granulating mechanism **40** through the outlet **55** to drop onto screening mechanism **60**, which is illustrated as having a fine screen portion **61**, and accepting coarser screen portion **62**, and an oversize reject discharge **63**. Also indicated at **64** is means for oscillating or shaking screen mechanism **60**.

FIGURE 5.3: COMPACTING AND GRANULATING PROCESS

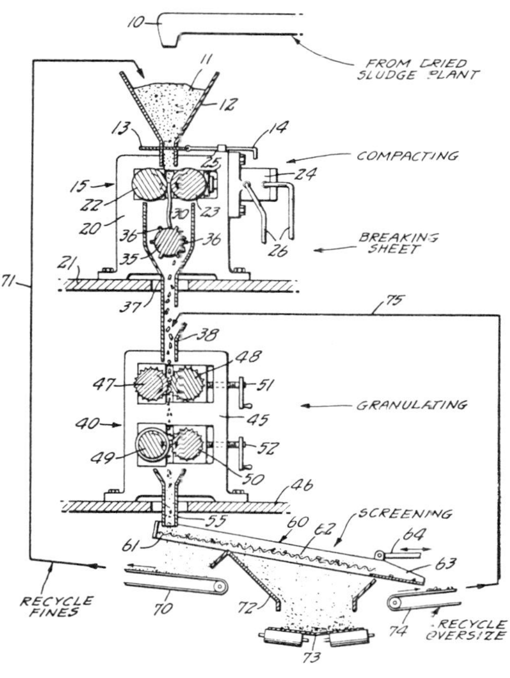

Source: U.S. Patent 2,977,214

The material emerging from outlet **55** of granulating unit **40** will include unacceptable pulverous fines, granules of an acceptable size, and oversize granules. Initially the fines will drop through the fine screen area **61** onto a conveying mechanism indicated at **70** to be recycled, as indicated by the arrows **71** back up to hopper **12** for mixture with the material **11** to be compacted.

The granules of acceptable size will fall through or pass the acceptance area **62** of the screening mechanism and fall through chute **72** to be received on conveying mechanism indicated at **73** for conveying to the ultimate storage or bagging or shipment facilities, while oversize granules will pass the acceptance area **62** of screen **60** and emerge at the oversize discharge **63** onto conveying mechanism indicated at **74** to be recycled, as indicated by arrow **75** to the inlet **38** of granulating mechanism **40**. As will be seen from the foregoing, highly pulverous material is interjected from a source **10** into hopper **12** and passes continuously through the apparatus until compacted, fractionated, and granulated to a specified or desired granule size for discharge of final conveyor for use as desired.

It may be noted that satisfactory results were achieved using, in compacting unit **15**, compacting rolls **22, 23** in the form of smooth rolls 18 inches in diameter by 6 inch face operating at 2 rpm. With a dried pulverous free flowing sludge having approximately 5% moisture, and with the nip pressure unit **24** set for a face-to-face initial pressure of 1,700 lb per linear inch, a satisfactorily hard sheet **30** was produced with a thickness of 0.055 inch at a rate of 1.8 lb of material per minute, although, during operation, the nip pressure actually went up to 3,170 lb per linear inch.

With pulverous material having a moisture content of approximately 10.6%, however, the sheet **30** produced under the foregoing conditions was somewhat softer and more crumbly than optimally desired. By increasing the initial setting of nip pressure between rolls **22** and **23** to approximately 2,500 lb a satisfactory compacted and frangible sheet **30** was produced from the higher moisture content material at a thickness of approximately 0.045 inch and at the rate of 1.8 pounds per minute, during which operation the recorded nip pressure rose to 4,800 lb per linear inch.

After formation of the compacted sheet, it was broken as by breaker **35**, and the large flakes or chunks fed through the granulating rolls in granulating unit **40**. Satisfactory results were obtained with granulating unit **40** when the first set of granulating rolls, **47** and **48**, consisted of one roll with four corrugations per inch and one roll with five corrugations per inch, both axial corrugations, and operating at somewhat differential speeds. The second set of rolls were of the Lepage type as illustrated at rolls **49, 50**, roll **50** having axial corrugations at ten to the inch, and roll **49** having circumferential corrugations at 8 to the inch.

The granulated product emerging from outlet **55** of corrugating unit **40** was screened on an 8 mesh and 30 mesh screen in a mechanism such as **60**, with the production of 80% of the original feed within the size range of -8+20 mesh. About 5% was retained on the 8 mesh screen and 15% went through the 30 mesh screen, with both these rejects being subject to recirculation through the appropriate steps, as noted, if desired.

Whereas the original density of the pulverous material **11** ran from approximately 25 to 28 pcf as poured loosely into a container, the 8 to 30 mesh granules produced as the acceptable fraction ran approximately 35 pcf, and even the 30 mesh fines were increased in density to from 30 to 35 pcf.

Microscopic examination of the interior of sheet **30** (or granules produced) reveals the inherent fibrous and cellular character of the initial dried sewage sludge starting material and that the compacted substance appears to have a texture similar to, for example, plug tobacco, with the microscopically visible cellular or fibrous particles being closely compressed and flattened parallel to the sheet surface and with both sheet surfaces coated with a thin, glossy black glaze, presumably, resulting from momentary liquefaction of tar-like organic components of the dried raw material under the influence of heat and pressure at the nip of the compacting rolls. It is believed, however, that the strength and frangibility of the formed sheet is a characteristic important for the ultimate granulation (as opposed to merely crumbling or pulverizing) and results mainly from the matting or felting or intertwining of the fibrous and lamellar fragments in the main body portion of the compressed sheets.

OTHER PROCESSES

Heat Conversion Using Magnesium Oxide and Sulfuric Acid

According to a process described by *H.P. Scheel; U.S. Patent 3,758,287; Sept. 11, 1973; assigned to Western Minerals, Inc.* conversion of organic waste into organic plant food is carried out by subjecting the waste to heat sufficient to sterilize, deodorize, and dewater the waste, the heat being generated by the chemical reaction of a mixture principally of magnesium oxide and sulfuric acid.

Although commercial magnesium oxide may be used, a preferable source is finely ground olivine, a less highly refined and less expensive material containing a large percentage of magnesium oxide. Olivine is a mineral composed essentially of the oxides of magnesium and iron but also containing aluminum, nickel, and other trace elements. When the olivine is mixed with concentrated sulfuric acid, a paste is formed which is then mixed with the waste. The exothermic reaction which takes place produces a compost-like material which is valuable as a plant food.

Example 1: Powdered olivine (200 mesh) was mixed with an equal amount, by weight, of 98% sulfuric acid, utilizing a gasoline powered Essick motor mixer of 5 cubic feet capacity. After being mixed for 2 to 3 minutes, the paste was allowed to set up for about 15 minutes. Sewage sludge was then added in the ratio of 3 parts, by weight, of sludge to 1 part, by weight, of paste. After being mixed an additional 5 minutes, the material was transferred to 55 gallon steel drums. When the drums were full, covers were placed on them equipped with ½ inch pipes for taking temperature readings.

Mercury thermometers were suspended in the pipes and blanket insulation packed about the steel drums. Two different types of sewage sludge were used: (1) digested, dewatered sewage sludge was obtained from the West Point treatment plant of the municipality of metropolitan Seattle (Metro); and (2) raw, dewatered sewage sludge was obtained from the treatment plant of Lynnwood, Washington. Analysis of the two sludge materials is shown in Table 1.

Temperature readings in the steel drums were taken at two locations, near the edge and in the center. A typical temperature versus time curve is shown in Figure 5.4.

Sawdust or sander dust was added to some of the batches to reduce moisture and quick or powdered limestone added to other batches to correct the pH. The results of bacteriological tests of the plant food end product from both the Metro and Lynnwood sewage are shown in Table 2.

FIGURE 5.4: OLIVINE-SULFURIC ACID TREATMENT OF ORGANIC WASTE

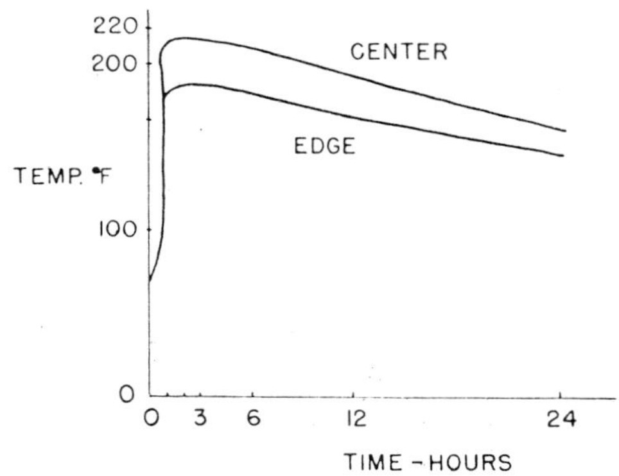

Source: U.S. Patent 3,758,287

TABLE 1: COMPOSITION OF SLUDGE MATERIALS

	Metro	Lynnwood
Moisture, percent	52.1	-
Protein (N x 6.25), percent	2.9	-
Bacteriological analysis:		
Std. plate count per gram[1]	120,000,000	240,000,000
Coliform count, MPN per gram[2]	>11,000	110,000,000
Fecal strep, MPN per gram[2]	>11,000	15,000,000
pH[3]	8.2	-

[1] Standard plate counts were reported in 1,000 as lowest dilution run was 1,000.
[2] Presumptive tests run on coliform counts and fecal strep.
[3] Run on a 10% solution.

TABLE 2: BACTERIOLOGICAL TESTS OF PLANT FOOD END PRODUCT

	Metro	Lynnwood
Standard plate count per gram[1]	<1,000	<10
Coliform count, MPN per gram[1]	<2	<2
Fecal strep, MPN per gram[1]	<2	<2
pH	3.5[2]	2.8[3]

[1] Same as in Table 1.
[2] Run on a 10% solution.
[3] Before addition of limestone. Run on a 50% solution.

The results show that, after treatment of the sewage, the standard plate count of bacteria was reduced to less than 1,000, which means essentially that no colonies of bacteria were observed. The coliform count was reduced to less than 2 and the fecal strep reduced to less than 2.

Example 2: Dead-burned magnesite (flue dust) was used in place of olivine. An analysis of the dead-burned magnesite was as follows:

	Percent
Silica	12.5
Fe_2O_3	2.9
Al_2O_3	4.8
CaO	3.0
MgO	63.4

Twelve hundred grams of damp magnesite was mixed with 900 grams sulfuric acid in a 2 to 3 gallon iron reaction vessel. Thirty-six hundred grams of sewage sludge was added. After reaction, the resulting plant food was submitted for bacteriological analysis. The results are given below.

Standard plate count	$<$10/g
Coliform count	Negative
Fecal coliform	Negative
Fecal strep	Negative

The plant food derived from the dewatered, digested sewage sludge of Metro (identified as A) and the plant food derived from the dewatered raw sewage sludge of Lynnwood, Washington (identified as B) were submitted for quantitative analysis to determine the percent by weight of nitrogen, phosphorus, potassium, and magnesium, with determination of other elements by semiquantitative spectrographic analysis of the ash. The results were as follows.

	Percent by Weight	
	A	B
Analysis on dry sample basis:		
Ash, residue on ignition	62.02	50.80
Total nitrogen (N)	0.49	0.97
Total phosphorus (P)	0.27	0.17
Total potassium (K)	0.04	0.05
Total magnesium (Mg)	5.7	6.3
Semiquantitative spectrographic analysis of ash:		
Silicon (Si)	23.0	24.0
Aluminum (Al)	5.2	2.1
Iron (Fe)	4.0	3.3
Strontium (Sr)	0.01	0.009
Sodium (Na)	1.9	Trace
Calcium (Ca)	4.3	6.6
Magnesium (Mg)	11.0	13.0
Titanium (Ti)	0.45	0.10
Lead (Pb)	0.06	0.05
Phosphorus (P)	0.41	0.44
Boron (B)	0.002	Nil
Manganese (Mn)	0.15	0.21
Chromium (Cr)	0.17	0.18
Nickel (Ni)	0.13	0.16

(continued)

	Percent by Weight A	B
Semiquantitative spectrographic analysis of ash:		
Tin (Sn)	0.02	0.03
Vanadium (V)	0.004	Nil
Copper (Cu)	0.03	0.04
Silver (Ag)	0.002	0.001
Zinc (Zn)	0.18	0.22
Zirconium (Zr)	0.007	Nil
Cobalt (Co)	0.01	0.02
Potassium (K)	Trace	Nil
Other elements	Nil	Nil

Some of the plant food material was applied to a patch of lawn; after 3 to 4 weeks the treated patch acquired a definitely greener color than the surrounding nontreated areas. Plant food produced by this method is particularly valuable because no gas or elements are lost in preparation of the plant food as is the case both when sewage is processed in digesters and when garbage is composted. Instead, a valuable plant food material including trace elements is added to the soil.

Using Brown Coal

G. Nettesheim, E. Petzold and H. Pruckner; U.S. Patent 3,607,212; Sept. 21, 1971; assigned to Union Rheinische Braunkohlen Kraftstoff AG, Germany have found that an agent for improving the fertility of soil can be obtained by adding to brown coal a small amount of a material which is capable of being decomposed by bacteria. Similar to peat, these materials provide better harvests.

Materials which are capable of being decomposed by bacteria are waste products of the cellulose industry containing α-cellulose, for example, so-called filter residues; sugar-containing waste products of the sugar industry, for example, lime residue; and public wastewaters, for example, mud of those waters or sewage sludge. In general, mixtures containing up to 15% by weight are sufficient to obtain an agent which is equal to and frequently even more effective than a corresponding mixture of peat. However, higher percentages of the material are sometimes useful.

Example: Pot tests with wheat as the experimental growing plant have been carried out in two test series, under different rates of irrigation. In these tests, a dry climate (Series A) was simulated by sprinkling to a water capacity of about 60% (sandy soil). By sprinkling to a water capacity of 70% and additionally by sprinkling during the period of growth once a week to a water capacity of 120%, a damp climate was simulated (Series B).

Except for the experiments with a lower rate of brown coal and with sewage sludge only, the amount of agent for improving fertility of soil, related to the dry substance was equal in all test series. In all series the following components served as fertilizer: 1.2 grams N, 1.2 grams P_2O_5 and 2 grams K_2O per pot. The fertility-improving agents tested are shown on the following page.

	water contents
Rhenish brown coal (RBK)	58.6%
Peat	74.8%
Cellulose-containing residue	7.6%
Sugar-containing lime residue	27.8%
Sawdust of spruces	16.4%
Sewage sludge	94.0%

As can be seen from Series A compared with the experiment carried out without adding an agent for improving soil fertility, peat delivers an increase in yield of about 5%, whereas the use of brown coal and sewage sludge, each added separately, results in distinctly less improvement. The combination according to this process, however, shows improvement in all cases.

Series B, fundamentally, shows the same results. Peat provides an increase in yield of about 15%; brown coal alone brought only about 4%. When employing the combination according to this process, the result is almost equal to the effect of peat. Only when adding sawdust was the improvement somewhat inferior.

AGENT	YIELD OF WHEAT GRAIN	
	(dry basis) g. per pot	% weight
SERIES A		
without	43.1	**100.0**
163 g. of peat	45.0	**104.5**
100 g. of brown coal	41.8	**97.0**
95 g. of brown coal + 2.2 g. of α-cellulose-residue	45.6	**106.0**
81 g. of brown coal + 11 g. of lime residue	44.2	**102.5**
5 g. of sewage sludge	42.0	**97.5**
95 g. of brown coal + 2 g. of sewage sludge*	44.9	104.0
88 g. of brown coal + 5 g. of sewage sludge*	46.0	**107.0**
76 g. of brown coal + 10 g. of sewage sludge	45.4	**105.5**

*(dry basis)

(continued)

AGENT	YIELD OF WHEAT GRAIN	
	(dry basis) g. per pot	% weight
SERIES B		
without	39.7	100.0
163 g. of peat	45.8	115.5
100 g. of brown coal	41.4	104.5
95 g. of brown coal + 2.2 g. of α-cellulose-residue	46.3	116.5
81 g. of brown coal + 11 g. of lime residue	46.9	118.0
60 g. of brown coal + 23 g. of lime residue	45.8	115.5
95 g. of brown coal + 2.4 g. of sawdust	43.7	110.0
95 g. of brown coal + 2 g. of sewage sludge*	45.7	115.0
88 g. of brown coal + 5 g. of sewage sludge*	44.3	111.5
76 g. of brown coal + 10 g. of sewage sludge *		
*(dry basis)	44.6	112.5

Treatment with N-Methylol Urea

J.M. O'Donnell; U.S. Patent 3,942,970; March 9, 1976; assigned to Orgonics, Inc. has found that acid-condensable, prepolymer solutions which contain a methylol monomer which sterilizes the sludge material may be reacted with the sludge material while being condensed, or polymerized under such conditions as to produce a solid compound of such moisture content as to produce granulation of the solid particles. The resulting granular product under such process conditions is particularly useful as a fertilizer or animal feed supplement. The preferred components of the process are domestic sewage sludge, commercially available urea (45 to 46% N) and formaldehyde (37%).

In the process the condensable methylol-urea solutions are preferred as prepared by the reaction of urea and formaldehyde to form a prepolymer N-methylol-urea solution, and the reaction of such solution with the sludge within a particular time period after the reaction, so as to provide a product with controlled decomposition or dissolution characteristics. Typically, the prepolymer solution contains a mixture of mono- and di-N-methylol urea.

While commercially available urea containing 45 to 46% nitrogen is the preferred reactant, it may be replaced by casein, melamine, phenol, or the water-soluble polymeric constituents thereof. Concentrated solutions of urea can be used with a corresponding increase in the concentration of formaldehyde. In place of formaldehyde (37%), so-called urea-formaldehyde concentrates can be used. These commercially available products usually contain about 60% formaldehyde, 25% urea and 15% moisture. In addition, other aldehydes, such as paraformaldehyde, crotonaldehyde, acetaldehyde, propionaldehyde and/or formaldehyde-generating agents, can be used. It has been found that a hydrogen ion is required to convert the HCOH to methylene (CH_2) groups in the resulting polymer linkage in the polymerization step. It is usually necessary to add this ion in the form of a strong mineral acid, such as sulfuric, hydrochloric or phosphoric acid, but organic acids, such as acetic, formic, or lactic acid or mixtures thereof, may be used.

Another and preferred way of controlling hydrogen-ion-concentration is to add an ammonium-acid salt to the solid materials prior to the addition of the prepolymer methylol-urea solution. Again a choice of such salts is available, such as ammonium nitrate, sulfate, borate, chloride or others. However, the preferred salts would be ammonium sulfate, phosphate (mono- or dibasic), since any of these will enhance the fertilizer value of the end product.

The use of ammonium salts to control pH has the dual effect of supplying sufficient hydrogen ion for the reactions and forming hexamethylenetetramine which greatly enhances the stability of the resulting polymer once formed. Using ammonium sulfate as a typical example, the reaction would be as follows:

$$\underset{(H^+\text{ donor})}{6CH_2O\text{ (aq)}} + 2(NH_4)_2SO_4 \longrightarrow C_6H_{12}N_4 + 2H_2SO_4 + 6H_2O$$

In previously described procedures for adding urea-formaldehyde to sewage sludge or other carriers, pH control during polymerization has typically been maintained between one and three for the best results. The resulting mixtures were then neutralized in order to prevent cross-polymerization of the polymer. For this process the pH of the reactants in the polymerization step should be maintained at between 3.0 and 5.0, and preferably between 4.0 and 4.5. The maintaining of such pH limits makes it unnecessary to neutralize the end product; thus avoiding an additional step and the use of a neutralizer. The physical and chemical natures of the prepolymers to be added to the sludge cake material are important, and are as follows:

They must be in a soluble or dispersed form so that they can be easily sprayed or sparged into the sludge.

They must be easily polymerizable at relatively low temperatures, for example, 40° to 70°C, and high pH's (4.0 to 4.5).

After polymerization, they must have the appearance of solids and have a solids-to-moisture ratio considerably higher than the sludge being treated. For example, if the solids-to-water ratio in the sludge cake is between 1 to 3 and 1 to 1, the solids-to-water ratio of the polymerizable material must be at least 3 to 2.

The resulting polymers must be of relatively low molecular weight, for example, not greater than 300, and preferably essentially a straight chain polymer notwithstanding the possibility of co-, branch- or graft-type polymerization with the constituents of the sludge. In any case, the number of CH_2 groups should not exceed 5, and preferably is 3 to 4, and a CH_2 radical should not be present as a terminal group.

In general, it has been found that the best method of preparing a prepolymer with the above properties is one in which 1 part of commercially available 37% formaldehyde is neutralized with triethanolamine to a pH of 8.0, and to this mixture is added 1 part of commercially available prilled urea containing 46%

nitrogen. A negative heat of solution ensues which will lower the temperature to 5°C. The solution is then gently heated to 30°C, at which point the urea is in solution with the formaldehyde. After 10 minutes, the temperature of the solution will rise to 60°C, at which point it is considered suitable for reaction with the sludge.

Activated Sludge Process

C.H. Carothers; U.S. Patent 3,728,254; April 17, 1973 describes a method and apparatus for ecologically recycling household organic waste, yard and garden trimmings and prunings, and storm sewerage along with sanitary sewage. An activated sludge treatment process is modified to increase its aeration efficiency while eliminating the need for preliminary sedimentation tanks, final sludge digester tanks and the utilization of the communities' natural water resources for the stabilization and the purification of the final effluent discharge by effecting on site utilization of the final effluent discharge.

This is accomplished by a separation of the sludge component and the nutrient-rich liquid component for respectively forming dried commercial fertilizer and/or commercially saleable compost and reducing the nutrient-rich liquor to nutrient-free water by conveying it as a feeder liquor through a suitably formed hydroponic garden containing commercially saleable plants arranged to extract the nutrients. The apparatus, in addition to the conventionally fitted activated sludge aeration tanks and a final sedimentation tank, provides a grinding unit for grinding the household organic waste and yard and garden trimmings and introducing them at suitable strategic points into the sanitary sewer lines to be conveyed by the sewage effluent to the plant site and delivered, after passage through a mixer to provide thorough mixing, to the aeration sludge tank fitted with a supplemental aeration system which adds to the conventional activated-sludge process a trickling filter effect substantially enhancing the aeration and bacterial growth ability of the aeration tank.

The process also requires a sludge removal pump having an outlet alternately adapted to deliver excess sludge to a dryer for conversion as an end product to commercial fertilizer or to a spray means for spraying the sludge as a composting catalyst over a chip bed produced by passing yard and garden prunings through a conventional chipper to produce as an end product a rich commercial compost. A hydroponic garden is used for extracting the nutrients from the separated nutrient-rich liquor to produce as an end product nutrient-free water.

MISCELLANEOUS PROCESSES

IN SITU ELECTROLYTIC PRODUCTION OF TREATMENT AGENTS FOR USE ON BOATS

R.W. Fenn, III and K.J. O'Leary; U.S. Patent 3,943,044; March 9, 1976; assigned to Diamond Shamrock Corporation describe a system and apparatus for disinfecting sewage water and removing suspended solids by in situ treatment with chemical agents and gases, each produced electrolytically. The process comprises a treatment chamber, disinfecting and solids conveying means including at least one pair of electrodes disposed in the chamber, solids removal means associated with the chamber, a source of aqueous alkali metal halide solution and optionally an ultraviolet light ray source and solids collection means.

Referring to Figure 6.1, the sewage water treatment apparatus is shown generally at **10**. The apparatus includes treatment chamber **11**, having top walls **6**, upwardly inclined from the top edges of the side walls to form a pyramidally-shaped chamber top, dimensionally stable anode **12a**, cathode **12b**, power lead connections **12**, a motor **13** having stirrer and disintegrator blades **15**, attached to the shaft **14** thereof, effluent pump **16**, salt bed **19**, water inlet **18**, brine conduit **20**, and sewage water inlet conduit **9**. The treatment chamber may be constructed of any material which is inert or not substantially affected by the environment of the treating chamber, for example, it may be constructed of a polymorphic material such as rubber or plastic, e.g., polyvinyl chloride and the like.

The structure material of the other elements of the apparatus should be selected so that they are also unaffected by the environment of the treatment chamber. The motor is a conventional motor and may be selected in accordance with the size of the treatment chamber and required rotation adequate for disintegrating the solids contained in the sewage water mixture. Optionally, the disintegrator may be used externally of the chamber for macerating the solids of the sewage water mixture prior to the introduction of the mixture to the chamber.

The electrode pair comprise at least one dimensionally stable anode and a cathode and power connections to leads, not shown. A source of current, not shown, is supplied to the anode for electrolysis of the halide solution of the sewage water mixture.

FIGURE 6.1: ELECTROLYTIC SEWAGE TREATMENT PROCESS

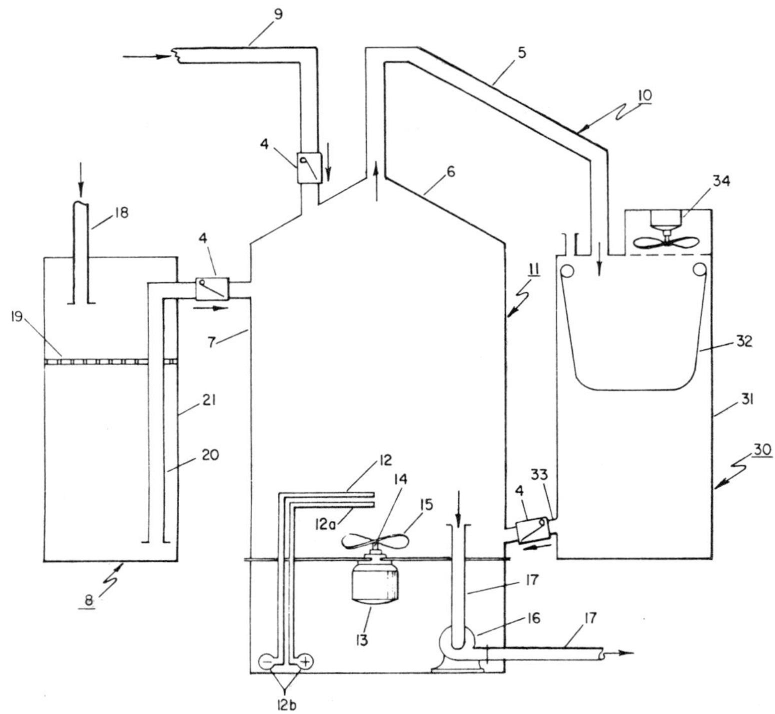

Source: U.S. Patent 3,943,044

The dimensionally stable anode **12a** comprises a valve metal electrically conductive substrate with a surface coating thereon of a solid solution of precious metal oxide and valve metal oxide. In these solid solutions an atom of valve metal in the characteristic rutile valve metal oxide crystal lattice is replaced with an atom of precious metal. This distinguishes them from mere physical mixtures of the oxides, pure valve metal oxides in fact being insulators. Such solid solutions are both electrically conductive and electrocatalytically active.

In the solid solutions the valve metals include titanium, tantalum, niobium and zirconium while the precious metals encompass platinum, ruthenium, palladium, iridium, rhodium and osmium. Titanium dioxide-ruthenium dioxide solid solutions are preferred. The mol ratio of valve metal:precious metal varies between 0.2 to 5:1, approximately 2:1 being presently preferred. The valve metals of the electrically conductive substrate are those defined above as included in the solid solutions.

If desired, these solid solutions may be modified by the addition of other components which may either enter into the solid solution itself or mix with the same to achieve a desired result. For instance, it is known that a portion of the precious metal oxide, up to 50%, may be replaced with tin dioxide without substantial detrimental effect on the overvoltage. Likewise, the solid solution may be modified by the addition of cobalt compounds particularly cobalt titanate.

It has been found that when the particularly preferred type of dimensionally stable anode is employed in the apparatus of this process gas bubbles develop at the anode and combine in a manner such that they cause rapid flotation of the suspended solids of the sewage water mixture to the surface of the vessel. While this phenomenon is not completely understood it has been found to be effective in providing rapid flotation of the solids.

The cathode, positioned in close relationship to the dimensionally stable anode should be constructed of a metal resistant to the cell environment, e.g., titanium. It is important that the anode and cathode be positioned as closely as practicably possible to provide maximum power efficiency since the larger the gap between cell electrodes the more voltage is required in overcoming the resistance of the electrolyte solution flowing between the electrodes. While only one pair of electrodes is shown, as many pairs may be utilized as are necessary in accordance with the size of the treatment vessel and the volume of sewage water mixture being treated.

The electrodes are preferably disposed in horizontal parallel position to facilitate the rapid ascent of the gas bubbles and concurrent conveyance of the solids to the surface of the mixture. The effluent pump **16**, disintegrator motor **13** and electrode leads **12** are preferably positioned externally of the treating chamber. Conduit **17** is positioned at the bottom of the chamber in order that sewage water mixture from which the suspended solids have been removed by flotation may be discharged from the treating chamber to a desired disposal site.

When the incoming sewage water mixture contains a saline solution from any source prior to being introduced into the chamber water inlet **18**, salt bed **19**, and conduit **20**, extending through wall or partition **21**, of brine supply tank **8** and wall **7** of chamber **11** and one-way valve **4**, are inoperative. However, when the entering sewage water mixture does not contain saline solution, fresh water is introduced through the inlet.

The inlet may be located in the top or upper sidewall of the treating chamber and salt bed **19** disposed at a position intermediate the top and bottom walls of the treating chamber. Conduit **20** extending from a point adjacent the bottom of the treating chamber through an opening in the partition is operative for delivering solution to chamber **11** only when the sewage water mixture introduced into the treating chamber at inlet **9** does not contain saline solution.

During operation of the apparatus a sewage water mixture, generally containing a saline solution, is introduced into the treating chamber, the disintegrator is operated for a short interval, generally from less than one minute to about ten minutes, preferably about two minutes and turned off. Suspended solids contained in the mixture are disintegrated by the stirrer and macerator blade **15**, connected to the shaft of pump **14** which blades have shapes and sharp edges designed to disintegrate the solid materials. Current is then supplied to the pair of electrodes. The sewage water mixture flowing through the gap between the dimensionally stable anode and cathode is electrolyzed. The chemical agents such as sodium hypochlorite provided by the reaction of a mixture of the chlorine, formed at the anode and sodium hydroxide formed at the cathode, permeate the sewage water mixture and sterilize or disinfect the liquid portion thereof.

The gas bubbles formed at the anode rise rapidly throughout the mixture causing the suspended solids to rise to the surface of the mixture. The solids conveyed to the surface are in the form of a foam-like substance which may readily be removed by various means such as skimming or scraping devices or a weir arrangement whereby the solids may be caused to flow from the chamber by influent sewage water and may be collected for convenient disposal. A solids collector which may be optionally utilized in combination with the other apparatus of the process is also shown in Figure 6.1.

The collector **30**, includes vessel **31**, with a disposable filter bag **32**, positioned therein for receiving solids from conduit **5**. The solids remain on the filter bag and the liquid filtrate which passes through the bag into the lower portion of vessel **31**, may be returned for further treatment to chamber **11** through conduit **32** and one-way valve **4**. A fan **34** may be arranged to assist the drying of the solids retained by the filter bag. The filter bag may be any material suitable for retaining the solids and preferably is made from material readily disposable along with the contained solids in any convenient and practical manner.

An especially convenient and preferred manner of disposing of the filter bag and collected solids is by burning after the materials are substantially dry. As collected about 90% of the water has been separated from the waste material and the residue has a high concentration of carbonaceous matter. After the water retained by the carbonaceous matter has been removed by flame contact or other drying means the residue may be completely converted to a small quantity of ash by flame contact or other heating means.

It should be understood that although halide solution tank **8**, chamber **11** and vessel **31** are shown as separate structures a unitary structure may be formed by combining the three units with the adjacent side walls then forming in combination compartments of a unitary structure.

One of the outstanding features of the process is the structure of the top portion of the chamber of Figure 6.1. The upwardly tapered sidewall, pyramidal and inverted cone designs have been found to greatly facilitate the removal of the solids from the chamber after the solids have been conveyed to the surface of the mixture. This particular design structure surprisingly and very unexpectedly provides removal of solids from the treatment chamber with no required auxiliary apparatus such as scrapers, skimmers and the like. In fact, the design may be considered to provide self-removal of the solids as movement of the solids proceeds rapidly and efficiently upwardly to the aperture at the apex of the upwardly tapered sidewall structure, pyramid or inverted cone and into the conduit connected to the aperture for convenient disposal.

Such designs provide substantially maintenance-free operation of the apparatus as all moving parts may be eliminated from the treatment chamber if the solids contained in the sewage water mixture are disintegrated prior to introduction of the mixture into the treatment chamber. Although the solids may be removed by overflow methods such as weir or baffle arrangements and the like, such techniques cause buildup and resultant delayed removal of the solids and interference with the other components of the apparatus. Consequently most available treatment devices necessarily utilize some type of mechanically operated means for solid removal.

Another very important and practical advantage of the above-described upwardly tapered sidewall, pyramidal and inverted cone top portion results when the process is utilized in combination with a marine toilet. Because of erratic movement of boats caused by water and wind conditions, sewage treatment devices installed on boats are subject to spillage and other disturbance or overflow of the contents. Such erratic movement also may result in interference with mechanically operated solids removal equipment.

The chamber top structural configuration of Figure 6.1 avoids such disadvantages in that if the chamber is tilted by erratic motion any resultant overflow will be contained by the closed top and discharged through the conduit connected at the chamber aperture for the usual normal removal of solids. Such beneficial operation is made possible not only by the structural design but also by the absence of any mechanically operated solids removal means at the surface of the mixture.

The effluent pump **16**, as noted above, is positioned on or proximate to the bottom wall of the treatment chamber for discharging the disinfected liquid. Since the liquid portion of the sewage water mixture being treated is substantially free of solid matter which has been conveyed by flotation to the surface of the chamber during treatment, the liquid portion may be discharged through effluent pump **16**, and conduit **17** which extends outside the treatment chamber. The effluent disinfected is sufficiently light in color, low in solid content and odor so that it may be disposed of without creating a health hazard or pollution problem, thus being in conformance with existing and probable future statutes intended to regulate the disposal of such materials.

When the sewage water mixture introduced into the treating chamber does not contain saline solution such solution may be added to the mixture through conduit **20**, positioned in tank **8**. Fresh water may be introduced through the inlet **18** so that it flows over the salt bed which may contain salt in the form of fine crystals, granules, blocks, or pellets The water passes over the salt bed to form a saline solution which because of its high density flows to the bottom of the tank and forms a saline solution reserve.

When additional fresh water is added through inlet **18**, saline solution from the reserve at the bottom of the tank is forced upwardly by hydrostatic pressure through conduit **20**, into treatment chamber **11** where it is mixed with the sewage water mixture and subjected to electrolysis. Water is added through inlet **18** in predetermined amounts to provide a saline solution having a sodium chloride content of about 1.0 g/l to about 100 g/l, preferably from about 5.0 to about 30.0 g/l.

In the examples presented in the table below the sewage water treatment was carried out batchwise by placing the wastewater in a treatment chamber of the type shown and described in Figure 6.1. In Examples 4 to 6, inclusive, water was added to the chamber to provide sufficient volume in the treatment chamber

Wastewater Treatment Results*

Ex. No.	Type of Wastewater	Dilution Factor (Ratio of Dilution Water to Sewage Water)	Time of Electrolysis (min.)	Percentage Reduction or Removal: Biochemical Oxygen Demand (BOD)	Percentage Reduction or Removal: Chemical Oxygen Demand (COD)	Percentage Reduction or Removal: Total Suspended Solids	Coliform/ 100 ml.	Biochemical and Chemical Oxygen Demand: Initial (ppm)	Biochemical and Chemical Oxygen Demand: Final (ppm)
1	Research Laboratory Building	Not diluted	10		83		100	COD 2890	492
2	Sewage Water Effluent	Not diluted	10		71		100	COD 2810	720
3		Not diluted	7		90		100	COD 2800	286
4	Toilet Effluent	2.25	9		83		100	COD 10740	1809
5		1.5	10	66	72	88	100	COD 7831 BOD 1200	2172 403
6		2.35	15	80	89	97	100	COD 5342 BOD 1045	594 205

*The data were obtained by analytical methods included in the textbook *Standard Methods for the Examination of Water and Wastewater,* 12th edition, New York, American Public Health Association, Inc. (1965)

for suitable operation of the equipment. The sodium chloride content of the mixture in the chamber was 10 g/l for all the examples except Example 2 which was 2 g/l. The sewage water mixture containing saline solution was subjected to disintegration after being placed in the treatment chamber by operation of the disintegrator for about one-half of a minute. After the disintegrator motor was turned off the mixture was electrolyzed for the time periods set forth in the table.

It will be readily observed from the examples in the table that the coliform bacteria are completely eliminated and the biological and chemical oxygen demand and total suspended solids significantly reduced in an extremely rapid treatment interval. The advantages of the apparatus and process over prior techniques are thus evidenced by the rapid and efficient cleaning of wastewater to meet disposal regulatory standard requirements. The large reduction in total suspended solids is particularly surprising and unexpected as extended periods have previously been required to accomplish such reduction.

Not only the lengthy time but also the expense and size of the equipment required to accomplish such results make known processes and apparatus unsatisfactory for wastewater treatment generally, and particularly for treatment at intermittent intervals and/or where small volumes must be treated. Where the influent of sewage-water mixture is continuous the solids are caused to overflow chamber **11** and enter conduit **5** for convenient disposal by influent displacement of the mixture being treated. Where operation and flow of the mixture into the treatment chamber is intermittent, the solids may be caused to overflow and exit through conduit **5**, by introducing water or an inflatable displacement device into the chamber.

COMPACT MECHANICAL TREATMENT FOR SMALL SHIPS

A process described by *J.J. Pizzo and M.F. McNeil; U.S. Patent 3,622,511; November 23, 1971; assigned to FMC Corporation* provides a simple and compact mechanical sewage treatment system for use in small ships while in harbors and inland waters to prevent pollution. One of the features of the process is to completely treat the sewage so that the end products are inactive and nonpolluting. This is achieved by concentrating the sewage in an evaporator to drive off water in the form of steam from which substantially pure water is recovered and can be reused in the sewage and boiler systems found on ships.

In the sewage treatment of the process as shown in Figure 6.2a, raw sewage coming from sanitary facilities, kitchen refuse, bilge water and other waste sources upon a ship are fed to a macerator pump **20** which reduces the size of the incoming refuse to a maximum of 5 mm. The sewage is then fed into a holding tank **22** which is provided to accommodate surges of raw sewage greater than the constant operating capacity of the system. From the holding tank the raw sewage is fed into an evaporator **24**. This evaporator includes a submersible burner **26** that is supplied externally by air at approximately 200 psi and by fuel such as diesel oil.

The submersed burner is effective to burn all combustibles in the raw sewage while raising the internal temperature in the evaporator to approximately 400°F at 15 psig. The evaporator is effective to concentrate the raw sewage such that the sludge reaches the level of 5% solids while water in the form of steam and other gases are driven off, as will be explained presently. The concentrated sludge leaves the evaporator **24** through line **28** and is supplied to a sludge incinerator **30** by means of a pump **32**. The incinerator may be of any well-known type and is supplied by fuel such as diesel oil by a pump **34**. This burner develops a temperature in the order of 1600°F at 0 psig and reduces the concentrated sludge to water vapor, carbon dioxide and a small amount of ash which exits the burner.

A water spray scrubber **37** is installed in line **36** to remove ash particles prior to discharging the incinerator exhaust from an exhaust stack or funnel via line **36**. The gases and water vapor driven off by the submersed burning in the evaporator pass through a demister **38** which causes some recondensation of the steam and prevents passage of solids. This demister may be made of such material as stainless steelwool or the like.

FIGURE 6.2: COMPACT MECHANICAL SEWAGE TREATMENT SYSTEM

a.

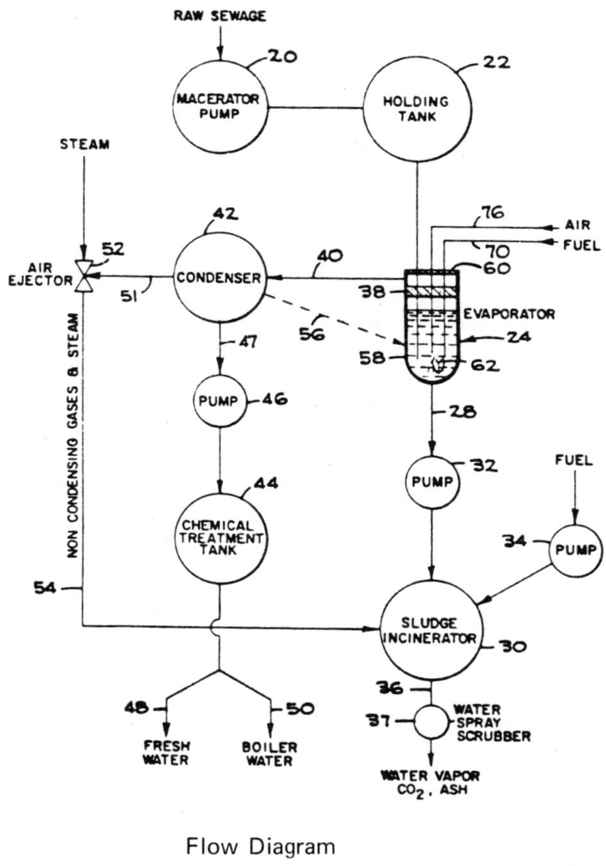

Flow Diagram

(continued)

FIGURE 6.2: (continued)

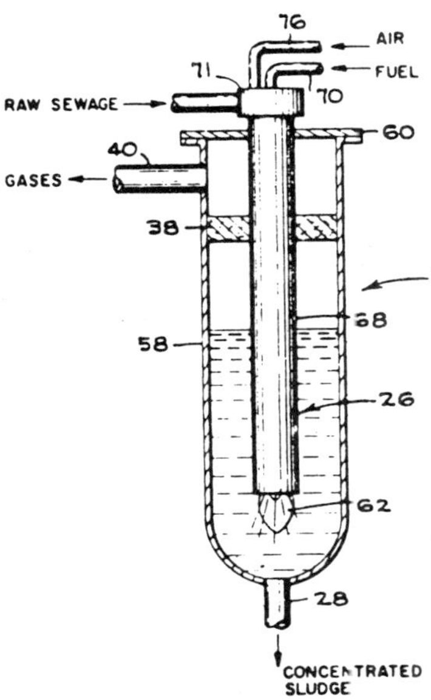

b.

Evaporator

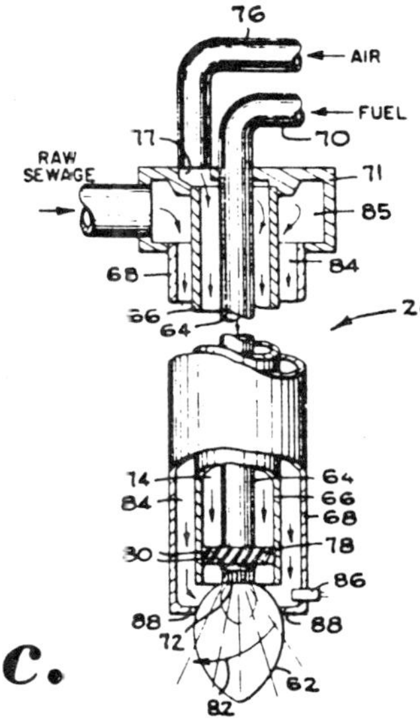

c.

Submersed Burner Nozzle

Source: U.S. Patent 3,622,511

The gases then pass out of the evaporator via line **40** into a condenser **42**. Here the steam and gases are condensed and if desired supplied to a chemical treatment tank **44** via a line **47** by means of a pump **46**. The employment of such chemical treatment tank in which the recovered gases now in the form of water may be treated with chlorine is optional and the condensed water may be directly reused as fresh water **48** or boiler water **50** in either the sanitary system or the steam generating boilers normally found on board ships.

Noncondensing gases are withdrawn from the condenser through line **51** by means of an air ejector **52**. The air ejector may be driven by a 150 to 200 psig source of steam. From the air ejector the noncondensing gases are routed to the sludge incinerator via line **54** where they are consumed in conjunction with the concentrated sludge. Some of these gases such as methane also assist in the burning of the sludge.

In order to improve the operating efficiency of the system and reduce heat loss, heat recovered from the gases in the condenser is returned to the evaporator **24**, by any well-known means such as indicated by line **56**. It is important, however, that such heat transfer means does not directly contact the raw sewage in the evaporator in order to prevent the buildup of organic deposits or scale which could eventually plug up the evaporator or seriously reduce its operating efficiency. The sewage treatment system may also be used with salt water instead of fresh. The efficiency is reduced somewhat because the salt is also removed with the sewage.

The evaporator **24** best seen in Figure 6.2b comprises a sealed tank **58** having a smooth interior to prevent the buildup of scale or sludge. Near the upper end of the tank, the steam and gases released are removed through line **40** while the concentrated sludge exits at the bottom via line **28**. A removable cover plate **60** at the top of the tank permits installation and removal of the submersible burner assembly **26** and the demister **38**.

As can be seen in Figure 6.2b, the submersible burner assembly extends beneath the surface of the liquid sewage in the tank and the combustion bubble **62** produced at its lower end transmits heat directly to the liquid sewage. The construction of the submersible burner assembly is best shown in Figure 6.2c. Here it can be seen that the burner is comprised of three concentric tubes **64, 66** and **68**. The innermost tube **64** serves to conduct fuel from an external supply line **70** through a cap **71** to the lower portion of the burner. At the bottom end of the tube **64**, a nozzle **72** serves to spray and distribute the fuel.

Tube **66** is mounted in the cap concentrically with the inner fuel supply tube forming an air passage **74** which communicates with an external air supply line **76** through a chamber **77**. A vortex generator **78** is positioned in the air passage between tubes **64** and **66** near their lower ends and includes multiple inclined vanes **80**. This vortex generator causes the combustion air to swirl as indicated by arrow **82**, thereby improving the mixing of fuel and air and improving the efficiency of combustion.

The outer tube **68** is supported in the cap concentrically with tube **66** forming a sewage supply passage **84**. A second annular chamber **85** in the cap formed between the cap and tube **66** connects the raw sewage line and the sewage supply passage. An ignitor **86**, positioned in the lower portion of the outer tube, is employed to ignite the air and fuel mixture. Formed at the end of the outer tube, is a circumferential lip **88**. This lip insures that the raw sewage passing through the passage is positively directed through the combustion bubble to insure consumption of all readily combustible matter (such as oil, paper and the like).

REFINERY WASTE SLUDGE TREATMENT

W.K. Lorenz, E.C. Sebesta and C.L. McClellan; U.S. Patent 3,835,021; Sept. 10, 1974; assigned to Sun Oil Company of Pennsylvania have found that refinery waste sludge materials may be efficiently disposed without pollution by dewatering and deoiling refinery sludges through a filter press at a temperature of from about 100° to about 200°F, washing the filter cake with hot water, recovering the oil removed from the sludge, and applying the solid filter cake to soil for biodegradation.

The process is particularly advantageous in several aspects. First of all, it permits high recovery of oil from waste materials. This efficient oil recovery is attributed to the combination of the filtration step which breaks the oil-in-water emulsions in the sludge materials and to carrying out the filtration at a temperature of from about 100° to about 200°F which tends to remove the maximum amount of oil from the solid materials. In addition, the step of washing the filter cake with hot water removes additional oil which is retained on the cake after the filtration and thereby permits maximum oil recovery and results in a filter cake low in oil content.

It is surprising that this hot water wash effectively removes additional oil from the filter cake, firstly, because the filtration at elevated temperatures would be expected to have removed essentially all of the oil that could be removed, and secondly, because the oil insolubility and immiscibility in water suggests that the water would flow through the filter cake without carrying any oil. Furthermore, since the filter cake obtained by the process is low in water and oil content, its transportation cost to the land spreading area is reduced, and incorporation in the soil requires no preliminary evaporation or other operations. Furthermore, since the solid filter cake is compact and low in oil content, a relatively small land area can accommodate the solid waste and storm water runoff does not carry off oil and cause stream pollution.

Example 1: To a water emulsion refinery oily sludge consisting by volume of 28% contaminated API separator sludge, 36% sludge decant pit material, and 36% tank bottoms, spent clay is added in an amount of 0.6 lb/lb of dry solids and lime is added in an amount of 0.12 lb/gal, thereby providing a waste composition at pH 11 of 12% total solids and 8.5% oil. This composition is fed to a Beloit-Passavant filter press precoated with diatomaceous earth having 3 plates of 24" diameter at a feed temperature of 175°F and filtration is completed in two hours.

The filter cake contains 49.7% solids and 17.5% oil, the balance being water. The filtrate exits from the press at 100°F and is observed to be no longer emulsified and oil and water form two layers after standing. Separation of the oil is readily accomplished by mechanical skimming. The filter cake is transported to a land farm where it is mixed into the soil. After weathering for 10 months, examination of the soil indicates essentially complete disappearance of the oil. When the above filtration is carried out at 58°F, the filter cake contains over 15% more oil, thus indicating the importance of filtering at the elevated temperature.

Example 2: A refinery oil sludge as used in Example 1 is similarly treated with 0.56 lb of clay per pound of dry solids and 0.07 lb of lime per pound of dry solids and is fed to a filter at 180°F. After one hour, the filter cake is washed with water under pressure at a temperature of 180°F for one hour, the maximum pressure reaching 220 psig. The filter cake after this treatment contains 60.5% total solids and 4.9% oil and the filtrate shows a sharp separation of oil and water which is readily separated. The solid filter cake is plowed into soil and results in essentially complete oil elimination after 12 months, at which time Bermuda grass seed previously sown on the soil germinated readily and matured to a good grass crop.

ORGANIC SOLVENT REMOVAL FROM ORGANIC SOLVENT-CONTAINING SLUDGE

G. Slikkers, Jr.; U.S. Patent 3,929,586; December 30, 1975; assigned to Organic Chemicals Company, Inc. describes a process for removing organic solvents from organic solvent-containing waste sludges such as organic solvent-containing sludges of various paints and resins which renders such sludges suitable for disposal in landfills and the like without a contaminating effect on our environment.

The process comprises subjecting a mixture of the organic solvent-containing waste sludge, water, and a source of alkali or alkaline earth metal ion to distillation, with the pH of the mixture being greater than 7. Preferably, a stoichiometric excess of the metal ion source is provided based on the nonorganic solvent constituents of the sludge, and preferably, about from 2 to 4 gallons of water are employed for every gallon of the sludge. The preferred metal ion sources are the hydroxides of potassium, sodium, calcium, and magnesium, with calcium hydroxide being especially preferred. For most efficient solvent removal, the mixture should be subjected to continuous agitation during the distillation and the pH should preferably be above about 12.

Example 1: 150 gallons of industrial waste sludge consisting of about 10% epoxy resin, 25% acetone, and 65% toluol are transferred to a mixing tank, to which 200 pounds of calcium hydroxide are added. This mixture is added over a 60-minute period to 400 gallons of boiling water in an agitated still and then cooked an additional 60 minutes, during which time the sludge changes to a granular form. During the addition period, the temperature drops to approximately 96°C and then returns to 100°C during the cooking period.

The vapors are passed through a heat exchanger, condensed, and the distillate collected in a storage tank. The lower aqueous layer is separated and can be used as part of the water required for a subsequent run. The upper layer (65 gal) is a mixture of toluol, acetone, and a very small quantity of water, and may be reserved as is or dried as by passing across a tower packed with a dehydrating agent, such as calcium chloride. The granular solids (250 lb on a dry basis) remaining in the distilland are filtered and discarded to a landfill and the aqueous filtrate is discarded.

Example 2: 150 gallons of industrial waste sludge consisting of about 1% ink pigments, 25% ethyl acetate, 39% toluol, and 35% propyl acetate are transferred to a mixing tank, to which 150 pounds of calcium hydroxide are added. This mixture is added over a 45-minute period to 400 gallons of boiling water in an agitated still and then cooked an additional 45 minutes, during which time the sludge changes to a granular form. During the addition period, the temperature drops to approximately 94°C and then returns to 100°C during the cooking period.

The vapors are passed through a heat exchanger, condensed, and the distillate collected in a storage tank. The lower aqueous layer is separated and can be used as part of the water required for a subsequent run. The upper layer (75 gal) is a mixture of ethyl acetate, toluol, propyl acetate, and a very small quantity of water, and may be reserved as is or dried as by passing across a tower packed with a dehydrating agent, such as calcium chloride. The granular solids (200 lb on a dry basis) remaining in the distilland are filtered and discarded to a landfill and the aqueous filtrate is discarded.

Example 3: 150 gallons of industrial waste sludge consisting of about 1% acrylic resin, 20% acetone, 20% methyl ethyl ketone, and 59% toluol are transferred to a mixing tank, to which 100 pounds of calcium hydroxide are added. This mixture is added over a 45-minute period to 500 gallons of boiling water in an agitated still and then cooked an additional 30 minutes, during which time the sludge changes to a granular form. During the addition period, the temperature drops to approximately 91°C and then returns to 100°C during the cooking period.

The vapors are passed through a heat exchanger, condensed, and the distillate collected in a storage tank. The lower aqueous layer is separated and can be used as part of the water required for a subsequent run. The upper layer (60 gal) is a mixture of acetone, methyl ethyl ketone, toluol, and a small quantity of water, and may be reserved as is or dried as by passing across a tower packed with a dehydrating agent, such as calcium chloride. The granular solids (175 lb on a dry basis) remaining in the distilland are filtered and discarded to a landfill and the aqueous filtrate is discarded.

CONVERSION OF SLUDGE TO SYNTHESIS GAS

According to a process described by *A. Brent and C.F. Teichmann; U.S. Patent 3,687,646; August 29, 1972; assigned to Texaco Development Corporation* sewage is converted into a useful nonpolluting product gas, e.g., synthesis gas or fuel gas by dewatering municipal sanitary sewage, producing a thickened sludge. The thickened sludge is mixed with a fluid selected from the group consisting of H_2O, liquid hydrocarbon fuel, particulate carbon slurries in water or in a liquid hydrocarbon fuel as produced subsequently in the process, and mixtures to produce a feed mixture stream.

Preferably the feed mixture is heated to a temperature in the range of about 212° to 600°F to vaporize volatile constituents in the feed mixture and to produce a feed dispersion stream. The feed dispersion is then reacted by partial oxidation with an oxygen-rich gas in an unpacked free-flow noncatalytic synthesis gas generator, at a temperature in the range of about 1500° to 3000°F and at a pressure in the range of about 1 to 250 atm to produce synthesis gas. Water separated from the sewage is purified and used for industrial applications or converted into by-product steam by indirect heat exchange with the product gas.

PAUNCH MANURE AND HUMUS DISPOSAL

J.W. Hatchel and J.C. Roberts; U.S. Patent 3,329,107; July 4, 1967; assigned to Pollution Eliminators, Inc. describe a sewage disposal plant and process in which moisture is removed from the sewage in successive steps followed by the pressing, compacting and comminuting of the sewage to a final drying step after which the sewage is combusted and the heat of combustion efficiently and effectively employed in the drying process.

Referring to Figure 6.3, the numeral **10** indicates an inlet conductor through which raw sewage may be admitted as indicated at **11**. The inlet conductor opens through one end of a water screen assembly **12** into the interior thereof, the assembly being mounted for rotation around a horizontal axis by a driving motor or other rotating means **13**.

As shown in Figure 6.3f, the water screen assembly is formed of a pair of concentric cylindrical screens **14** and **15**, the screens being spaced apart from one another and confining in the annulus between the screens a body of relatively loosely packed filtering material **16** which may be any suitable type of inorganic filter media. The loose packing of the filter material allows some movement of the particles so that the filter is substantially self-cleansing.

FIGURE 6.3: PAUNCH MANURE AND HUMUS DISPOSAL PROCESS

a.

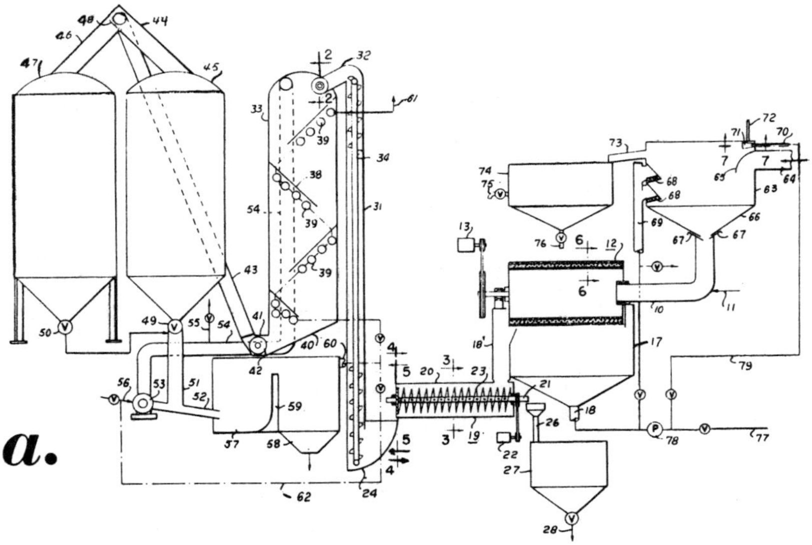

Process Schematic

b.

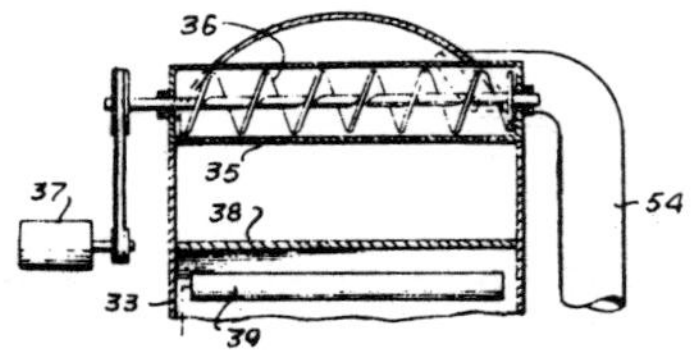

Cross-Sectional View Along Line 2–2 of Figure 6.3a

c.

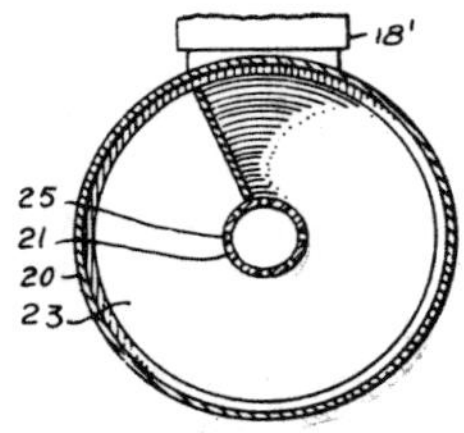

Cross-Sectional View Along Line 3–3 of Figure 6.3a

d.

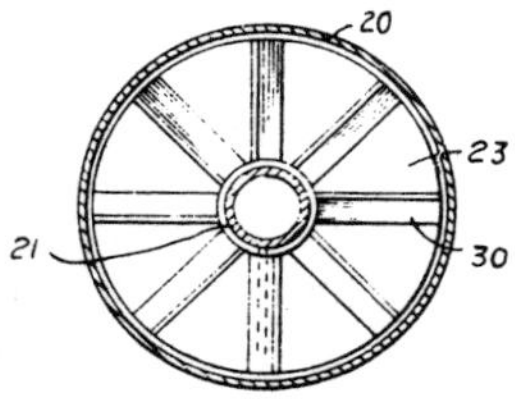

Cross-Sectional View Along Line 4–4 of Figure 6.3a

e.

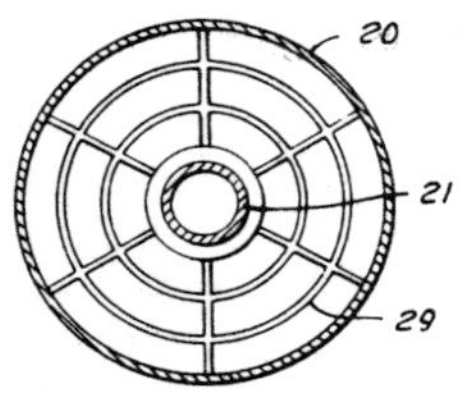

Cross-Sectional View Along Line 5–5 of Figure 6.3a

(continued)

FIGURE 6.3: (continued)

f.

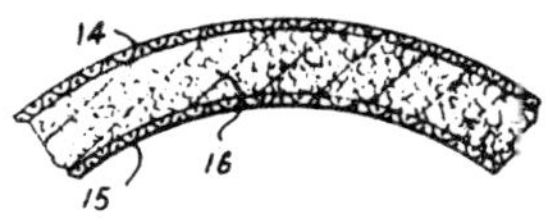

Cross-Sectional View Along Line 6–6 of Figure 6.3a

g.

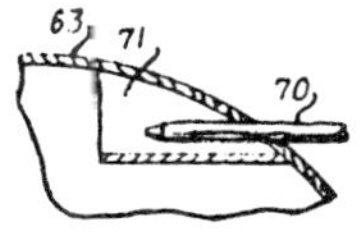

Horizontal Cross-Sectional View Along Line 7–7 of Figure 6.3a

Source: U.S. Patent 3,329,107

The filter screen assembly operates through gravity, excess water being filtered from the raw sewage through the lower half of the filter and being received in a storage tank **17** underlying the screen assembly and having a discharge conductor **18** leading from its lower end through which separated water and liquids may be removed for subsequent use or disposal.

The end of the water screen assembly opposite the inlet end is open and overlies a discharge chute **18′** into which partially dewatered raw sewage passes from the screen assembly downwardly to a dewatering press **19**. The press includes an elongated, horizontal, cylindrical housing **20** having mounted therein for revolving about the horizontal axis of the housing, a tubular perforated shaft **21** driven by a motor or other motive means **22**. A screw conveyor **23** is carried upon the shaft, having its periphery extending nearly to the inner wall of the housing, the convolutions of which may be progressively more closely spaced from the inlet end of the press to the outlet end.

Chute **18′** opens into one end of the housing, the opposite end of the housing being open and overlying a discharge bin **24**. As shown in Figure 6.3c, the tubular shaft **21** is provided with a multiplicity of radial perforations **25**, the end of the shaft adjoining the outlet end of the housing being plugged. The opposite end of the shaft is open, extending from the housing and overlying the open upper end of a discharge flume **26** which leads downwardly into a storage vessel **27** having a valved outlet conductor **28** at its lower end.

Thus, as the partially dewatered raw sewage passes through the press, it is squeezed or compressed and compacted by the revolving screw conveyor, quantities of water being extracted therefrom and removed through the perforations for delivery through the tubular shaft into the vessel **27**.

The housing carries in its outlet end a plurality of fixed, circular, concentric knife blades **29**, as shown in Figure 6.3e, through which the discharged solids from the raw sewage are forced and which cut and break up the solids into discrete particles. For further cutting and breaking up the solids, the shaft carries immediately adjacent the knife blades or in close proximity to the knife blades a plurality of radial, angular knife or cutting blades **30** which revolve with the shaft and, being in close proximity to the blades **29**, cause a very effective chopping and cutting action to take place. In this manner, the semidried solids from the raw sewage are discharged as discrete particles into the bin for further processing.

Vertical conveyor enclosure **31** extends upwardly from bin **24**, terminating at its upper end in a laterally directed branch **32** which opens downwardly and laterally into the upper end of drying tower **33**. Vertical bucket conveyor **34** is housed in **31** and extends from the bottom of the bin adjacent with **32** for lifting and conveying the comminuted raw sewage solids from **24** into **32** and then into the upper end of the drying tower. For distributing the solids uniformly across the width of the drying tower, the branch opens within the tower into a horizontal, perforated tube **35** extending transversely of the tower and enclosing a screw conveyor **36** revolved by a motor or other suitable driving means **37**. Obviously, the screw conveyor functions to spread the solids uniformly through the length of the tube so that they may pass downwardly therefrom through the perforations of the tube.

The drying tower is provided below the distributor tube with a plurality of spaced, staggered, downwardly inclined baffles **38** over which the comminuted solids are spread and cascaded for relatively complete drying. A plurality of heating tubes **39** are positioned beneath the baffles, these tubes being connected in series and receiving a fluid heating medium for drying the solids. The bottom wall **40** of the tower is inclined and adjoins a lateral discharge passage **41** having a suitably driven blower **42** for blowing the dried solids upwardly through a conduit **43** into the apex of a Y-shaped distributor conduit having one branch **44** depending into the upper end of a first storage tank **45** and a second branch **46** depending into a second storage tank **47**. A conventional flapper vane **48** determines into which storage tank the solids are directed.

The tanks **45** and **47** have discharge valves **49** and **50**, respectively, at their lower ends which communicate with a feed pipe **51** depending into a solids burner tube **52**. The feed pipe may enclose suitable feed means such as a screw conveyor, and a blower **53** at the inlet end of the burner supplies combustion air, and possibly some additional gaseous or vaporous fuel to the burner.

In the heating of the solids in the drying tower, certain combustible vapors and gases will be evolved, and these are withdrawn from the upper end of the tower through a conduit **54** for suitable discharge as indicated at **55** or for conveyance to the blower as additional fuel to that provided by the dried sewage solids. Further, additional supplementary fuel, such as natural gas, may be supplied to the blower and the burner as indicated at **56**.

Burner **52** opens into a combustion chamber **57** of generally rectangular configuration having an ash discharge outlet **58** at its end remote from the burner with a vertical transverse baffle **59** between the burner and the outlet for causing the burning or combusting solids to be blown upwardly and over the baffle so as to travel in a circuitous path. The hot flue gases or vapors from the combustion chamber are withdrawn therefrom through a flue **60** and are passed in series fashion through the heating tubes **39** of the drying tower, being ultimately discharged at **61**. If desired, and since most flue gases still retain some combustible materials, a portion of the flue gases may be diverted through a branch flue **62** and conducted to the blower as supplemental fuel.

In the treatment of paunch manure, it is found that some fats and oils will be present, and it is desirable to separate these initially both to facilitate the subsequent treatment of the sewage material as well as to recover the fats and oils which may be further processed for the recovery of by-products having commercial value. For this purpose, there is provided an initial fat separator tank **63** having at its upper end an inlet conductor **64** terminating within the tank in a downwardly facing diverter baffle **65** for diverting the incoming material downwardly into the interior of the tank. The tank has a hopper bottom **66** opening through a double slide valve **67** into the inlet conduit **10**. A plurality of screening plates or grids **68** open upwardly through one wide wall of **63** above water outlet conductor **69** through which some water may be extracted from the sewage present in the tank.

A water-operated jet apparatus is provided in the upper end of **63** above the baffle and includes a water inlet jet nozzle **70** opening into a deflector enclosure **71** provided with an air inlet conductor **72**. This directs a high velocity jet of aerated water across the surface of the sewage in **63** to aerate and float the fats and oils and drive them across the tank to an outlet conductor **73** positioned above the screens **68** and discharging into a stratification vessel **74** from which the fats and oils may be withdrawn through an upper outlet **75**, and the stratified water withdrawn through a lower or bottom outlet **76**.

The water discharge conductor **69** extends downwardly into the outlet conductor **18** of the vessel **17** and liquids from both may be discharged through a valved outlet conductor **77**, and by means of a suitable pump **78**, a portion of these liquids may be, and desirably are, driven upwardly through a conductor **79** to furnish a water supply for the jet nozzle.

In the operation of the plant, the paunch manure enters the vessel **63** and undergoes some degree of stratification and water separation, the fats and oils, together with some water, passing to the vessel **74** for stratification into fatty and aqueous layers. As previously noted, some portion of water is drawn off through the screens or grids, the remainder of the material, in more or less aqueous dispersion, passing downwardly to the inlet conductor **10** at a rate governed by the setting of the valve plates **67**.

The material is additionally dewatered in the screen assembly **12**, compressed and further dried in the press **19**, dried in the tower **33** and stored in the tanks **45** and **47** for regulated supply to the burner **52** and combustion chamber **57**, in which the now relatively dry material is combusted and discharged as ash material.

In the event humus is to be treated rather than paunch manure, the vessel **67** is not utilized, but the raw sewage is instead admitted directly to the inlet conductor **10**, the subsequent steps remaining as previously described. Experimentation has shown that the dried solids from the sewage have a heating value of approximately 10,000 Btu/lb, and hence, one pound of this material would furnish sufficient heat to evaporate at atmospheric pressure approximately ten pounds of water. Thus, adequate heat is provided for final drying of dewatered solids or sludge materials from the raw sewage, and there are possibilities of using the dried solids as a fuel.

It is contemplated that the dried solids, supplemented by the gases and vapors withdrawn from the upper end of the tower through the conductor **54**, will furnish the sole fuel required in the process, but this, of course, is subject to variation. If necessary or desirable, additional fuel of any nature can be supplied as needed or desired through the conductor **56** and/or conductor **62**.

CEMENT MADE FROM SLUDGE

In a process described by *N.S. Davis; U.S. Patent 3,442,498; May 6, 1969; assigned to North American Rockwell Corporation* activated sludge sewage treatment is combined with a cement making process for making cement economically from normally waste materials. Activated sludge is flocculated with sequential additions of carbon dioxide and calcium oxide, filtered, and fed to a cement kiln. The organics in the sludge reduce fuel requirements in the kiln and the resultant clinker is in the Portland cement composition range.

Referring to Figure 6.4, raw sewage has primary sludge separated in a primary settler **10** and the effluent is converted to activated sludge by microorganisms in a tank **11**. Settled and concentrated sludge has carbon dioxide and calcium oxide sequentially mixed in a tank **14** to cause flocculation. After composition blending the sludge is filtered in a filter **16** and the filter cake is burned in a cement kiln **17**, thereby producing cement clinker which is ground with gypsum in a grinder **19** to produce Portland cement.

FIGURE 6.4: COMBINED WASTE TREATMENT AND CEMENT MAKING PROCESS

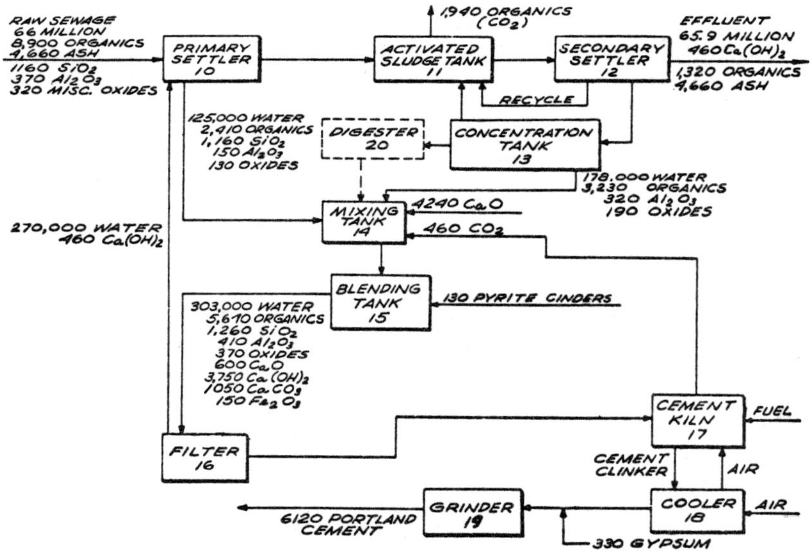

Source: U.S. Patent 3,442,498

One of the substantial costs in a cement manufacturing operation comprises the fuel required for firing the kiln to the elevated temperatures required. Since the filter cake fed to the cement kiln comprises a substantial amount of combustible organic materials a substantial fuel economy results since less fuel is required in the kiln. Thus, by combining activated sludge treatment with cement production, otherwise wasted organic materials are burned to produce heat for making cement and simultaneously are reduced in volume to avoid disposal problems.

BOD REDUCTION USING IONIZING RADIATION IN PRESSURIZED OXYGEN ATMOSPHERE

A process as described by *F.N. Case and D.E. Smiley; U.S. Patent 3,650,926; March 21, 1972; assigned to U.S. Atomic Energy Commission* relates to a method of increasing the light transparency and reducing the biological oxygen demand of organic-containing waste solutions. The process comprises the step of exposing such solutions to penetrating ionizing radiation under pressure of an oxygen-containing atmosphere of at least 1,000 psi until the desired degree of optical and ultraviolet transmittancy as well as biological oxygen demand have been reached. The following examples illustrate the process.

Example 1: This example is designed to demonstrate the effects achieved by irradiating lignin-bearing wastes and textile plant waste to a pressure of at least 1,000 psi in an oxygen-containing atmosphere. Effluent from the sewage disposal plant of the city of Knoxville, Tenn., having color introduced from lignin, and textile plant effluent were irradiated at a gamma flux of 1.2 x 10^7 r/hr.

For purposes of comparison, a first set of samples of 125 ml each were irradiated in air at atmospheric pressure for periods of 10 to 120 minutes, one sample being set aside as an unirradiated control sample. The samples were analyzed for BOD, pH and percent light transmission. The results are shown below.

Run No.	Time, min.	Exposed dose, γ	pH	BOD	Light transmission, percent, at—			
					625 mμ	500 mμ	400 mμ	280 mμ
1	0	0	7.0	202	58	48	22	0
2	10	2×10^6	7.1	128	52	43	15	0
3	20	4×10^6	6.8	101	51	41	12	0
4	30	6×10^6	6.7	85	41	26	5	0
5	60	1.2×10^7	6.4	76	48	36	8	0
6	120	2.4×10^7	6.1	53	37	26	5	0

A second set of 15-ml samples were irradiated in air at 2,000 psi. The samples were analyzed for BOD, pH, and percent light transmission. The results are shown below.

Run No.	Time, min.	Exposed dose, γ	pH	BOD	Light transmission, percent, at—			
					625 mμ	500 mμ	400 mμ	280 mμ
7	0	0	7.0	202	58	48	22	0
8	10	2.3×10^5	6.6	166	54	50	26	0
9	20	4.6×10^5	6.1	140	62	62	38	1
10	30	7×10^5	6.0	122	65	68	54	8
11	60	1.4×10^6	4.3	100	71	74	65	25
12	120	2.8×10^6	3.4	28	87	88	83	58

Several major points of difference can be noted by comparing runs 1 through 6, irradiated at atmospheric pressure, with runs 7 through 12, which were irradiated at 2,000 psi air pressure. Although the samples irradiated at atmospheric pressure were exposed to a higher gamma intensity, the drop in BOD and pH was relatively small compared to the samples irradiated at 2,000 psi air pressure. Irradiation at 2,000 psi pressure, in addition, yielded an essentially optically clear solution within 30 minutes with increasing ultraviolet transparency. In marked contrast, irradiations conducted at atmospheric pressure in air actually reduced light transmission in the optical region and had no effect in improving light transmission in the ultraviolet region, even at total dosages of nearly an order of magnitude higher as compared to samples irradiated under pressure.

Example 2: An opaque paper mill effluent was irradiated to a dose of 2.4 x 10^7r under 2,000 psi oxygen. The opaque solution was converted to a pale yellow solution containing a white fibrous material. Apparently, the white fibrous material was pulp contained in the waste, which was also bleached by the process.

Example 3: An aqueous slurry of raw sulfite, wood pulp, brown in color, was irradiated in the presence of sodium hypochlorite under an oxygen pressure of 1,600 psi for 2 hours. The dose rate was 2 x 10^7 r/hr. The brown fibrous material was converted to white fibrous material. The combination of radiation, hypochlorite, and high pressure oxygen gave better decolorization than did hypochlorite alone or irradiation and hypochlorite in the absence of high pressure oxygen. The benefits of the oxidizing power of hypochlorite ion to the high-pressure radiolysis can be achieved with any water-soluble hypochlorite salt.

ADDITION OF POWDERED COAL TO CONTROL ODOR

A.J. Kehoe; U.S. Patent 3,300,403; January 24, 1967 has found that if powdered coal, the major portion of which is rice size or smaller, is added to sewage so as to be substantially uniformly distributed in the sludge, the sludge remains fresh for an extended period of time. Rather than deteriorating within several hours, such sludge has been held for periods of time that exceed 48 hours without giving off undue odors. Additionally, when the sludge so treated commences to become acid or septic (due to acid digestion) from remaining static for too long a period of time, it is possible to recirculate this sludge and add it to the incoming sewage to allow it to resettle with the fresh solids and such treatment freshens the deteriorating sludge so that it will keep for an additional extended period of time.

The addition of powered coal enhances the sedimentation or settling of the sewage solids, so that the need for using thickener tanks or holding tanks is largely eliminated. However, where such use is necessary, the resistance of powdered-coal-treated sludge to septic action or acid digestion makes it possible to employ these devices while reducing the offensive odors. Further, it has been found that by adding powdered coal to the sludge, the sludge itself exhibits considerably more body and is of greater porosity than the nontreated sludge so that it is more readily and completely chemically reacted with flocculating agents. As a consequence, this sludge filters far more efficiently than the nontreated sludge, and the yield of filter cake, or sludge cake per gallon of sewage treated is greatly increased.

CHLORINATION

In a process described by *E.R. Grich and J.W. Hood; U.S. Patent 3,300,402; January 24, 1967; assigned to Purifax, Inc.* chlorine, preferably in the gaseous form, is introduced through a conventional metering device into the sludge and waste in a zone undergoing rapid agitation. The object is to provide for ready access of the chlorine to all particles of the sludge or waste. As is well known, in the presence of readily oxidizable matter chlorine also reacts directly with water releasing nascent oxygen and hypochlorous acid, both of which are powerful oxidants.

Introduction of the chlorine may be made into the inlet side of a suitable centrifugal pump. An alternative is to introduce the chlorine adjacent a high-speed impeller disposed to agitate the material.

In the process, immediately following introduction of the chlorine and subsequent to being thoroughly mixed, the sludge or waste is caused to undergo rapid circulation in an unobstructed vessel or vessels, the rate of circulation being sufficient to maintain in suspension in the waste insoluble or nitrogenous matter, carbonaceous matter, sand or bits of ground bone in the case of sewage, and titanium dioxide and clay or other filler in the case of industrial waste from a plant converting waste paper into paperboard.

In addition it has been found that continuance of the centrifugal action after oxidation has taken place causes the fine air bubbles attached to the solid particles to separate from the solid particles and coalesce. The air or gas bubbles upon leaving the reactors and progressing into a zone where the pressure is lower, expand and separate from the sludge mass. As a result, sedimentation of the solid particles can take place in the receiving clarifier. Of further importance it has been found that in such instances where dewatering of the sludge from the reactors is desirable on mechanical dewatering equipment or sand beds, removal of the fine air bubbles attached to the solid particles increases the rate of separation of the solids from the liquids by threefold or more.

In the case of sanitary sewage the improved chemical oxidation process may be utilized in a number of ways. One arrangement would be in combination with a conventional aerobic-anaerobic sewage treatment installation, whereby the chemical oxidation operation would be carried out on the supernatant liquor from the anaerobic digester.

AERATION AND FLOCCULATION TREATMENT OF A POLLUTED BODY OF WATER

W.E. Budd and T.J. Powers; U.S. Patent 3,470,091; September 30, 1969; assigned to Dorr-Oliver Incorporated and The Dow Chemical Company describe a method for combating the pollution of streams or rivers or canals, by treating the polluted water collectively in the flowing stream itself, instead of separately treating the individual contributing sources of pollution in order that the construction of numerous shore-located individual expensive sewage treatment plants may be forestalled or avoided.

This system provides for applying aeration and flocculation treatment to a predetermined length of the stream itself, so that the resulting floc structures from that treatment, weighted by entrapped inert solids, will settle in a downstream adjoining sedimentation zone as sludge which may be removed or pumped away periodically.

With the flowing stream carrying the polluting matter to a treatment station located in the stream channel itself, the treatment preferably comprises the steps of oxygenating the polluted stream by either pure oxygen or air to convert the pollutants into biologic floc structures, or increasing the settleability of the floc structures by the introduction to the stream of an agglomerating flocculating material preferably of the polymeric class, causing the thus-conditioned organic solids weighted by sand, silt, or clay particles present in the stream, to settle on the bottom of the stream bed as sludge, and of removing the resulting sludge from the settling zone in the stream.

A portion of the sludge thus removed may be recirculated to the oxygenation zone and/or to the flocculation zone in the stream for seeding, while waste sludge is subjected to solids concentrating treatment for disposal. The weighting particles are present in the polluted stream from naturally occuring dust or run-off reaching the stream.

Preferably, the oxygenating medium or fluid is introduced in a first conditioning zone or station as by means of diffuser tubes located at or near the bottom of the stream, thus taking advantage of the depth of the stream or river to attain optimum absorption or dissolution and maximum utilization of the oxygen supply in the formation of the relatively fragile biologic floc structures. The flocculation agent in solution is similarly introduced into the stream, for instance in a second conditioning zone or station which may be located downstream from the flocculation zone. The floc structures are thus further coalesced or agglomerated as well as weighted by trapped particles of inorganic particles, so they will settle on the stream bottom in the third or settling zone downstream from the treatment zone.

The sludge is allowed to drift into this sedimentation zone, and may collect for instance, in a depression or pocket of the river bottom, and which may overlap with the preceding flocculation zone. The sludge is pumped up from the pocket of the settling zone, and is subjected to solids concentration preferably to the extent of producing a wet cake material which may be subjected directly to incineration producing but a mixture of sand, silt and ash suitable for use as fill.

Features of this process are concerned with various ways of introduction into the stream of the oxygenating fluid and of the flocculating agent, as well as with various modes of sludge withdrawal, all depending more or less upon the nature of the stream and the degree of its pollution, and particularly also upon the depth and width of the stream and upon the profile of the river bed.

For example, in the case of a relatively narrow (nonnavigable) stream one way of introducing the oxygenating fluid is by way of a gas pump or compressor forcing pure oxygen or air through diffuser pipes or the like located at the bottom of the stream. Flocculating solution is then introduced by a similar distributing pipe arrangement in the next downstream treatment zone. Subsequent sludge withdrawal downstream from the flocculating zone may be carried out through a stationary sludge withdrawal pipe system located at the bottom of the stream, or a bridge structure across the stream may carry a depending sludge withdrawal pipe movable thereon back and forth across the stream, while the bridge structure itself may be movable parallel to itself.

In the case of wide navigable streams or rivers oxygenation may be provided through the diffuser pipes at the river bottom more economically by the supply of previously dissolved oxygen. That is to say, a relatively small auxiliary stream of water preferably derived from the river itself is passed through a shore-based oxygen absorption column. This provides a supply of water highly saturated with dissolved oxygen, to be forced through the diffuser pipes or the like into the stream.

Inasmuch as this mode of oxygen introduction is characterized by a relative absence of turbulence as compared with the direct introduction of a gaseous oxidizing medium, the introduction of the flocculating agent or solution may be effected within the oxygenation zone or even in a zone preceding it, since undue turbulence would not interfere with the initial formation of the fragile biologic floc structures.

OXIDATION USING PHOTOSYNTHESIS OF ALGAE

P.J. Gresham; U.S. Patent 3,565,797; February 23, 1971 describes apparatus and a process for the treatment of sewage by oxidation and by contacting the sewage with living filamentous algae affixed to a suitable supporting framework. Several stages of treatment may be used, but a single stage of treatment through moving contact of the sewage with an algae bed or screen of sufficient size will provide complete or almost complete removal of suspended solids, and the biochemical oxygen demand of the effluent water can be reduced to acceptable levels for disposal in streams and lakes.

The algae utilized in this process grow rapidly, trapping suspended solids and removing dissolved organic matter to utilize both types of materials as food. Through photosynthesis the organic material is converted to new compounds by the algae and oxygen released. The oxygen released by the algae oxidizes sewage solids. One aspect of this process is the discovery that oxidation of sewage solids causes sludge separation and compaction.

Moreover, the coagulating effect of algae filaments together with the effect of oxidation produces a heavy, dense sludge which can be readily compacted by the compression created by scraping mechanisms moving such sludge up an incline to a point above the water line. The compacted sludge can be further compressed to remove a substantial portion of its water content, producing a relatively dry sludge which is in a form suitable for incineration or other disposition.

BENZYL BROMOACETATE TO CONTROL PAPER MILL SLIME

A process described by *S.J. Lederer; U.S. Patent 3,658,700; April 25, 1972; assigned to Merck & Co., Inc.* relates to a method for inhibiting the formation and growth of slime in paper mill white water systems by introducing into paper mill water systems an antibacterial and antifungal quantity of benzyl bromoacetate.

In general, effective control and inhibition of the formation of slime is achieved by adding to the paper mill water system sufficient benzyl bromoacetate to provide a concentration of from about 0.1 to 25.0 ppm based on the total system volume with concentrations between 0.1 to 10.0 ppm being generally preferred. In mills having severe slime formation problems, much higher concentrations (i.e., as high as 1,000 ppm) can be employed.

The benzyl bromoacetate in appropriate concentrations may be added to the mill water at any point in the system from the pump preparation stages up to machine head box. It will be preferred usually to add the benzyl bromoacetate at the point or points of greatest slime accumulation. In general, areas of maximum slime accumulation will be found at, or just before, the paper-making machine. Conveniently, therefore, the benzyl bromoacetate may be added at the fan pump where good mixing is assured, or into the tray white water. The toxicant then is distributed throughout the system by recirculation of the white water to earlier stages. The toxicant may be added either continuously or by intermittent slugging techniques.

Benzyl bromoacetate, being liquid, may be incorporated neat into paper mill water systems by any simple mixing technique which will insure good distribution. Alternatively, however, the benzyl bromoacetate may be added in the form of an adsorbate of benzyl bromoacetate on finely powdered solid dispersants. Typical dispersants which may be employed for this purpose will include, for example, calcium magnesium silicate, silicon dioxide, clays and talcs. From 10 to 50% by weight of benzyl bromoacetate may be adsorbed on such dispersants by thoroughly blending the benzyl bromoacetate and the powdered dispersant in the suitable mill. The adsorbate then may be added to the paper mill water system at any desired point in sufficient quantity to give the desired final concentration of benzyl bromoacetate in the total mill system.

PRESSURE SEWAGE SYSTEM AND MACERATOR

J. Burkholder; U.S. Patent 3,897,600; August 5, 1975; assigned to Robintech, Incoporated describes a method of directing sewage from a holding tank, septic tank, or the like, into a pressure sewage system. The process comprises providing a receptacle or storage area for receiving sewage by gravity from a residential or similar source, which receptacle has a downwardly converging or conical bottom region, whereby heavier solid matter in the sewage fluids will be directed into a relatively confined area at the bottom of the reservoir.

A sewage-treating unit comprising a macerator and pump is disposed in the apex portion of the bottom of the receptacle or storage area for comminuting the solids, agitating the sewage mixture and transferring the effluent into a relatively small discharge conduit and moving the fluid through the conduit for discharge under a predetermined hydraulic head into the pressure sewage system.

Referring to Figure 6.5, a fluid receiving tank **10** has outer sidewalls **12** which may be of substantially any configuration, such as cylindrical, square, rectangular, or the like, which tank is normally at least partially buried beneath the surface of the earth. The upper end **14** of the tank is open and may be provided with a suitable manhole extension member complementary to the configuration of the sidewalls and extending upwardly therefrom for receiving a suitable manhole cover member and lid, thus providing access to the interior of the buried tank **10** from the surface of the ground, as is known. The bottom **14** of the tank preferably is downwardly and inwardly tapered as particularly shown in Figure 6.5a, and preferably is provided with a relatively flat or horizontal floor portion **16** at the central converging point of the bottom.

An inlet pipe or conduit **18** of conventional size (typically 4" nominal diameter) is provided in the upper portion of the tank spaced downwardly from the open upper end thereof, and extends through an aperture in the sidewall **12** to an elbow or angled conduit **20** for directing effluent into the tank and discharging the fluid in a downward direction therein. An overflow conduit **22** extends through a second aperture provided in the sidewalls to provide communication between the interior of the tank and the exterior.

The overflow conduit is spaced below the inlet conduit in order to permit emergency discharge of any accumulating fluids in the tank and to forestall any inadvertent back-up of the fluids through the inlet conduit as will be hereinafter set forth. In addition, a discharge conduit or line **24**, which as shown is quite small in diameter relative to the size of the inlet pipe, extends through the sidewalls in any well-known manner for providing communication between the interior of the tank and the main sewer system **106** (Figure 6.5c).

FIGURE 6.5: PRESSURE SEWAGE SYSTEM

a.

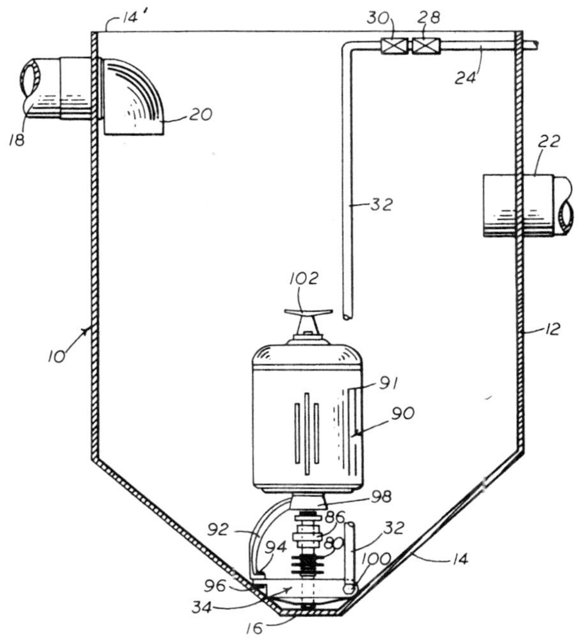

Elevational View Partly in Section of a Holding Tank and Pump Unit

(continued)

FIGURE 6.5: (continued)

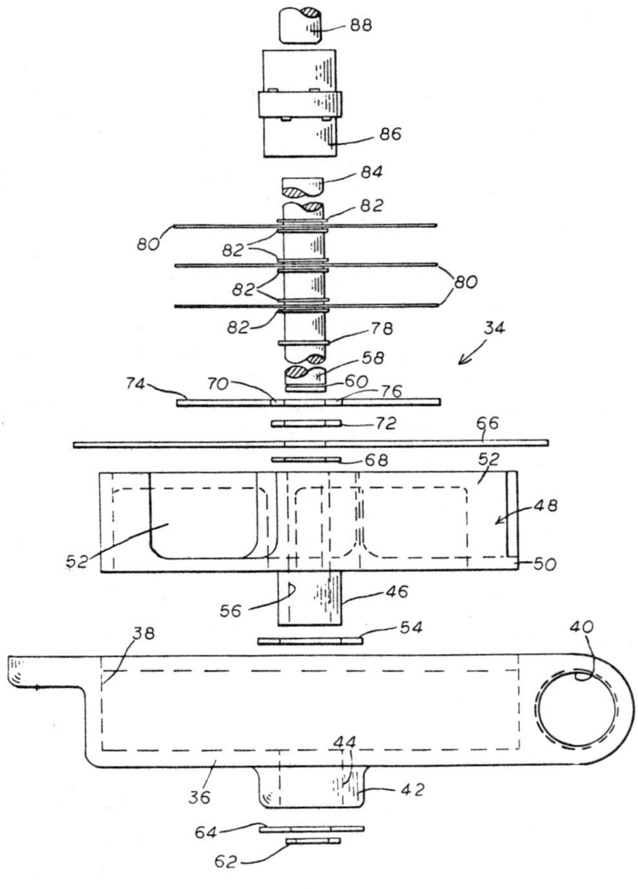

b.

Enlarged Exploded Side Elevation View of a Pump

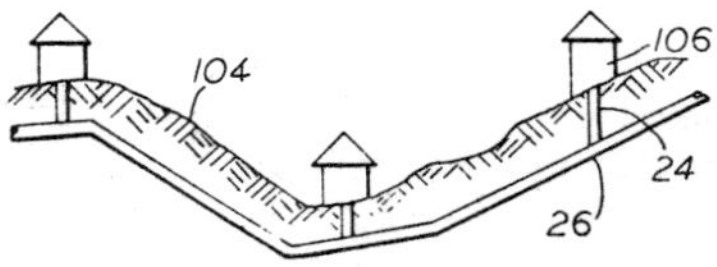

c.

Sectional Schematic View of a Pressure Sewage System

Source: U.S. Patent 3,897,600

The discharge line **24** is provided with a plurality of suitable valves, such as schematically shown check valve **28**, and a suitable compression end valve **30**, as is well known, which may be threaded or otherwise attached to a vertically disposed portion **32** extending downwardly into the proximity of the bottom **16** of the tank **10**, as particularly shown in Figure 6.5a. As small as practicable diameter discharge pipes may be employed. In practice 1" diameter discharge conduits **24** and **32** have been found entirely satisfactory for use with this system. The economic advantages in furnishing and burying such a small pipe in preference to the much larger tile or cast-iron pipe of conventional size are manifest.

A macerator-type impeller pump generally indicated at **34** is disposed in the proximity of the lower portion **16** of the tank and as shown in Figure 6.5a is secured to the lower end of the immediately adjoining, generally centrally located and vertically disposed pipe portion **32**. As best seen in Figure 6.5b, the pump **34** comprises a body portion **36** having an interiorly disposed convulate-type recess **38** terminating in a discharge port **40**, as is well known. A centrally disposed boss **42** extends axially from the body and is provided with a central bore **44** for journaling one end of a hub **46** of an impeller **48**.

The impeller comprises a substantially circular plate **50** intergral with or secured around the hub and axially spaced from one end thereof. A plurality of circumferentially spaced fins or blades **52** are secured to or integral with the plate and are disposed substantially perpendicular with respect thereto. The blades are preferably of a substantially spiral configuration and are integral with or secured around the hub and extend radially outwardly. The hub is rotatably disposed or journaled in the bore, and a shim member **54** is preferably disposed between the plate and the body for facilitating rotation of the impeller member with respect to the body.

A central bore **56** extends through the hub for receiving a shaft **58**. The shaft may be keyed or otherwise secured within the bore for transmitting rotation to the impeller, and an annular groove **60** is provided on the outer end of the shaft for receiving a retainer ring **62** for locking the shaft against longitudinal movement in one direction with respect to the hub. In addition, it is preferable to interpose an annular shim member **64** between the retainer ring and the body for facilitating rotation of the shaft.

A stationary grinder plate or disc **66** is suitably secured around the shaft and is disposed on the upper surface of the blades. It is preferable however to provide a suitable annular shim member **68** between the plate **66** and the fins. The grinder plate may be of any suitable type, such as perforate plate, or the like which screens the sewage material that flows through the holes of the plate enroute to the pump, as is known.

In addition, a grinder blade **70** is secured around the shaft in any suitable manner for rotation therewith and is spaced from the grinder plate by a suitable annular shim member **72**. The grinder blade may be of any suitable type, and comprises one or more radially extending blades **74** (preferably three) integral with or secured to a central disc member **76** which in turn is secured to the shaft in any known manner. An annular retainer ring **78** may be suitably secured to the shaft for cooperating with the shim **72** to retain the grinder blade. A plurality of suitable macerator blades **80** are secured on the shaft in any known manner for rotation simultaneously therewith and are spaced longitudinally by suitable retainer rings **82**.

The upper end of the shaft **58** is adapted to connection preferably through suitable flexible-type coupling **86** to the drive shaft **88** of an electric motor **90**. The drive shaft of the motor is suitably keyed or otherwise connected with the coupling for transmitting rotation. The outer housing or casing **91** of the motor is reciprocally supported on the motor for limited vertical movement.

The motor housing is secured to the pump **34** by means of an arcuate arm **92** which has one end thereof secured to the body **36** by a bolt **94** and locknut **96**. The opposite end of the arm is provided with a flared sleeve member **98** adapted for receiving and supporting the mounting hub and associated parts of the motor. The motor is provided with a float switch whereby the motor is inoperative while the housing is in a lowered position, and operative when the housing becomes buoyant and assumes a raised position.

In addition, the port **40** of the pump is secured to the lower end of the pipe portion **32** by a suitable fitting member **100** in such a manner as to secure the body to the pipe, and provide open communication between the port and the interior of the pipe. A suitable handle member **102** is provided on the uppermost end of the motor for facilitating installation in and removal from the interior of the tank **10**.

In a pressure sewer system, such as schematically depicted in Figure 6.5c, the main sewer conduit **26** is normally buried beneath the surface of the ground **104** at a substantially uniform depth, so it generally follows the contour of the ground or earth. The individual buildings **106**, or the like, feed sewage into their respective tanks and are independently connected with the main sewer conduit by the lines **24**, and the pump of each tank provides adequate pressure for operation of the main sewer system.

The line **24** is provided with the valves **30** and **28** in order that fluid may be pumped from line **24** into the line **26**, but back-flow of the fluids from line **26** into the pipe portion **32** is precluded. Raw sewage from the building is directed to the tank by gravity through the usual tile conduit or pipe **18**, and is discharged into the interior of the tank through the elbow member **20** for accumulation in the lower portion.

Assuming the motor and pump are in a nonoperative condition following a pumping cycle, sewage fluid routinely accumulates in the tank, and the motor casing floats or rises slightly to actuate the float switch, thereby activating the entire sewage treating and pumping unit, comprising motor **90**, macerator blades **80**, grinder blades **74**, and impeller **48**.

When the pump is operating, the agitated sewage is swirled into a vortex, as might be expected, particularly when the adjoining tank walls are conical and convergent, and the exposed macerator blades above the pump apparently assist in producing a substantial degree of aeration indicated by extensive bubbling when viewed through an observation window in test units, which undoubtedly encourages chemical action through oxidation. This reduces odor, and in fact field reports consistently indicated that the unit is odorless.

The solids in the sewage are accumulated in the proximity of the bottom **16** of the tank by gravity, and are thus efficiently directed to the pump and macerator blades and grinding blades thereof; thus, the solids are efficiently macerated and mixed with the liquids as they descend and are drawn into the pump cavity **38**. The pump action forces the mixture centrifugally upward through the discharge pipe **32**, valves **30** and **28**, and the relatively small line **24** replacing the usual large tile conduit and leading up and down with the terrain into the main sewer conduit **26**, under sufficient hydraulic head to pressurize the main sewer conduit. After a predetermined quantity of the sewage has been pumped from the tank **10**, the motor casing **91** will have moved downwardly to its lowermost position, and the operation of the motor **90** and pump ceases, so the liquid level in the tank is maintained within predetermined limits.

In the event of any pump or motor failure, the sewage accumulating in the tank will flow through the discharge pipe **22** prior to any back-up of the sewage into the pipe **18** since the discharge pipe is disposed at a lower elevation than the inlet pipe **18**. Of course, if desired, a suitable alarm system may be provided for indicating any pump failure in order to preclude damage from overflow of the sewage.

Removal of the entire treating and pumping unit for servicing or cleaning is readily accomplished by disconnecting a union, conventional ground joint, or similar threaded connection normally associated with valves of the type of valve **30**, whereby, if the handle **102** is submerged, the pipe **32** may serve as a convenient handle means located well above any normal fluid level for lifting the entire operating assembly from the tank. Likewise on reassembly, the unit may be lowered by gravity to its operating position, the pump being automatically located and concentrically aligned by the convergent side walls of the tank, and the unit resting on the frustoconical floor section, all as shown in Figure 6.5a.

COMPANY INDEX

The company names listed below are given exactly as they appear in the patents, despite name changes, mergers and acquisitions which have, at times, resulted in the revision of a company name.

INVENTOR INDEX

U. S. PATENT NUMBER INDEX

NOTICE

Nothing contained in this Review shall be construed to constitute a permission or recommendation to practice any invention covered by any patent without a license from the patent owners. Further, neither the author nor the publisher assumes any liability with respect to the use of, or for damages resulting from the use of, any information, apparatus, method or process described in this Review.

HOW TO DISPOSE OF TOXIC SUBSTANCES AND INDUSTRIAL WASTES 1976

by Philip W. Powers

Environmental Technology Handbook No. 4

This book discusses all recognized ultimate disposal methods in detail and contains a long list of specific recommendations for specific substances plus alternative disposal or recovery methods.

Ultimate waste disposal implies the final disposition of non-degradable, persistent, harmful, and cumulative wastes that may be solid, liquid, or gaseous. Workable solutions to ultimate disposal problems as described in this book include conversion to harmless end products, subsurface storage in ponds or landfills, and disposal in the ocean. This condensed information will enable anyone to establish a sound background for action towards disposal of toxic and hazardous materials with safety.

The five general categories of hazardous wastes are:

1. Toxic Chemical
2. Radioactive
3. Flammable
4. Explosive
5. Toxic Biological

Many toxic or hazardous wastes contain valuable materials. Whenever this is the case, recovery and reuse is one of the most desirable methods of hazardous waste avoidance.

The book is based on government-sponsored reports (often elaborated by highly qualified personnel from industrial companies), U.S. patents, and on pertinent articles in authoritative journals. A partial and condensed table of contents follows here. Chapter headings are given in full whenever possible.

ISBN 0-8155-0615-5 **500 pages**

MUNICIPAL SOLID WASTE MANAGEMENT 1975

by David Rimberg

Pollution Technology Review No. 26

This book has been prepared with the intention to provide sufficient technological and administrative guidance for designing practical and economical solid waste management systems in accordance with the *Federal Solid Waste Disposal* and *Resource Recovery Acts.*

The reader is given ample background and current information pertinent to the subject as obtained from recent studies performed mostly under the auspices of the EPA (U.S. Environmental Protection Agency). Armed with this information the reader can endeavor to plan or modify a local government's solid waste management system.

The book is logically arranged into three sections. The first section is devoted to the technological aspects of solid waste management and thus comprises the first four chapters. Discussed in these chapters are the sources and characteristics of solid waste, methods of storage, collection and disposal, and a survey of the newest equipment and architectural designs. The next two chapters (the second section of the book) deal with the financial, economic, and labor aspects of solid waste management. The last part of the book is a description of several prototype municipal installations in various parts of the United States as built or modified rather recently. A partial and condensed table of contents follows here. Chapter headings are given, and some of the more important subtitles are included also.

ISBN 0-8155-0596-5 **380 pages**

Date Due

Due	Returned	Due	Returned
APR 2 2 1982	APR 5 1982		
DEC 0 1 1983	NOV 2 3 1983		
DEC 1 8 1986	DEC 1 8 1986		
OCT 3 0 1989	DEC 1 5 1989		
OCT 0 1 1991	JUL 1 0 [illegible]		
NOV 1 4 [illegible]	NOV 1 5 1994		
MAR 1 8 1996			
APR 0 8 1996	APR 0 8 1996		
DEC 0 8 1996	NOV 2 6 1996		
DEC 3 0 1996			
FEB 2 0 1997	DEC 3 0 1996		
OCT 1 2 1999	SEP 0 7 1999		
JAN 1 0 2001	JAN 1 0 2001		
JUN 0 1 2001	MAY 1 5 2001		